Principles of Biofuels and Hydrogen Gas

Production and Engine Performance

Ahindra Nag, Ph.D.

New York Chicago San Francisco
Athens London Madrid
Mexico City Milan New Delhi
Singapore Sydney Toronto

Library of Congress Cataloging-in-Publication Data

Names: Nag, Ahindra, author.
Title: Principles of biofuels and hydrogen gas : production and engine performance / Ahindra Nag, Ph.D., editor.
Description: New York : McGraw-Hill Education, 2020. | Includes bibliographical references and index. | Summary: "Principles of Biofuels explores the chemistry of biofuels, the latest refining processes, and major performance characteristics-discussing the theory and experimental procedures used to prepare biofuels economically. Unlike most books on biofuels that broadly cover non-conventional energy sources, this book goes into detail about engine performance, making it a highly valuable resource for students, researchers, and practitioners. Principles of Biofuels is grounded in professional relevance and expertise. It covers the derivation of hydrogen gas from biomass and water media, as well as biofuels derived from algae and biomass. End-of-chapter questions reinforce student comprehension"—Provided by publisher.
Identifiers: LCCN 2020005816 | ISBN 9781260456424 | ISBN 9781260456431 (ebook)
Subjects: LCSH: Biomass energy. | Biodiesel fuels. | Hydrogen as fuel.
Classification: LCC TP339 .N345 2020 | DDC 662/.88—dc23
LC record available at https://lccn.loc.gov/2020005816

McGraw Hill Education books are available at special quantity discounts to use as premiums and sales promotions, or for use in corporate training programs. To contact a representative please visit the Contact Us page at www.mhprofessional.com.

Principles of Biofuels and Hydrogen Gas: Production and Engine Performance

Sponsoring Editor		**Proofreader**	
Robin Najar		Arbind Tiwari	
Editorial Supervisor		**Indexer**	
Donna M. Martone		Jerry Ralya	
Acquisitions Coordinator		**Production Supervisor**	
Elizabeth Houde		Lynn M. Messina	
Project Manager		**Composition**	
Poonam Bisht		MPS Limited	
Copy Editor		**Art Director, Cover**	
Theresa L. Kay		Jeff Weeks	

Contents

*Hajra Javed, Muhammad Naveed Anwar, Abdul Sattar Nizami,
and Mujtaba Baqar*

Contributors

Mujtaba Baqar *Sustainable Development Study Centre, Government College University, Lahore, Pakistan* (CHAP. 4)

Sebastian Bojanowski *Department of Math, Natural Science, and Information Technology, Laboratory for Waste Treatment Processes, University of Applied Sciences Giessen-Friedberg, Giessen, Germany* (CHAP. 8)

L. M. Das *Mechanical Engineering Department, Indian institute of Technology, Delhi India* (CHAP. 10)

Jayita Datta *Guru Nanak Institute of Technology, Electronics Instrument, Kolkata, West Bengal* (CHAP. 1)

M. P. Dorado *Department of Physical Chemistry and Applied Thermodynamics, EPS, University of Cordoba, Cordoba, Spain* (CHAP. 3)

B. B. Ghosh *Department of Mechanical Engineering, Indian Institute of Technology, Kharagpur, India* (CHAP. 7)

Hajra Javed *Sustainable Development Study Centre, Government College University, Lahore, Pakistan* (CHAP. 4)

Keikhosro Karimi *Department of Chemical Engineering, University of Technology, Isfahan, Iran* (CHAP. 9)

Gerhard Knothe *National Center for Agricultural Utilization Research, Agricultural Research Service, U.S. Department of Agriculture, Peoria, Illinois* (CHAP. 5)

P. Manchikanti *Agriculture Engineering Department, Indian Institute of Technology, Kharagpur, India* (CHAP. 2)

Ahindra Nag *Department of Chemistry, Indian Institute of Technology, Kharagpur, India* (CHAPS. 1, 2, 6, 7)

Muhammad Naveed Anwar *Sustainable Development Study Centre, Government College University, Lahore, Pakistan* (CHAP. 4)

Behzad Satari *Department of Food Technology, College of Aburaihan, University of Tehran, Tehran, Iran* (CHAP. 9)

Abdul Sattar Nizami   *Center of Excellence in Environmental Studies (CEES), King Abdulaziz University, Jeddah, Saudi Arabia* (CHAP. 4)

M. Sedano   *Department of Physical Chemistry and Applied Thermodynamics, EPS, University of Cordoba, Cordoba, Spain* (CHAP. 3)

Ernst A. Stadlbauer   *Department of Math, Natural Science, and Information Technology, Laboratory for Waste Treatment Processes, University of Applied Sciences Giessen-Friedberg, Giessen, Germany* (CHAP. 8)

Mohammad J. Taherzadeh   *School of Engineering, University of Borås, Borås, Sweden* (CHAP. 9)

Preface

The continuous use of the world's crude oil reserve and a corresponding escalation in its price together with the limited coal reserves have stimulated the hunt for renewable sources of energy. The main sources of renewable energy are biomass, biogas, methanol, ethanol, and biodiesel; solar active (photovoltaic), solar passive (preheating of water), wind, mini hydel, and mini tidal are important sources that produce less pollution and protect the environment.

Much attention has been given to biomass and its modifications as a substitute for fossil fuels in the Western world. Among the modifications are biogas, alcohol, biodiesel, and manure. Presently, electrical power is attractive in many respects, and the search is on for renewable and nonfinite resources to produce and supplement electrical energy.

The first chapter discusses energy and its biological sources. If biofuel is one of the expected solutions, we must know where is the beginning of the crisis and its solution. This chapter reviews the background story along with an optimistic outlook for a safe energy resource on our green earth. The second chapter discusses energy from photosynthetic plants and their inherent recycling nature, as well as the environmental benefits involved. These sources of energy are the solution for energy management. The third chapter discusses bioethanol, which is now one of the main actors in the fuel market. Its market grew from less than a billion liters in 1975 to more than 39 billion liters in 2006, and is expected to reach 100 billion liters in 2015. The chapter discusses the variety of raw materials, such as sugars, starch, and lignocellulosic substances, that produce bioethanol and also covers some of the market issues. To extend the use of biodiesel, the main concern is the economic viability of producing biodiesel. Edible oils are too valuable for human feeding to run automobiles. So, the emphasis must be on low-cost oils, i.e., nonedible oils, animal fats, and used frying oils. There are many nonedible feedstock crops growing in underdeveloped and developing countries; biodiesel programs here would give multiple social and economic benefits.

The fourth chapter discusses different plant sources used for production of biodiesel, properties of biodiesel, and processing of vegetable oils as biodiesel, and compares engine performance with different biodiesels.

Biodiesel is the methyl or other alkyl esters of vegetable oils, animal fats, or used cooking oils. Biodiesel also contains minor components such as free fatty acids and acylglycerols. Important fuel properties of biodiesel that are determined by the nature of its major and minor components include ignition quality and exhaust emissions, cold flow, oxidative stability, viscosity, and lubricity. The fifth chapter discusses how the major and minor components of biodiesel influence the mentioned properties.

Different techniques of biodiesel preparation and resulting engine performance are discussed in detail in Chap. 6. The seventh chapter discusses ethanol and methanol as fuel in the internal combustion engine and emphasizes their advantages (such as a higher octane number) over gasoline. Cracking of lipids turns polar esters into nonpolar hydrocarbons. This is accompanied by a fundamental change in physical and chemical properties. Products formed give rise to new applications in the fuel sector and for chemical commodities, e.g., detergents. The eighth chapter explores routes to provide these alternative hydrocarbons from lipids. It concentrates on substrates (seeds, vegetable oils, animal fat) and conversion pathways as well as analytical tools.

The ninth chapter discusses the fuel cell, an electrochemical device and nonpolluting alternative energy source that converts the chemical energy of a fuel (hydrogen, natural gas, methanol, gasoline, etc.) and an oxidant (air or oxygen) into electricity with water and heat as by-products.

The book is organized in a manner to cater to the needs of students, researchers, managerial organizations, and readers at large. We welcome the reader's opinions, suggestions, and added information, which will improve future editions and help readers in the future. Readers' benefits will be the best reward for the authors.

This revised edition of *Biofuels Refining and Performance* is based on constructive and helpful suggestions from my beloved teachers and students. The book covers advanced techniques of nonconventional energy refining and engine performance and could be used to educate undergraduate and postgraduate students. It should also be useful for researchers and practicing professionals. The main features of this new edition include the following:

1. Different types of nonconventional energy sources and applications such as solar, wave, wind, and fuel cell system have been added to reflect the rapid development of energy sources and their impact on the global economy.

2. Hydrogen gas is a promising energy option that provides eventual freedom from the difficulties caused by the depletion of natural (fossil) resources and environmental pollution. The most distinctive feature of hydrogen energy is that it can be produced from water; a wealth of fossil and nonfossil sources of energy are covered in the text.

3. Questions to review and test comprehension have been added at the end of all chapters.

I gladly invite constructive suggestions from fellow teachers, students, scientists, and engineers for further improvement of this book.

Ahindra Nag

About the Editor

Ahindra Nag, Ph.D., FIC, MIC, has 35 years of teaching and research experience in the Department of Chemistry at the Indian Institute of Technology, Kharagpur. He is presently a professor in this department. He holds three patents, and has published 90 research papers in major national and international journals. He is also the author of 10 other books, including *Biofuels Refining and Performance; Biosystem Engineering; Solid Waste Management; Computer Aided Drug Design and Delivery System; Asymmetric Synthesis of Drugs and Natural Products;* and *Analytical and Instrumental Techniques in Agriculture, Environmental and Food Engineering*. He has visited the United States, Italy, Taiwan, and Spain as a visiting professor. He is on the editorial board of different renowned journals. He was the key note speaker in different Green Chemistry Conferences in Rome (2017), Dublin (2018), and Zurich (2019) and also one of the organizing members in that conferences.

CHAPTER 1

Introduction to Nonconventional Energy Resources

Ahindra Nag and Jayita Datta

1.1   Introduction

The rapid development of the global economy and increase in the world's population threaten to exhaust fossil fuels at a very fast rate. A challenge for today's scientists is to find nonconventional energy sources that are infinite, natural, and restorable, such as solar energy, wind energy, biofuels, etc.

The energy received on earth from the sun is used in different ways: 50 percent to 60 percent of solar energy is absorbed by the atmosphere, 10 percent heats land and earth, and only 8 percent falls on plants, which use only 0.02 percent during photosynthesis. The availability of energy sources are divided into two groups: (a) **renewable** sources of energy and (b) **nonrenewable** sources of energy. The differences between these two sources of energy are described in Table 1.1.

At present, high oil prices show that energy produced from nonconventional sources, such as biofuel, can be a cost–competitive alternative. Their market development has been taken seriously in recent years, particularly by industrialized countries. However, the competitive edge for liquid biofuel production seems to lie with developing countries that have favorable climatic and environmental conditions for plant growth, low labor costs, low input in agriculture production, and low production costs for energy crops. Interest in such production goes beyond the satisfaction of domestic transport fuel demand. It includes the international trade with liquid biofuels as well as potential by-product usage to support development.

1

Renewable sources	Nonrenewable sources
1. They are nonconventional sources	1. They are conventional sources.
2. These sources can be regenerated such as flowing water (hydroelectricity), biomass (wastes), nuclear fuel, geothermal, etc.	2. These sources cannot be regenerated, such as coal, petroleum, and natural gas.
3. These are the natural sources that are not exhausted with time.	3. These are the natural sources that are exhausted with time.
4. Energy from sources can be obtained continuously over a long period of time.	4. Energy from sources cannot be obtained continuously over a long period of time.

TABLE 1.1   Comparison between renewable and nonrenewable sources

Advantages of Nonconventional Energy

1. Nonconventional energy reduces the dependence on deplorable conventional energy resources, such as oil, natural gas, coal, etc. Renewable resources are replaced through natural processes at a rate equal or greater than the rate in which they are used.

2. Nonconventional energy contributes strength to the energy and security of supply at the national level.

3. Nonconventional sources are pollution-free and inexhaustible sources of energy (sun, wind, rivers, organic matter, animal and farm wastes). They are environment- and human-friendly in nature.

4. Nonconventional sources do not require heavy expenditure since they have low operating costs and are not influenced by the fluctuation of international markets.

5. Nonconventional energy meets a wide range of users' energy needs. They may be used for the regeneration of economically and socially depressed areas through investment such as biogas, geothermal energy, etc. This type of energy creates new jobs at the local level.

Disadvantages of Nonconventional Energy

1. Natural conditions such as storms, high rainfall, cloud cover, and erratic winds may hamper energy development.

2. When the supply is inconsistent and unreliable, large quantities of energy might not be generated.

3. Industrial sectors are still struggling to invest money for the long term. Land acquisition will be a problem for non-conventional energy setups where access to land sources is limited.

Judicious Uses of Energy

The world population will have reached 7.8 billion in 2020 and according to the Organization Petroleum Exporting countrird report (October 2015) that there are 1.5 trillion barrels of crude oil of reserves left in the world. Half of the population will have only 20 GJ/y. Wastage of energy should be avoided. A "firewood crisis" has created a "firewood catastrophe," and the forest cover is diminishing globally at the rate of 250,000 km^2/yr. The cutting of trees must be banned and more trees should be planted.

1.2   Solar Energy

Solar energy is received by the earth from the sun in the form of solar radiation; drying is one of the oldest applications of solar energy. It was used since the dawn of humankind mainly for food preservation but also for drying clothes, construction materials, etc. The first installation for drying by solar energy was found in South France at about 8000 BC, and now this energy utilization is prevalent around the world. The surface of earth receives about 47 percent of total solar energy, and this amount is usable. Generally, the earth absorbs radiation in the visible region and emits radiation in the infrared region. The solar flux density (same as power density) varies between 250 to 2500 kilowatt hours per meter squared per year (k Whm^{-2} per year). The actual solar radiation depends on geographical location, which is highest at the equator, especially in the desert areas.

1.2.1   Measurement of solar radiation

a. **Pyranometer** is a Greek word in which "pyr" means "fire" and "ano" means "above sky." It is designed to measure global radiation (see Fig. 1.1), usually on a horizontal surface and also inclined surface. When shaded from beam radiation by using a shaded ring, a pyranometer measures diffused radiation. There is a thermopile sensor (1) with black coating that absorbs all solar radiation covering 300 to 5000 nanometer range. The black coating on the thermopile sensor, which absorbs the ration, is converted to heat, and this heat flows thermopile sensor generates a voltage output signal that is proportional to solar radiation. The instrument has a voltage output of approximately 9 μ V/W m^2 and has an output independence of 650 Ω. A glass dome shields the thermopile sensor from convection. Pyranometers are used in metrological stations, climatology, and solar energy studies.

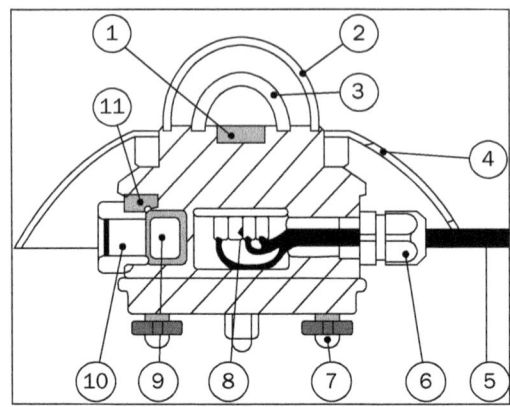

(1) sensor, (2,3) glass dome, (5) cable, standard length 5m, (9) desiccant

FIGURE 1.1   Pyranometer.

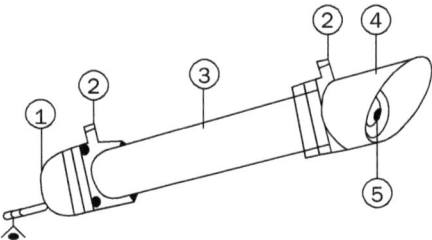

1) Humidity indicator 2) sight 3) body
4) protection cap 5) windows with heater
6) connector 7) cable

FIGURE 1.2   Pyrheliometer.

b. **Pyrheliometer** measures beam radiation by using a long, narrow tube to collect only beam radiation from sun at normal radiation. Both pyranometers and pyrheliometers are used at metrological research centers. A pyrheliometer has a voltage output of approximately 8μ V/W m^2 and has an output independence of $200\,\Omega$. Pyrheliometer (Fig 1.2) measurement applications include scientific meteorological and climate observations, material testing research, and assessment of the efficiency of solar collectors and photovoltaic devices.

c. Sunshine recorder (Fig. 1.3) measures the hours of sunshine in a day. The results provide information about the weather and climate as well as the temperature of a geographical area, which is useful in meteorology, science, agriculture, tourism, and other fields. For the specific purpose of sunshine duration recording, Campbell–Stokes recorders are used, which use a

Figure 1.3 Campbell–Stokes recorder.

spherical glass lens to focus sunrays on a specially designed tape. When the intensity exceeds a predetermined threshold, the tape burns. The work of the Campbell–Stokes recorder (sometimes called a Stokes sphere) is a sunshine recorder. Stokes's refinement was to make the housing out of metal and to have a card holder set behind the sphere. The unit is designed to record the hours of bright sunshine that will burn a hole through the card.

1.2.2 Applications of solar energy

Applications include (i) solar drying of food in agriculture and green house heating; (ii) architecture and urban planning; (iii) cooking of food; (iv) heating, cooling, and ventilation; (v) solar desalination; (vi) solar ponds that combine sensible energy collection and sensible heat storage; (vii) industrial process heat system—even in office buildings, roof areas can be covered with solar panels; (viii) production of high temperature; (ix) pumping of water; (x) solar telephones along highways; (xi) photovoltaic cells; (xii) cars and calculators; and (xiii) electricity generation in remote area where it is expensive to extend an electricity power grid.

1.2.3 Advantages of solar energy

Solar energy is renewable energy and can replace fossil fuel. It is pollution free and does not emit greenhouse gas after installation. Electricity generation by solar panels can be done in remote areas and is safer than traditional electric currents. The technology advancement in solar energy systems makes it cost effective, and most of the systems do not require any maintenance during their life span. It takes less space than an average power station and can be installed anywhere in a field or onto a building. Solar energy can also be stored in batteries for use at night. Solar energy installation creates jobs by employing solar panel manufacturers and solar installers, which, in turn, helps alleviate unemployment and improve economies.

1.2.4 Disadvantages of solar energy

Obviously, solar energy can only be generated during daytime. For night, there is a need for a large battery bank. Cloudy days do not produce high energy, and there is also lower production in winter months. Solar energy depends on geographical location as the size of solar panel varies for the same power generation. There is often high initial cost, and it takes a lot of solar panels initially to be efficient. Devices that run on DC power directly are more expensive, and due to lack of material and cost-lowering technology, solar panels are not being massed produced.

1.2.5 Structure of solar cell

A photovoltaic cell (Fig. 1.4) produces both a current and a voltage to generate electricity. In that process the device's material absorbs the light, which makes the electron create higher energy, and the movement of this higher energy electron from the solar cell into external circuit and the electron dissipates its energy in the external circuit and returns to the solar cell. A typical solar cell is a multilayered unit consisting of the (i) **cover,** which is a clear glass or plastic and transparent adhesive that holds the glass to the rest of the body and provides outer protection; (ii) **antireflective coating** that is designed to prevent the light that strikes the cell from bouncing off so that maximum energy is absorbed in the cell; (iii) **front contact,** which transmits the electric current; (iv) **N-type semiconductors layers,** a layer of silicon mixed with phosphorous by doping process; (v) **P-type semiconductors layers** in which silicon is doped with boron; and (vi) **back contact,** which transmits the light.

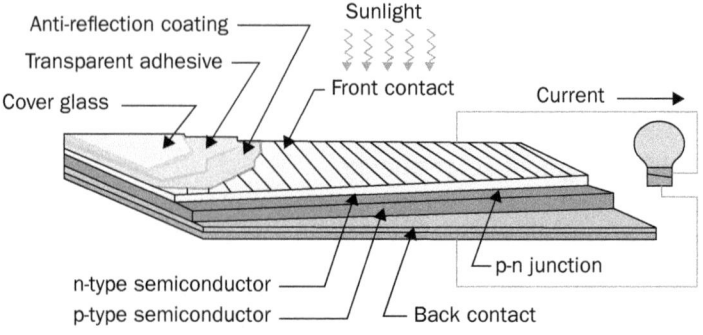

FIGURE 1.4 Structure of solar cell from sunlight energy to electrical energy conversion.

The efficiency of a solar cell is determined as the fraction of incident power, which is converted to electricity and is defined as:

$$P_{max} = V_{OC}I_{SC}FF$$

where:

V_{oc} is the open-circuit voltage; I_{sc} is the short-circuit current; FF is the fill factor; and η is the efficiency.

The input power for efficiency calculations is 1 kW/m² or 100 mW/cm². Thus the input power for a 100 × 100 mm² cell is 10 W. Generally solar cells developed by industries are four types:

   a. 100 mm diameter, round single crystalline

   b. 100 cm² of single crystalline

   c. 100 mm × 100 mm square multi crystalline

   d. 125 mm × 125 mm square multi crystalline

1.2.6   Solar panel and solar thermal collector

a) *Solar panel*

A solar panel is composed of a package of photovoltaic cells in which solar radiation that falls directly is converted into direct current. The electric energy output of each panel varies from 100 to 320. Its main applications are solar-powered radios, solar-powered fans, solar flashlights, solar nightlights, and charging batteries. The efficiency of the panels is also affected by the orientation of the panel, tilt of the roof and panel, and temperature and shade of the roof. The advantages of solar panels include being ecofriendly, low maintenance cost, easy to install, operating efficiently with beam or diffuse solar radiation, and being noise-free with no moving parts (Fig. 1.5).

FIGURE 1.5   Solar panel.

FIGURE 1.6   Solar thermal collector.

Solar panel efficiency (expressed as a percentage) quantifies a solar panel's ability to convert sunlight into electricity, and the more efficient panel will produce more electricity than the less efficient panel.

b) *Solar thermal collector*

A solar thermal collector collects heat by direct absorption of sunlight and is employed in solar power plants for heating water to produce steam, which in turn drives a turbine connected to an electric generator for generating electricity. It has special application in solar-assisted cooling, and the main use of this technology is in residential buildings where the demand for hot water has a large impact on the energy bill. Other uses are in pool heating and for domestic purpose such as heating hot water for washing clothes and showers. Advantages of solar thermal collector are that it has high efficiency, no extra parts except the collector itself, it is sustainable to high temperature, heat can be stored to power generation during overnight or during a cloudy sky, and it can utilize larger areas by inexpensive mirrors (Fig. 1.6).

Disadvantages of using solar collectors are the inability to produce electricity under diffused light conditions and the requirement of sun tracking for maintaining sunlight focus.

The *efficiency* (η) of the *collector* is defined as the ratio of the rate of useful thermal energy leaving the *collector* to the useable solar irradiance.

1.3   Geothermal Energy

Thermal energy is defined as the rise in temperature of a substance due to faster vibration of molecules and atoms. The three types of thermal energy are: (a) conduction, which involves direct contact of atoms;

FIGURE 1.7   Geothermal energy production.

(b) convection, which takes the movement of warm particles; and (c) radiation, which involves the movement of electromagnetic waves.

Geothermal energy comes from the Greek as *geo* (earth) and *therm* (heat), which has been the source of electricity or hot water in the early part of the 21st century. Other recent uses are air conditioning, industrial purposes, greenhouses, aqua culture, hot water resort and pool, and melting snow. Twenty-four countries had 10,715 megawatts (MW) of geothermal power plants, and the United States led the world in geothermal electricity production with 3,086 MW of the installed capacity from 77 power plants.

At the core of earth (4000 miles below the surface) temperatures may reach more than 9000 degrees Fahrenheit. Heat conducts from the core to surrounding rock. Extremely high temperature and pressure cause some rock to melt, which is commonly known as magma and is lighter than solid rock. Magma then rises to the surface. During drilling in geothermal power plants, cooling water gets converted into steam as it comes into contact with magma and rises to the surface. It is relatively simple to capture the steam and use it to drive electric generators (Fig. 1.7).

1.3.1   Advantages of geothermal energy

Some advantages of geothermal energy are as follows: (i) no atmospheric pollution; (ii) the operation process is simple and can be operated continuously for years; (iii) cost is half of that for a hydro-electric power station; (iv) geothermal plants do not have transport fuel like most power plants; and (v) unlike solar energy it does not depend on weather.

1.3.2   Disadvantages of geothermal energy

Geothermal energy also has some disadvantages: (i) it creates noise pollution; (ii) not all areas are suitable for production of electricity as it depends on structural rocks; (iii) earthquakes may occur during drilling or there may be danger of volcano; (iv) even though operating cost is low, cost of setting up and scaling up of geothermal plant is high.

1.3.3   Application of geothermal energy

Use of geothermal energy in power plants to make energy is as follows: (i) **Dry steam** where steam is produced directly from the reservoir to run the turbine, which powers the generator. Here no separation is needed because drilling wells only produce steam. (ii) **Flash steam** where heated water is separated in a vessel called a separator into the steam and hot water (called brine). Steam is delivered to the turbine, which powers the generator. The liquid is injected back into the reservoir. (iii) **Binary cycle** is a well accepted power plant worldwide. Here geothermal water heats another liquid such as isobutene or pentafluropropane, which boils at a lower temperature than water. The two liquids are completely separated through a heat exchanger, and heat exchange will take from hot water to the working fluid, which expands into gaseous vapor. The force of expanding steam turns the turbine that powers the generator. All the produced geothermal water is injected back into the reservoir. No harmful gas is emitted to the atmosphere as the underground water is never disclosed outside. Here initial cost is high but in the long run the cost of electricity becomes comparable to other energy resources.

1.4   Tidal Energy

Tidal energy is a form of hydropower that converts the energy of tides into power. Tides are waves caused by the gravitational pull of moon and sun. There are two tides as high tides and low tides during the rotation of earth. This building up and receding of wave happens twice a day. In any month at the full moon and new moon, the line joining the centers of the earth, sun, and moon are in straight line. On these days, the sun's and moon's attraction act in directly additive manners and result in tides particularly higher than normal, which are called spring tides. During high tides water flows to a reservoir and during low tides this water flows from the reservoir. This flowing water can be used to turn a turbine to produce electricity. The mean tidal range varies from place to place (Fig. 1.8).

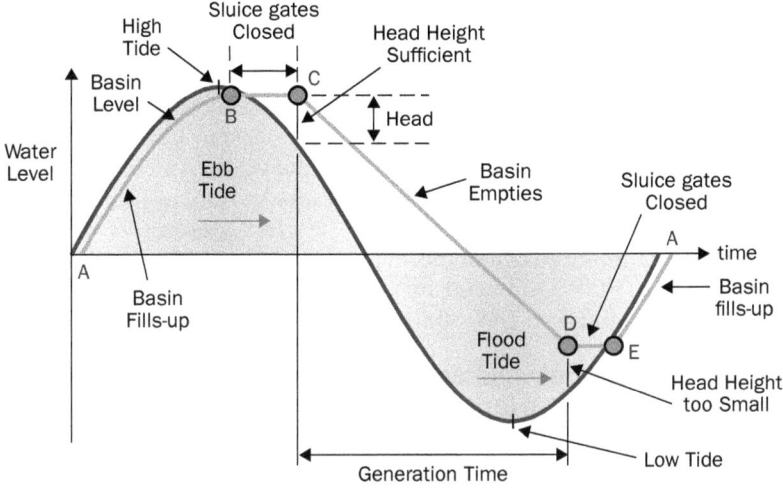

Figure 1.8   How high and low tides are formed.

1.4.1   Advantages of tidal energy

There are several advantages to tidal energy: (i) it is ecofriendly energy and does not produce greenhouse gases; (ii) there is scope to generate the energy in large scale as it is a source of inexhaustible energy; (iii) efficiency of tidal energy is greater as compared to coal, solar, and wind energy, which is around 80 percent efficiency; and (iv) its maintenance cost is very low.

It has the following uses: (i) **tidal electricity,** where tidal energy is used for generation of electricity; (ii) **grain mills,** where this energy has been used for mechanical crushing of grains; and (iii) **energy storage,** where this energy can be used as store of energy. Tidal energy barrages with their reservoir can be modified to store energy.

1.4.2   Disadvantages of tidal energy

Disadvantages of tidal energy include the following: (i) initial cost of construction is very high; (ii) intensity of sea waves is unpredictable and may damage the power generation; (iii) to find out the ideal location for construction is difficult and too localized to coastal regions only; (iv) it may influence aquatic life adversely, especially fish; (v) generation of tides are twice a day and hence electricity can be produced that time; (vi) sometimes frozen sea, low and weak tides are some of its obstruction; (vii) waves are not perfectly sinusoidal; and (viii) design to withstand storms from the substation as control section in order to detect and resolve any defect. Insufficient of wind flow the backup is taken in the form of (i) battery storage, (ii) connection with the local electricity distribution system, or (iii) a standby generator powered by liquid or gaseous fuels.

1.5   Wind Energy

Wind energy is another form of renewable energy where the kinetic energy of wind is transformed into mechanical or electrical energy that can be harnessed for practical uses. Seashore windmills set up in Holland made the country become one of the world's most industrialized countries by the 17th century. Global wind power generation amounted to 950 TWh in 2015, nearly 4 percent of total global power generation. China has the largest wind energy installed capacity with 145 GW, followed by the United States with 73 GW, Germany 45 GW, India 25 GW, Spain 23 GW, and the UK 14 GW.

1.5.1   Advantages of wind energy technology

Wind energy technology has several advantages: (i) it reduces utilization of fossil fuels and saves energy imports; (ii) it is renewable and inexhaustible; (iii) it is a clean form of energy—wind power could be supplying up to 19 percent of the world's electricity and avoiding more than three billion tons of CO_2 a year, and by 2050, 25 percent to 30 percent of global power could come from harnessing the wind (the wind industry has grown at around 26 percent per year over the past 18 years; Europe and China have been solid wind markets for over a decade); (iv) its maintenance costs after initial construction are very low as wind also has relatively low operations; (v) it can be used as hybrid energy source, e.g., wind–diesel, wind–photovoltaic; (vi) electricity can be produced at lower costs after installation, which are economically competitive; (vii) windmills are highly economical in the rural areas far from existing grids, which may be useful for power supply at offshore and onshore (such as hilly regions, etc.) sites; and (viii) it creates new local employment system and contributes sustainable development. Around 600,000 people currently work in the wind power industry. That figure could rise to around 1.5 million by 2020 and exceed 2 million jobs by 2030.

1.5.2   Disadvantages of wind energy technology

There are also disadvantages to wind energy technology: (i) generally, site selection is a problem as an average wind speed of 14 mph is needed to convert wind energy into electricity economically; (ii) wind speed varies from season to season causing intermittency issues of power grids; (iii) due to the fluctuating nature of wind blowing, the produced mechanical energy has to be stored by some means; e.g., battery storage; (iv) wind power is not steady and consistent, which results in complications while designing the whole plant; (v) there are often problems regarding land use and noise—a wind plant's impact on bird populations and its visual impact on the surrounding landscape are environmental drawbacks, as well; and (vi) wind plants also need a lot of land. One wind machine needs about two acres of land to call its own.

FIGURE **1.9**   Wind energy plant.

1.5.3   Output determination of windmill

Three factors determine the output from a wind energy converter: (i) wind speed; (ii) cross section of wind speed of rotor; and (iii) overall conversion, efficiency of the rotor, transmission system, and generator or pump (Fig. 1.9).

1.5.4   Different systems in wind energy

a) *Wind turbine*

Wind turbines can be connected to a utility power grid or even combined with a solar cell system. A wind turbine is a device that converts kinetic energy from wind to electricity. In that process, the blades of a wind turbine turn between 13 to 20 revolutions per minute depending on the technology set up at a constant or variable velocity, where the velocity of the rotor varies in relation to the velocity of the wind to reach a greater efficiency. Wind turbines are mounted on a tower to capture the wind energy. Usually blades are mounted on a shaft to rotors. A single 1 MW windmill operating at a 45 percent reduction rate will generate about 3.9 million KW of electricity in a year, which would enough for about 500 households per year.

b) *Wind turbine blades*

Generally, wind turbine blades are shaped to generate more efficient blade design and the maximum power from the wind at the minimum construction cost. A blade acts much like an airplane wing. When the wind blows, a pocket of low-pressure air forms on the down side of the blade and the low pressure air pocket then pulls the blade toward it, causing the rotor to turn, which is called lift. The force of the lift is actually much stronger than the wind's force against the front side of the blade, which is called drag. The combination of lift and drag causes to rotor to spin like a propeller, which turns the shaft of a generator to make electricity.

There are two types of blades: flat rotor blades and curved rotor blades. The advantage of a curved rotor blade compared to a flat blade is that lift forces allow the blade tips of a wind turbine to move faster than the wind is moving, generating more power and higher efficiencies. As a result, lift-based wind turbine blades are becoming more common now. There are two types of forces acting on blades and thrust.

(i) Circumferential force, which is acting in the direction of wheel rotation that provides the torque. The circumferential force can be obtained from

$$T = \frac{P}{\omega} = \frac{P}{\pi\,DN}$$

where T = torque kgf or Newtone (N)
ω = angular velocity f wheel, m/s
D = diameter of turbine wheel
N = wheel revolution per minute time s^{-1}

$$\text{The real efficiency} = \frac{\eta P}{P_{total}}$$

$$P = h P_{total} = \frac{\eta.1\rho AV3^{i}}{2g_c}$$

For a turbine operating at power P, the expression for torque becomes

$$T = \frac{\eta.1\rho AV3^{i}}{2g_c\,DN}$$

$$\text{For D} = \sqrt{\frac{4}{\pi}}\ \text{A.m}$$

$$T = \frac{\eta.1\rho AV3^{i}}{8g_c\,DN}$$

(ii) Axial force, which is acting in the direction of the wind stream, provides an axial thrust that must be counteracted by mechanical proper design.

The axial force on a turbine wheel operating at a maximum efficiency:

$$Fx = \frac{\pi \rho D^2 V_i^2}{9g_c}$$

The axial forces are proportional to the square of the diameter of turbine wheel; this limits turbine wheel diameter of large size.

c) Rotor

The rotor is connected to a gear box that lifts the turning velocity from 13 to 1,500 revolutions per minute. A wind's turbine rotor may have any number of blades from wood, metal, or composites of several materials. Generally the rotor is mounted on a horizontal shaft and is further connected either to a generator or some mechanical devices such as a water pump or heat generators. The whole assembly is mounted on a tower so that it turns clear of the ground and away from surface drag and interference. Generally performance increases with the increase of tower because wind speed increases with height. All the critical functions of wind turbines are monitored from the substation as control section to detect and resolve any defect. In sufficient of wind flow the backup is taken in the form of (i) battery storage, (ii) connection with the local electricity distribution system, or (iii) a standby generator powered by liquid or gaseous fuels.

1.6 Biofuels

Biofuels are a challenging arena in which academia, industry, and governments are joining efforts to find alternative and sustainable sources of energy to replace fossil fuels and reduce waste, pollution, and the greenhouse effect and global warming. Energy can be produced from biomass, which is a natural resource from animals and plants such as crop residue, fuel wood urban refuse, municipal sewage, aquatic plants, and livestock manure. Biofuels production should also lead to some benefits for local communities (including food availability, new job opportunities, biodiversity perseverance, quality of water, and soils used). The European Union (EU) has adopted a resolution that 10 percent of the fuels used in the transportation sector should be of renewable sources by 2020. Solid state fermentation, waste valorization, and biogas are all topics of fundamental importance in the field of biofuels.

The prospects of ethanol and biodiesel as substitutes for conventional fuels will not be discussed here; these two aspects are presented in sufficient detailed in Chaps. 3, 4, 5, 6, 7, and 8. One of the promising approaches for future fuel is, perhaps, hydrogen and methane, both of which could be obtained from living, particularly microbial resources, which is discussed in Chap. 10.

1.6.1   Advantages of biomass as fuel

Biomass as fuel can do the following: (i) it reduces the environmental hazards and has the potential to reduce the greenhouse gas emission; (ii) it provides an expensive fuel, at times with an effective low sulphur fuel, and is very effective for poor rural people with plenty of feed stock; and (iii) production of biological fuel may be coupled to the synthesis of proteins.

1.6.2   Disadvantages of biomass as fuel

There are also disadvantages of biomass as fuel: (i) a high amount of moisture is present, which requires expensive collection and transport; (ii) high skill and design are required as the energy conversion is expensive; (iii) scarcity of additional land is required for suitable growing plants; (iv) there is relatively low concentration of biomass per unit area of land and water. The burning of biomass causes atmospheric pollution and a lot of carbon monoxide is produced.

1.7   Biogas

Biogas is a renewable energy source and is using nature's elegant tendency to recycle substances into productive resources. This process is known as anaerobic digestion. Biogas is principally a mixture of methane (CH_4 containing 55 percent to 90 percent by volume biogas) and carbon dioxide (CO_2) along with other trace of gases. Biogas, like natural gas, (i) has a low volumetric energy density compared to the liquid biofuels, ethanol, and biodiesel; (ii) is comparatively simple and can be produced easily; (iii) reduces deforestation; (iv) burns without smoke and leaving no residue; (v) household waste and biowastes can be disposed in a healthy manner; and (vi) slurry from the biogas plant is excellent manure. It can be used in trains and buses (Fig. 1.10).

Biogas can be produced by degradation of organic matter (from animals and plants) in an anaerobic environment. Plant materials with high lignin content are an inferior type of feed for such reactions. Lignocelluloses biomass mainly contains cellulose, hemicelluloses, and lignin and has the potential to be used as raw material for biogas production. However, the compact crystalline structure, the fact that lignin physically shields the cellulose and hemicellulose parts, makes these materials more resistant to anaerobic digestion and thus lower the efficiency of the process.

Applications

A biogas bus, Sweden

The Biogas Train "Amanda"
Sweden

Figure 1.10 Examples of biogas in use in Sweden.

$$
\begin{array}{ccc}
\textbf{Waste} & \rightarrow & \textbf{Biogas plant} \rightarrow \textbf{Biogas} \\
\textbf{(collected from} & & \downarrow \\
\textbf{plants and animal feed stock)} & & \textbf{Sludge manure}
\end{array}
$$

The results of the investigations showed that by using a pretreatment method, the (enzymatic) hydrolysis of cellulose and hemicelluloses could be noticeably accelerated, providing greater yields during the following anaerobic digestion. A suitable pretreatment method is needed to break down the structure, remove the lignin, and reduce the crystallinity, which will enhance the solubilization of the material resulting in improved methane production during the subsequent anaerobic digestion process. Generally the process is three steps:

(1) Hydrolytic stage → (2) Acetogenic stage → (3) Methanogenic stage

(Fermentive stage)   (Thermophilic)   (Mesophilic)

35°C, pH 5-6   45°C, pH 4-6

Organic matter → Acetic acid, H_2, CO_2 → Methane, Co_2

An oversimplified mass balance may be written as:

$$ C_6H_{12}O_6 \quad \rightarrow \quad 3CH_4 + 3CO_2 $$

The technical values of yield coefficient, biological efficiency, chemical/biological oxygen demand (COD/BOD), biological efficiency in productivity/ecologic efficiency rate (BEP/EER) ratios, and so forth are yet to be determined for each setup or system.

(1) Hydrolytic stage

Chemical acid treatments and enzymatic methods are the most common methods for hydrolyzing cellulose. For acid treatment concentrated acids such as sulphuric or hydrochloric acid (in the range of 10 percent to 30 percent) at temperatures of about 140 to 160°C and pressures about 10 atmosphere are conducted. H_2SO_4 is the usual acid employed although HCl, HNO_3, and H_3PO_4 are also used. The mechanism involved in the acid hydrolysis is that it catalyzes the breakdown of long hemicellulose chains to form shorter chain oligomers and then to sugar monomers. In this process, acid concentration, temperature, and pressure are controlled to avoid sugar and lignin degradation into by-products.

Enzymatic hydrolysis is performed under mild conditions around pH 5 and temperature less than 50°C generally by cellulose enzymes, which correspond to a mixture of several enzymes. Specific microorganisms or hemicellulases produced by many species of bacteria and fungi, as well as by several plants, can also promote the hemicellulose hydrolysis. Examples of a number of hemicellulases, including xylanases and mannanases, have been identified in *Trichoderma reesei* or *Aspergillus* strains. Enzymatic treatment is better than acid treatment because of its high specificity, low energy consumption, and no chemical requirement.

(2) Acetogenic stage

In this process organic monomers can also be degraded to acetate via acetogenesis, which refers to the synthesis of acetate, which includes the formation of acetate by the reduction of CO_2 and the formation of acetate from organic acid. Acetogens that oxidize organic acids obligatory use hydrogen ions and carbon dioxide as electron acceptor. Acetogenic bacteria can only derive energy for growth from these conversions if the concentration of the products is kept low and this results in an obligate dependence of acetogenic bacteria on methanogenic archaea. In a typical acetate fermentation reaction, glucose obtained from first-stage reaction is converted to two moles of acetate and two moles of CO_2. However, acetogens can reduce the two CO_2 molecules to acetate yielding a total of three moles of acetate. Acetogenic bacteria are widespread in nature and are an essential link in the anaerobic mineralization of organic matter. These bacteria are nutritionally versatile and can grow heterotrophically as well as lithoautotrophically. Mostly obligate anaerobes and a few facultative microbes contributing to these conversions belong to different genera. A few may be mentioned: *Actinomyces, Aerobacter, Aeromonas, Arthrobacter, Bacillus, Bacteroides, Cellulomonas, Citrobacter, Clostridium, Corynebacterium, Enterobacter, Escherichia, Klebsiella, Lactobacillus, Laptospira, Micrococcus, Nocardia, Peptococeus, Proteus, Pseudomonas, Ruminococcus, Sarcina, Staphylococcus, Streptococcus, Streptomyces*, and many others.

(3) Methanogenic stage

Methanogens are microorganisms that produce methane as ameta-bolic by-product in hypoxic conditions (low oxygen conditions). A few methanogenic species are also known: *Methanobacterium bryantii, Methanococcus vanniellii, Methano-genum aggregans, Methanomicro-bium mobile, Methanosarcina barkeri, Methanothrixconcillii,* usually eukary-otic organisms, and blue-green algae are incapable of performing such bioconversions. Morphologically, the organisms belong to wide groups: coccus, sarcina (flower-like), rod, filamentous, and other shapes. G_C (guanine-cytosine) values of DNA of these organisms also suggest that they all have varied origin and hence are likely to have different metabolic patterns. Khan (1980) found that *Acetivibrio cellulolyticus* producing acetic acid and hydrogen from cellulose are readily utilized by *M. Barkeri* to produce methane and carbon dioxide. It has been established beyond doubt that the process is chemolithotrophic metabolism, favored by strict anaerobic condition and facilitated by the absence of sulfates, abundance of moisture, approximate temperature range of 25–40°C (37°C), and pH 6.2–8.0 (pH 6.8). The organic materials on which these organisms survive and grow are usually cellulose in nature. Crop residues, agricultural residues, animal excreta, municipal sewage, and other organic mate-rials derived from terrestrial and aquatic origin are also considered as good substrates. But the organic components of wastewater vary vastly. Chemical structures of the organic matter select for specific methanogens to perform anaerobic digestion is very difficult. In this respect, animal excreta appear to be a ready-made substrate.

The important factors of biogas plant design are (a) the tank where biomass is mixed with water (mixing tank design), (b) the tank where biomass undergoes decomposition (digester), (c) the tank where digested biomass (slurry) is collected (outflow tank), and (d) the storage tank of gas. However, biogas may be purified to a natural gas equivalent fuel for pipeline injection and further compressed for use as a transportation fuel.

Biogas plant may be (1) batch type, (2) continuous type, or (3) fixed type digester or dome type.

1. In **batch type** plant the biomass in batches is fed in a tank with large interval time between two batches and the retention time may vary from 30 to 50 days to complete the fermentation. Biogas is derived from the digester from 10 to 15 days intermittently. In batch type digester needs initial seeding for fermentation. There may be several digesters fed sequentially and discharged in sequential manner to get biogas continuously. After completion of digestion the digester residue is emptied and fresh biomass is added. In this process initial cost is high and maintenance is more complex.

2. **Continuous type** of biogas plant are two types: (a) single stage continuous type biogas plant and (b) two stage continuous type biogas plant where biogas is continuously delivering.

a. Single stage continuous type biogas plants: In these type of plants acid formation and methanation are carried out in the same chamber. This process is preferred for a medium gas plant in which the plant is simple, low cost, and easy to operate.

b. Two stage continuous type biogas plants have two chambers where acid formation is carried out in one chamber and methanation is carried out in another chamber. The gas is collected from the second chamber after digestion of biomass.

c. A fixed-dome plant is comprised of a closed, dome-shaped digester (Fig. 1.11) with no moving parts and no steel rusting parts and the plant is constructed underground, protecting it from physical damage. No day/night fluctuations of temperature in the digester positively influence the bacteriological processes. Biogas is collected in the upper portion of the digester in a dome-shaped cavity. The excess slurry in the digester gets accumulated in the displacement chamber. The advantages of these plants have low initial costs and long, useful life-span; no moving or rusting parts are involved. The construction of fixed dome plants is labor-intensive, thus creating

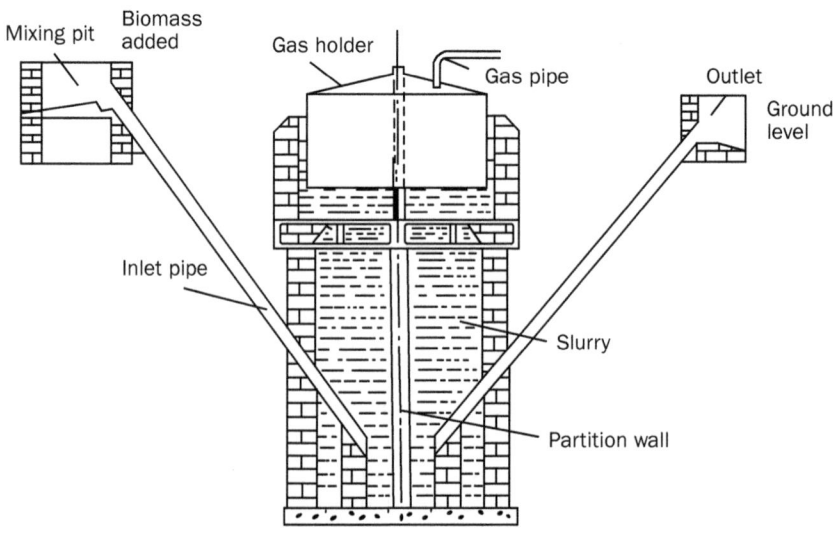

FIGURE 1.11 Fixed dome type digester.

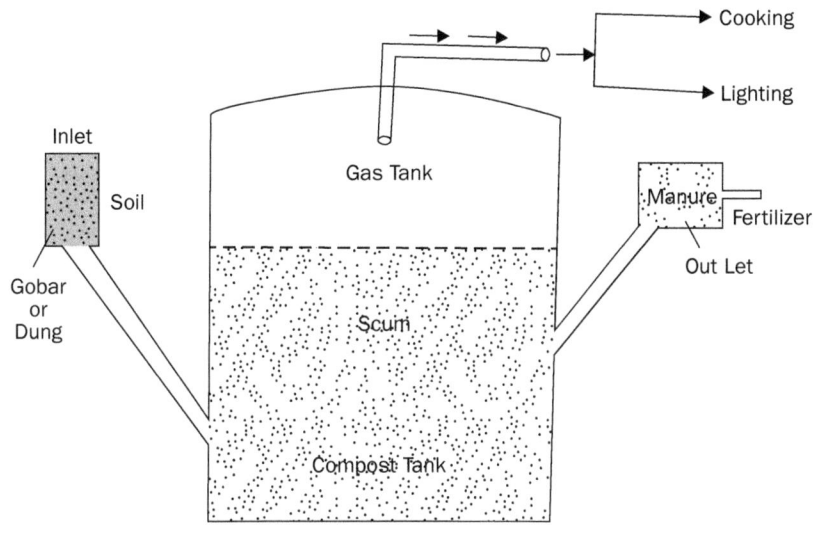

FIGURE **1.12**   Gobar gas.

skilled local employment. The underground construction saves space and protects the digester from temperature change. But disadvantages include tight construction and frequent gas leaks. The gas pressure fluctuates substantially depending on the volume of the stored gas. It complicates gas utilization and amount of gas produced is not immediately visible.

1.8   Gobar gas

The art of producing gaseous fuel out of cattle excreta is well known in the Indian subcontinent as the **gobar gas** (see Fig. 1.12). The ecologic role of biogas is manifold. Chemical anoxic transformation reduces the BOD value of the organic residues, which in turn are enriched proportionately in C, N, P, and mineral ratios. In lignocellulosics, after the anoxic process, enrichment of lignin occurs and may lead to peat formation. This may be the origin of coal; natural gas and coal deposits are likely to be found within a reasonable stretch. This is a built-in machinery of nature for BOD and pollution control.

Usually, when cattle excreta (gobar) is the starting material for anoxic fermentation to flammable gas, it is called **gobar gas**. Before a scientific and technical approach was given to this promising field, the technique was developed with dehydrated animal excreta, which when ignited produces fumes and burn for a short duration with a partially sooty flame a little above the solid fuel. Slurred excreta, when stored in closed earthen vessels for a while, produced flammable gas.

Based on these observations, villagers developed techniques of producing gas similar to illicit brewing.

The important factors of gobar gas projects to consider include

a. Dehydrated cow dung is a popular fuel and does not need special or expensive containers for keeping throughout the year.

b. Untended herds make the collection of dung laborious and cost intensive.

c. Installation of community biogas plants is not easy. Due to the fragmentized small households, individual plants are also difficult to erect. Most families cannot provide the minimum 50-kg average dung put into the plant. About 50 L of water should also go with it. Fifty percent of the settlements are located in drought-prone areas. The remaining 50 percent face water shortage during the 5 months of dry season.

d. Temperature fluctuations throughout the year are significant and affect the rate of biogas production.

The daily output of dung from an average of five cattle (a minimum of four) may suffice for a household with a miniature gobar gas plant. When underground ambient conditions (30° C) are favorable, at least 2.7 m^3 of gas (50 m^3/ton of wet dung) per day is expected out of the plant. This gas has a minimum of 9500 kcal (3500 kcal/m^3) of heat value (equivalent to 1.5 L of kerosene), which may serve the daily need of a five-member family. It is estimated that the average daily requirements of the gas per adult per day are 0.3 m^3 for cooking and 0.2 m^3 for lighting purposes.

Installation of a 3 m^3 digester (gobar gas plant) will require about 50–60 kg (4 buckets) of raw wet cattle dung and an equal amount of water. If the dung is slurried prior to feeding the digester plant, stirring may not be needed. Initially, a 15-day incubation is necessary and combustible gas starts coming out after about 3 weeks, when stabilized, and will continue to produce a gas mixture that is satisfactorily flammable. The average retention time of the materials in the digester is 3–7 weeks (average 5 weeks). The temperature is between 15–65°C with pH varying between 6.5 and 7.

1.9   Fuel Cell

Global primary energy consumption (i.e., energy used for space heating, transportation, generating electricity, etc.) is expected to triple from about 400 exajoules (EJ = 10^{18} joules) per year in 2000 to about 1200 EJ/yr in 2050 at the present rate of increase in consumption. However, due to increased energy efficiency of the devices, the actual increase is expected to be about 800–1000 EJ.

More than 80 percent of the present primary energy requirements are met by fossil fuels. The consequences of burning hydrocarbons at such a large scale for our energy needs are already evident in the form of global warming and its disastrous environmental effects. To permit stabilization of anthropogenic greenhouse gases, fossil fuel consumption will have to be limited to about 300 EJ/yr by 2050. Hopefully, the concern about global warming, limit on fossil fuel supplies, and rise in their prices will force us to gradually decrease the use of fossil fuels in the future. Reducing hydrocarbon consumption to 300 EJ requires carbon-free energy sources to supply the difference ~700 EJ/yr. This shortfall is a problem that requires immediate attention and proactive action for sustainable development.

The need for an efficient, nonpolluting energy source for transportation, large-scale generation, and portable devices has spurred the development of alternative energy sources. Fuel cells are a promising alternative energy source that fits the above requirements. A fuel cell is an electrochemical device that converts the chemical energy of a fuel (hydrogen, natural gas, methanol, gasoline, etc.) and an oxidant (air or oxygen) into electricity, with water and heat as by-products. Since no combustion is involved in the hydrogen fuel cell process, no NO_x are generated. Since sulfur is a poison to fuel cells, it has to be removed from fuel before feeding it to a fuel cell; therefore, no SO_2 is generated in the fuel cell.

The trend toward portability and miniaturization of computing and communication devices has created a requirement for small and lightweight power sources that can operate for long periods of time without any refill or replacement. Also, advances in the medical sciences are leading to an increasing number of electrically operated implantable devices such as pacemakers, which need power supplies to operate for an extremely long duration (years) without maintenance, as any maintenance would necessitate surgery. Ideally, implanted devices would be able to take advantage of the natural fuel substances found in the body. The idea of a biofuel cell that can generate electricity based on various metabolic processes occurring in our own cells is appealing. A biofuel cell converts chemical energy to electrical energy by the catalytic reaction of microorganisms. Most microbial cells are electrochemically inactive, and electron transfer from microbial cells to the electrode requires mediators such as thionine, methyl viologen, methylene blue, humic acid, and neutral red. In recent years, mediatorless microbial fuel cells have also been developed; these cells use electrochemically active bacteria (*Shewanella putrefaciens*, *Aeromonas hydrophila*, etc.) to transfer electrons to the electrode. A major advantage of the biofuel cell over the hydrogen fuel cell is the replacement of expensive and precious platinum (Pt) as a catalyst by much cheaper hydrogenase enzymes. A brief description of the development and state of the art of hydrogen and biofuel cells is presented in this chapter.

1.9.1   Fuel cell basics

Although fuel cells have been around for more than a century (William Grove first discovered the principle of the fuel cell in 1839), it was not until the National Aeronautics and Space Administration (NASA) demonstrated its potential applications in providing power during space flights in the 1960s that fuel cells became widely known and the industry began to recognize the commercial potential of fuel cells. Initially, fuel cells were not economically competitive with existing energy technologies; but with advancements in fuel cell technology, it is now becoming competitive for some niche applications.

The main components of a fuel cell are anode, anodic catalyst layer, electrolyte, cathodic catalyst layer, and cathode, as shown in Fig. 1.13.

The anode and cathode consist of porous gas diffusion layers, usually made of high-electron-conductivity materials such as thin layers of porous graphite. The most common catalyst is platinum for low-temperature fuel cells. Nickel is preferred for high-temperature fuel cells. Some other materials (Pt-Pt/Ru, Perovskites, etc.) are also used, depending on the fuel cell type .

The electrolyte is made up of materials that provide high proton conductivity and zero or very low electron conductivity. The charge carriers (from the anode to the cathode or vice versa) are different, depending on the type of fuel cell. A fuel cell stack is obtained by connecting such fuel cells in series/parallel to yield the desired voltage and current outputs (Fig. 1.14). The bipolar plates (or interconnects) collect the electrical current and also distribute and separate reactive

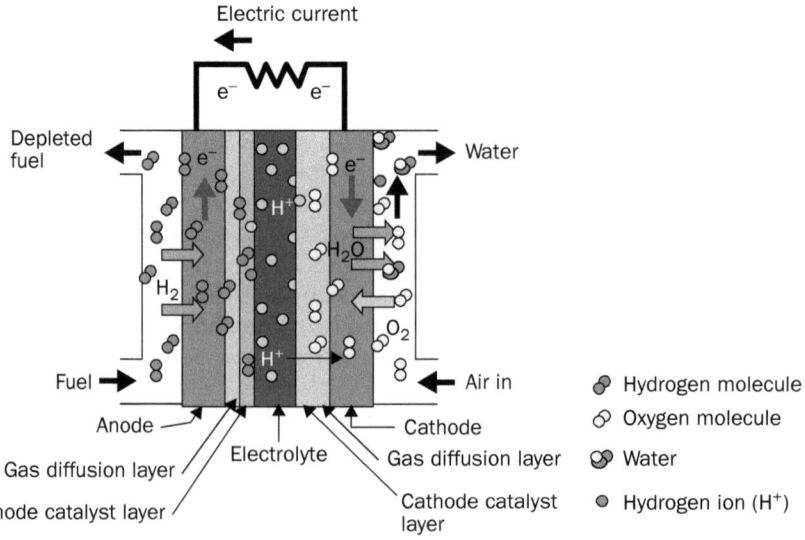

FIGURE 1.13   Generic H_2-O_2 fuel cell.

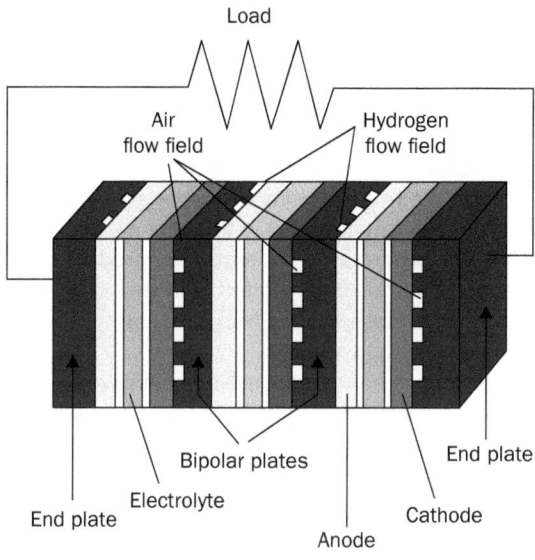

FIGURE 1.14   A fuel cell stack.

gases in the fuel cell stack. Sometimes, gaskets for sealing/preventing leakage of gases between anode and cathode are also used.

The anode reaction in hydrogen fuel cells is direct oxidation of hydrogen. For fuel cells using hydrocarbon fuels, the anodic half reaction consists of indirect oxidation through a reforming step.

In most fuel cells, the cathode reaction is oxygen (air) reduction. The overall reaction for hydrogen fuel cells is

$$H_2 + \tfrac{1}{2}O_2 \rightarrow H_2O \text{ with } \Delta G = -237 \text{ kJ/mol}$$

where AG is the change in Gibbs free energy of formation. The reaction product is water released at the cathode or anode, depending on the type of fuel cell.

For an ideal fuel cell, the theoretical voltage E_0 under standard conditions of 25°C and 1 atm pressure is 1.23 V, whereas typical operating voltage for high-performance fuel cells is ~0.7 V. Stack voltage depends on the number of cells in a series in a stack. Cell current depends on the cross-sectional area (the size) of a cell.

Fuel cell systems are not limited by Carnot cycle efficiency. Therefore, a fuel cell system with a combined cycle and/or cogeneration has high efficiency (55 percent to 85 percent) as compared to the efficiency of about 30 percent to 40 percent of current power generation systems. In a distributed generation system, fuel cells can reduce costly transmission line installation and transmission losses. There are no moving parts in a fuel cell and few moving parts (compressors, fans, etc.) in a fuel cell system. Therefore, it has higher reliability compared to an internal combustion or gas turbine power plant.

Fuel cell-based power plants have no emissions when pure hydrogen and oxygen are used as fuel. However, if fossil fuels are used for generating hydrogen, fuel cell power plants produce CO_2 emissions. Compared to a steam power plant, a fuel cell plant has very low water usage; water/steam is a reaction product in a fuel cell. This clean water/steam does not require any pretreatment and can be used for reactant humidification and cogeneration. Another advantage of the fuel cell power plant is that it does not produce any solid waste and its operation is very silent compared to a steam/gas turbine power plant. The noise generated in a fuel cell power plant is only from the fan/compressor used for pumping/pressurizing the fuel and the air supply to the cathode.

A fuel cell power plant has good load-following capability (it can quickly increase or decrease its output in response to load changes). The modular construction of fuel cell plants provides good planning flexibility (new units can be added to meet the growth in electric demand when needed), and its performance is independent of the power plant size (efficiency does not vary with variation in size from W to MW size).

The major technical challenges in fuel cell commercialization at present are (1) high cost, (2) durability, and (3) hydrogen availability and infrastructure. For fuel cells to compete with contemporary power generation technology, they have to become competitive in terms of the cost per kilowatt required to purchase and install a power system. A fuel cell system needs to cost ~$30/kW to be competitive for transportation applications and for stationary systems; the acceptable price range is $400–$750/kW for widespread commercial application. Fuel cell technology needs a few breakthroughs in development to become competitive with other advanced power generation technologies.

1.9.2   Types of fuel cells

Fuel cells are classified primarily on the basis of the electrolyte they use. The electrolyte is the heart of the fuel cell as it decides the important operating parameters such as the electrochemical reactions that take place in the cell, the type of catalysts required, the temperature range of cell operation, and the fuel (reactants) to be used, and therefore the applications for which these cells are most suitable. There are several types of fuel cells currently under development; a few of the most promising types include

- polymer electrolyte membrane fuel cells (PEMFCs)
- direct methanol fuel cells (DMFCs)
- alkaline electrolyte fuel cells (AFCs)
- phosphoric acid fuel cells (PAFCs)

- molten carbonate fuel cells (MCFCs)
- solid oxide fuel cells (SOFCs)
- biofuel cells

1.9.3 Polymer electrolyte membrane fuel cell (PEMFC)

The PEMFC uses a solid polymer membrane as an electrolyte. The main components of this fuel cell are an electron-conducting anode consisting of a porous gas diffusion layer as an electrode and an anodic catalyst layer, a proton-conducting electrolyte, a hydrated solid membrane, an electron-conducting cathode consisting of a cathodic catalyst layer and a porous gas diffusion layer as an electrode, and current collectors with the reactant gas flow fields (Fig. 1.15).

In the PEMFC, platinum or platinum alloys in nanometer-size particles are used as the electrocatalysts with NafionTM (a DuPont trademark) membranes. The polymer electrolyte membranes have some unusual properties: In a hydrated membrane, the negative ions are rigidly held within its structure and are not allowed to pass through. Only the positive ions contained within the membrane are mobile and free to carry a positive charge through the membrane. The proton exchange membrane (PEM) is a good conductor of hydrogen ions (protons), but it does not allow the flow of electrons through the electrolyte membrane. As the electrons cannot pass through the membrane, electrons produced at the anode side of the cell must travel through an external wire to the cathode side of the cell to complete the electrical circuit in the cell.

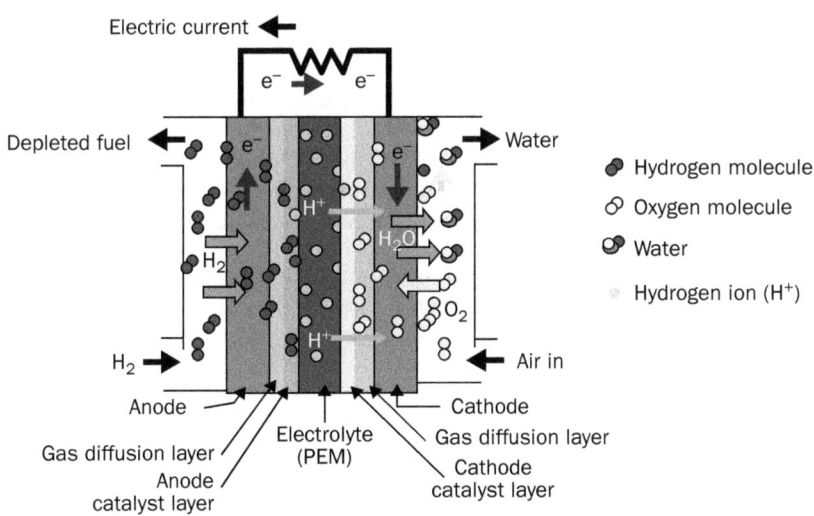

FIGURE 1.15 Polymer electrolyte membrane fuel cell.

In the PEMFC, the positive ions moving through the electrolyte are hydrogen ions, or protons. Therefore, the PEMFC is also called a proton exchange membrane fuel cell. The polymer electrolyte membrane is also an effective gas separator; it keeps the hydrogen fuel separated from the oxidant air. This feature is essential for the efficient operation of a fuel cell.

The heart of a PEMFC is the membrane electrode assembly (MEA), consisting of the anode–electrolyte–cathode assembly that is only a few hundred microns thick.

Electrochemistry of PEM fuel cells

All electrochemical reactions consist of two separate reactions: an oxidation half reaction occurring at the anode and a reduction half reaction occurring at the cathode.

$$\text{Oxidation half reaction: } 2H_2 \longrightarrow 4H^+ + 4e^-$$
$$\text{Reduction half reaction: } O_2 + 4H^+ + 4e^- \longrightarrow 2H_2O$$
$$\text{Overall cell reaction: } 2H_2 + O_2 \longrightarrow 2H_2O$$

The H_2 half reaction.   At the anode, hydrogen (H_2) gas molecules diffuse through the porous electrode until they encounter a platinum (Pt) particle. Pt catalyzes the dissociation of the H_2 molecule into two hydrogen atoms (H) bonded to two neighboring Pt atoms; here each H atom releases an electron to form a hydrogen ion (H^+). These H^+ ions move through the hydrated membrane to the cathode while the electrons pass from the anode through the external circuit to the cathode, resulting in a flow of current in the circuit.

The O_2 half reaction.   The reaction of one oxygen (O_2) molecule at the cathode is a four-electron reduction process that occurs in a multistep sequence. The catalysts capable of generating high rates of O_2 reduction at relatively low temperatures (~80°C) appear to be the Pt-based expensive catalysts. The performance of the PEMFCs is limited primarily by the slow rate of the O_2 reduction half reaction, which is many times slower than the H_2 oxidation half reaction.

Electrolyte.   The polymer electrolyte membrane is a solid organic polymer, usually poly-[perfluorosulfonic] acid. A typical membrane material used in the PEMFC is Nafion. It consists of three regions:

1. The teflon-like fluorocarbon backbone, hundreds of repeating $-CF_2-CF-CF_2-$ units in length
2. The side chains, $-O-CF_2-CF-O-CF_2-CF_2-$, which connect the molecular backbone to the third region
3. The ion clusters consisting of sulfonic acid ions, $SO_3^- H^+$

The negative ion SO^- is permanently attached to the side chain and cannot move. However, when the membrane becomes hydrated by absorbing water, the hydrogen ion becomes mobile. Ion movement occurs by protons (H^+) bonded to water molecules, hopping from one SO_3^- site to another within the membrane. Because of this mechanism, the solid hydrated electrolyte is an excellent conductor of hydrogen ions.

Electrodes. The anode and the cathode are separated from each other by the electrolyte, the PEM. Each electrode consists of porous carbon to which very small Pt particles are bonded. The porous electrodes allow the reactant gases to diffuse through each electrode to reach the catalyst. Both platinum and carbon are good conductors, so electrons are able to move freely through the electrode.

Catalyst. The two half reactions occur very slowly under normal conditions at the low operating temperature (~80°C) of the PEMFC. Therefore, catalysts are needed on both the anode and cathode to increase the rates of each half reaction. Although platinum is a very expensive metal, it is the best material for a catalyst on each electrode.

The half reactions occurring at each electrode can occur only at a high rate at the surface of the Pt catalyst. A unique feature of Pt is that it is sufficiently reactive in bonding H and O intermediates, as required to facilitate the electrode processes, and is also capable of effectively releasing the intermediate to form the final product. The anode process requires Pt sites to bond H atoms when the H2 molecule reacts; next, these Pt sites release the H atoms, as follows:

$$2H^+ + 2e^- \longrightarrow H_2$$
$$H_2 + 2Pt \longrightarrow 2(Pt\text{-}H)$$
$$2(Pt\text{-}H) \longrightarrow 2Pt + 2H^+ + 2e^-$$

This optimized bonding to H atoms (neither very weak nor very strong) is a unique property of the Pt catalyst. To increase the reaction rate, the catalyst layer is constructed with the highest possible surface area. This is achieved by using very small Pt particles, about 2 nm in diameter, resulting in an enormously large total surface area of Pt that is accessible to gas molecules. The original MEAs for the Gemini space program used 4 mg of platinum per square centimeter of membrane area (4 mg/cm^2). Although the technology varies with the manufacturer, the total platinum loading has decreased from the original 4 mg/cm^2 to about 0.5 mg/cm^2. Laboratory research now uses platinum loadings of 0.15 mg/cm^2. For catalyst layers containing Pt of about 0.15 mg/cm^2, the thickness of the catalyst layer is ~10 μm; the MEA with a total thickness ~200 μm can generate more than half an ampere of current for every square centimeter of the MEA at a voltage of 0.7 V between the cathode and the anode. Recently, scientists at Los Alamos National Laboratory, USA have developed a new class of hydrogen fuel cell catalysts that exhibit promising activity and stability. The catalysts, cobalt-polypyrrole-carbon (Co-PPY-XC72) composite, are made of low cost metals entrapped in a heteroatomic-polymer structure.

1.9.4   The cell hardware

The hardware of the fuel cell consists of backing layers, flow fields, and current collectors. These are designed to maximize the current that can be obtained from an MEA. The backing layers placed next to the electrodes are made of a porous carbon paper or carbon cloth, typically 100–300 μm thick. The porous nature of the backing material ensures effective diffusion of the reactant gases to the catalyst. The backing layers also assist in water management during the operation of the fuel cell; too little or too much water can halt the cell operation. The correct backing material allows the right amount of water vapor to reach the MEA and keep the membrane humidified.

Carbon is used for backing layers because it can conduct the electrons leaving the anode and entering the cathode. A piece of hardware, called a plate, is pressed against the outer surface of each backing layer. The plate serves the dual role of a flow field and current collector. The side of the plate next to the backing layer contains channels machined into the plate. The plates are made of a lightweight, strong, gas-impermeable, electron-conducting material; graphite or metals are commonly used, although composite material plates are now being developed. Electrons produced by the oxidation of hydrogen move through the anode, through the backing layer, and through the plate before they can exit the cell, travel through an external circuit, and reenter the cell at the cathode plate. In a single fuel cell, these two plates are the last of the components making up the cell.

In a fuel cell stack, current collectors are the bipolar plates; they make up over 90 percent of the volume and 80 percent of the mass of a fuel cell stack.

1.9.5   Water and air management

Although water is a product of the fuel cell reaction and is carried out of the cell during its operation, it is necessary that both the fuel and air entering the fuel cell be humidified. This additional water keeps the polymer electrolyte membrane hydrated. The humidity of the gases has to be carefully controlled, as too little water dries up the membrane and prevents it from conducting the H^+ ions and the cell current drops. If the air flow past the cathode is too slow, the air cannot carry all the water produced at the cathode out of the fuel cell, and the cathode "floods." Cell performance deteriorates because not enough oxygen is able to penetrate the excess liquid water to reach the cathode catalyst sites. Cooling is required to maintain the temperature of a fuel cell stack at about 80°C, and the product water produced at the cathode at this temperature is both liquid and vapor.

1.9.6   Performance of the PEM fuel cell

Energy conversion in a fuel cell is given by the relation:

Chemical energy of the fuel = electric energy + heat energy

Power is the rate at which energy (E) is made available ($P = dE/dt$, or $AE = PAt$). The power delivered by a cell is the product of the current (I) drawn and the terminal voltage (V) at that current ($P = IV$ watts). To compute power delivered by a fuel cell, we have to know the cell voltage and load current. The ideal (maximum) cell voltage (E) for the hydrogen/air fuel cell reaction ($H_2 + 1/2O_2 \rightarrow H_2O$) at a specific temperature and pressure is calculated from the maximum electrical energy:

$$W_{el} = -\Delta G = nFE \quad or \quad E = -\frac{\Delta G}{nF}$$

where ΔG is the change in Gibbs free energy for the reaction, n is number of moles of electrons involved in the reaction per mole of H_2, and F (Faraday's constant) = 96,487 C (coulombs = joules/volt). At a constant pressure of 1 atm, the change in Gibbs free energy in the fuel cell process (per mole of H_2) is calculated from the reaction temperature (T) and from changes in the reaction enthalpy (H) and entropy (S).

$$AG = AH - T\,AS$$

$$= -285{,}800 \text{ J} - (298 \text{ K})(-163.2 \text{ J/K})$$

$$= -237{,}200 \text{ J}$$

For the hydrogen–air fuel cell at 1 atm pressure and 25°C (298 K), the cell voltage

$$E = -\frac{\Delta G}{nF}$$

$$= -\left(-\frac{237{,}200\text{J}}{2 \times 96{,}487\text{J / V}}\right) = 1.23\text{V}$$

As temperature rises from room temperature to the PEM fuel cell operating temperature (80°C or 353 K), the change in values of H and S is very small, but T changes by 55°C. Thus the absolute value of AG decreases. Assuming negligible change in the values of H and S,

$$AG = -285{,}800 \text{ J/mol} - (353 \text{ K})(163.2 \text{ J/mol K})$$

$$= -228{,}200 \text{ J/mol}$$

Therefore,

$$E = -\left(-\frac{228,200\,J}{2 \times 96,487\,J/V}\right) = 1.18\,V$$

Thus, for standard pressure of 1 atm, the maximum cell voltage decreases from 1.23 V at 25°C to 1.18 V at 80°C. An additional correction is needed for using air instead of pure oxygen, and also for using humidified air and hydrogen instead of dry gases. This further reduces the maximum voltage from the hydrogen–air fuel cell to 1.16 V at 80°C and 1 atm pressure. With an increase in load current, the actual cell potential is decreased from its no-load potential because of irreversible losses, which are often called polarization or overvoltage (h). These originate primarily from three sources:

- Activation polarization (h_{act})
- Ohmic polarization (h_{ohm})
- Concentration polarization (h_{conc})

The polarization losses result in a further decrease in actual cell voltage (V) from its ideal potential E ($V = E -$ potential drop due to losses). The activation polarization loss is dominant at low current density. This is because electronic barriers have to be overcome prior to current and ion flows. Activation polarization is present when the rate of an electrochemical reaction at an electrode surface is controlled by sluggish electrode kinetics. Therefore, activation polarization is directly related to the rates of electrochemical reactions. In an electrochemical reaction with $h_{act} > 50 - 100$ mV, activation polarization is described by a semiempirical equation known as the Tafel equation:

$$h_{act} = \left(\frac{RT}{\alpha nF}\right)\ln\left(\frac{i}{i_0}\right)$$

where α is the electron transfer coefficient of the reaction at the electrode (anode or cathode), and i_0 is the exchange current density. The Tafel slope for the PEMFC electrochemical reaction is about 100 mV/decade at room temperature. There is an incentive to develop electrocatalysts that yield a lower Tafel slope.

Ohmic losses occur because of the resistance to the flow of ions in the electrolyte and resistance to the flow of electrons through the electrode materials. Decreasing the electrode separation and enhancing the ionic conductivity of the electrolyte can reduce the ohmic losses. Both the electrolyte and fuel cell electrodes obey Ohm's law; the ohmic losses can be expressed by the equation: $h_{ohm} = iR$, where i is the current flowing through the cell and R is the total cell resistance, which includes ionic, electronic, and contact resistance (see Fig. 1.16).

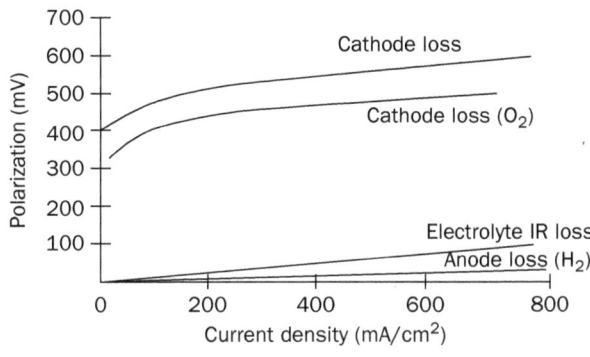

FIGURE 1.16   Activation losses in a PEM fuel cell.

Due to the consumption of reactants at the electrode by an electrochemical reaction, the surrounding material is unable to maintain the initial concentration of the bulk fluid and a concentration gradient is formed, resulting in a loss of electrode potential.

Although several processes contribute to concentration polarization, at practical current densities, slow transport of reactants and products to and from the electrochemical reaction site is a major contributor to concentration polarization. The effect of polarization is to shift the potential of the electrode:

For the anode,

$$V_{anode} = E_{anode} + |h_{anode}|$$

and For the cathode,

$$V_{cathode} = E_{cathode} + |h_{cathode}|$$

The net result of current flow in a fuel cell is to increase the anode potential and to decrease the cathode potential. This reduces the cell voltage. The cell voltage includes the contribution of the anode and cathode potentials and ohmic polarization:

$$V_{cell} = V_{cathode} - V_{anode} - iR; \text{ or}$$
$$V_{cell} = E_{cathode} - |h_{cathode}| - (E_{anode} + |h_{anode}|) - iR; \text{ or}$$
$$V_{cell} = E_{cell} - |h_{cathode}| - |h_{anode}| - iR$$
$$\text{where } E_{cell} = E_{cathode} - E_{anode}$$

The goal of fuel cell developers is to minimize the polarization losses so that the V_{cell} approaches the E_{cell} by modifications to the fuel cell (Fig. 1.17) design by improvement in the electrode structures, better electrocatalysts, more conductive electrolytes, thinner cell components, and so forth. It is possible to improve the cell performance by modifying the operating conditions such as higher gas pressure,

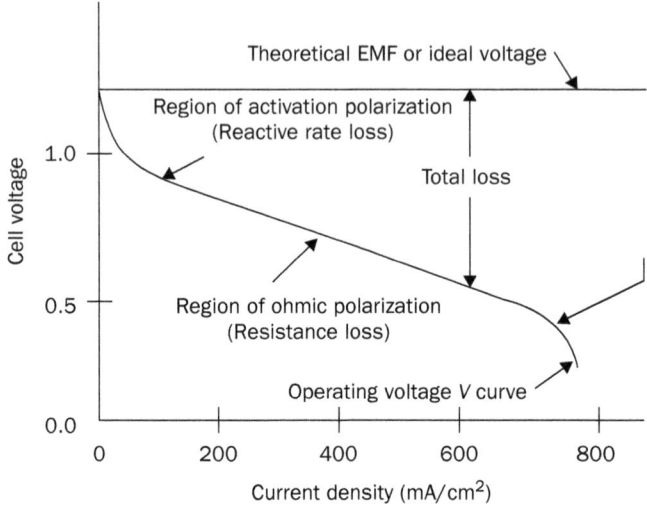

FIGURE 1.17 PEM fuel cell voltage versus current density curve.

higher temperature, and a change in gas composition to lower the gas impurity concentration.

1.9.7 Direct methanol fuel cells (DMFCs)

Direct methanol fuel cells are similar to the PEMFC as they also use a polymer membrane as the electrolyte. However, it produces power by direct conversion of liquid methanol to hydrogen ions on the anode side of the fuel cell. In the DMFC, the anode catalyst draws hydrogen directly from the liquid methanol, thus eliminating the need for a fuel reformer. All the DMFC components (anode, cathode, membrane, and catalysts) are the same as those of a PEMFC. A DMFC system is shown in Fig. 1.18. Methanol diluted to a specified concentration is fed to the fuel cell stack. During operation, the concentration of the methanol solution exiting the stack is reduced. Therefore, pure methanol is added in the feed cycle to restore the original concentration of the solution. A gas–liquid separator is used to remove carbon dioxide from the solution loop, and a compressor feeds air to the DMFC stack. Water and heat are recovered by passing the outlet air through a condenser. A portion of the recovered water is returned to the fuel circulation loop. The stack temperature is maintained by removing the excess heat from the fuel circulation loop using a heat exchanger. The DMFC can attain high efficiencies of 40 percent with a Nafion-117 membrane at 60°C, with current density in the range of 100–120 mA/cm^2. Studies have shown that DMFC efficiency decreases with increasing methanol concentration. Therefore, operating a fuel cell to maintain the maximum efficiency needs close control of methanol concentration and temperature. An online concentration

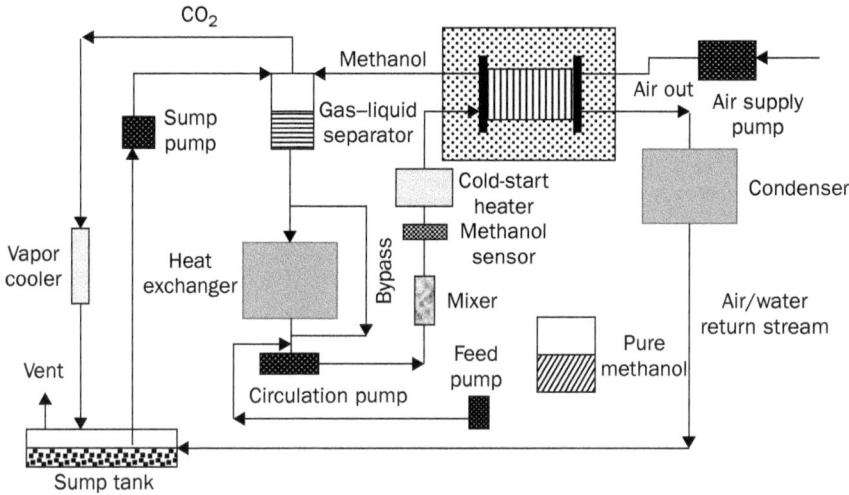

Figure 1.18   A DMFC system.

sensor is used in the feedback loop for this purpose. Some of the advantages of this system, relative to the hydrogen systems, are that the liquid feed (methanol) helps in attaining the uniform stack temperature and maintenance of membrane humidity; it is also easy to refill since the fuel (methanol) is in liquid form.

As compared to the PEMFC, the DMFC has a very sluggish electrochemical reaction (significant activation over voltage) at the anode. It therefore requires a high surface area of 50:50 percent Pt-Ru (a more expensive bimetal) alloy as the anode catalyst to overcome the sluggish reaction and an increase in catalyst loading of more than 10 times that for the PEMFC. Even then, the output voltage on the load is only 0.2–0.4 V with an efficiency of about 40 percent at operating temperatures between 60°C and 90°C. This is relatively low, and therefore, the DMFC is attractive only for tiny to small-sized applications (cellular phones, laptops, etc.). Another potential application for the DMFC is in transport vehicles; as it operates on liquid fuels, it would greatly simplify the onboard system as well as the infrastructure needed to supply fuel to passenger cars and commercial fleets and can create a large potential market for commercialization of fuel cell technology in vehicle applications.

1.9.8   Alkaline-electrolyte fuel cells (AFCs)

Alkaline-electrolyte fuel cells (see Fig. 1.19) are one of the most developed fuel cell technologies. They have been in use since the mid-1960s for Apollo and space shuttle programs [3, 6, 18, 19]. The AFCs onboard these spacecraft provide electrical power as well as drinking water. AFCs are among the most efficient electricity-generating fuel cells with an efficiency of nearly 70 percent. The electrolyte used in the AFC is an alkaline solution in which an OH^- ion can move freely across the electrolyte.

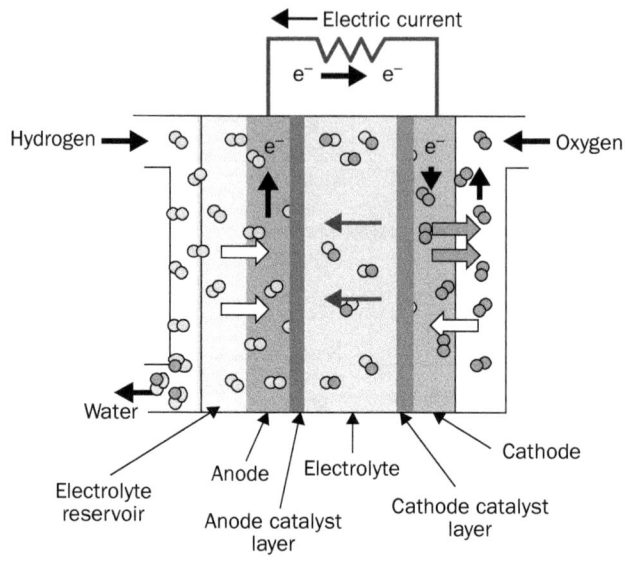

FIGURE **1.19** An alkaline-electrolyte fuel cell.

Electrochemistry of AFCs. The electrolyte used in the AFC is an aqueous (water-based) solution of potassium hydroxide (KOH) retained in a porous stabilized matrix. The concentration of KOH can be varied with the fuel cell operating temperature, which ranges from 65 to 220°C.

The charge carrier for an AFC is the hydroxyl ion (OH⁻) that migrates from the cathode to the anode, where they react with hydrogen to produce water and electrons. Water formed at the anode migrates back to the cathode to regenerate hydroxyl ions.

$$\text{Anode reaction: } 2H_2 + 4OH^- \longrightarrow 4H_2O + 4e^-$$
$$\text{Cathode reaction: } O_2 + 2H_2O + 4e^- \longrightarrow 4OH^-$$

Hydroxyl ions are the conducting species in the electrolyte. Overall cell reaction: $2H_2 + O_2 \rightarrow 2H_2O + \text{heat} + \text{electricity}$

In many cell designs, the electrolyte is circulated (mobile electrolyte) so that heat can be removed and water eliminated by evaporation. Since KOH has the highest conductance among the alkaline hydroxides, it is the preferred electrolyte.

Electrolyte. Concentrated KOH (85 wt.%) is used in cells designed for operation at a high temperature (~260°C). For lower temperature (<120°C) operation, less concentrated KOH (35–50 wt.%) is used. The electrolyte is retained in a matrix (usually asbestos), and a wide range of electrocatalysts can be used (e.g., Ni, Ag, metal oxides, and noble metals). A major advantage of the AFC is the lower activation polarization at the cathode, resulting in a higher operating voltage (0.875 V). Another advantage of the AFC is the use of inexpensive electrolyte materials. The electrolyte is replenished through a reservoir on the anode side. The typical performance of this AFC cell is 0.85 V at a current density of 150 mA/cm^2. The AFCs used in the space shuttle orbiter have a rectangular cross-section and weigh 91 kg. They operate at an average power of 7 kW with a peak power rating of 12 kW at 27.5 V. A disadvantage of the AFC is that it is sensitive to CO_2 present in the fuel or air. The alkaline electrolyte reacts with CO_2 and severely degrades the fuel cell performance, limiting their application to closed environments, such as space and undersea vehicles, as these cells work well only with pure hydrogen and oxygen as fuel.

Electrodes. A significant cost advantage of alkaline fuel cells is that both anode and cathode reactions can be effectively catalyzed with nonprecious, relatively inexpensive metals. The most important characteristics of the catalyst structure are high electronic conductivity and stability (mechanical, chemical, and electrochemical). Both metallic (typically hydrophobic) and carbon-based (typically hydrophilic) electrode structures with multilayers and optimized porosity characteristics for the flow of liquid electrolytes and gases (H_2 and O_2) have been developed. The kinetics of oxygen reduction in alkaline electrolytes is much faster than in acid media; hence, AFCs can use low-level Pt catalysts (about 20% Pt, compared with PEMFCs) on a large surface carbon support.

Performance. The AFC development has gone through many changes since 1960. To meet the requirements for space applications, the early AFCs were operated at relatively high temperatures and pressures. Now the focus of the technology is to develop low-cost components for AFCs operating at near-ambient temperature and pressure, with air as the oxidant for terrestrial applications. This has resulted in lower performance. The reversible cell potential for an H_2 and O_2 fuel cell decreases by 0.49 mV/°C under standard conditions. An increase in operating temperature reduces activation polarization, mass transfer polarization, and ohmic losses, thereby improving cell performance. Alkaline cells operated at low temperatures (~70°C) show reasonable performance.

Pure hydrogen and oxygen are required to operate an AFC. Reformed H_2 or air containing even trace amounts of CO_2 dramatically affects its performance and lifetime. There is a drastic loss in performance when using hydrogen-rich fuels containing even a small amount of CO_2 from reformed hydrocarbon fuels and also from the presence of CO_2 in the air (~350 ppm CO_2 in ambient air). The CO_2 reacts with OH^- ($CO_2 + 2OH- \longrightarrow CO_3 2- + H_2O$), thereby decreasing their concentration and thus reducing the reaction kinetics. Other ill effects of the presence of CO_2 are:

- Increase in electrolyte viscosity, resulting in lower diffusion rate and lower limiting currents

- Deposition of carbonate salts in the pores of the porous electrode

- Reduction in oxygen solubility

- Reduction in electrolyte conductivity

A higher concentration of KOH decreases the life of O_2 electrodes when operating with air containing CO_2. However, operation at higher temperatures is beneficial because it increases the solubility of CO_2 in the electrolyte. The operational life of air electrodes polytetrafluoroethylene (PTFE) bonded carbon electrodes on porous nickel substrates at a current density of 65 mA/cm^2 in 9-N KOH at 65°C ranges from 4000 to 5500 h with CO_2-free air, and their life decreases to 1600–3400 h when air (350-ppm CO_2) is used. For large-scale utility applications, operating times >40,000 h are required, which is a significant hurdle to commercialization of AFC devices for stationary electric power generation. Another problem with the AFC is that the electrodes and catalysts degrade more on no-load or light-load operation than on a loaded condition, because the high open-circuit voltage causes faster carbon oxidation processes and catalyst changes. The AFC with immobilized KOH electrolyte suffers much more from this as the electrolyte has to stay in the cells causing residual carbonate accumulation, separator deterioration, and gas cross leakage during storage or unloaded periods if careful maintenance is not carried out. In circulating an electrolyte-type AFC, the electrolyte is emptied from the cell during nonoperating periods. Shutting off the H_2 electrodes from air establishes an inert atmosphere. This shutdown also eliminates all parasitic currents and increases life expectancy. The exchangeability of the KOH in a circulating electrolyte-type AFC offers the possibility to operate on air without complete removal of the CO_2.

1.9.9 Phosphoric acid fuel cells (PAFCs)

Phosphoric acid fuel cells (see Fig. 1.20) operate at intermediate temperatures (~200°C) and are developed and commercially available today. Hundreds of PAFC systems are working around the world in hospitals, hotels, offices, schools, utility power plants, landfills and wastewater treatment plants, and so forth. Most of the PAFC plants are in the 50- to 200-kW capacity ranges, but large plants of 1- and 5-MW capacity have also been built; a demonstration unit has achieved 11 MW of grid-quality ac power [3]. PAFCs generate electricity at more than 40 percent efficiency and if the steam produced

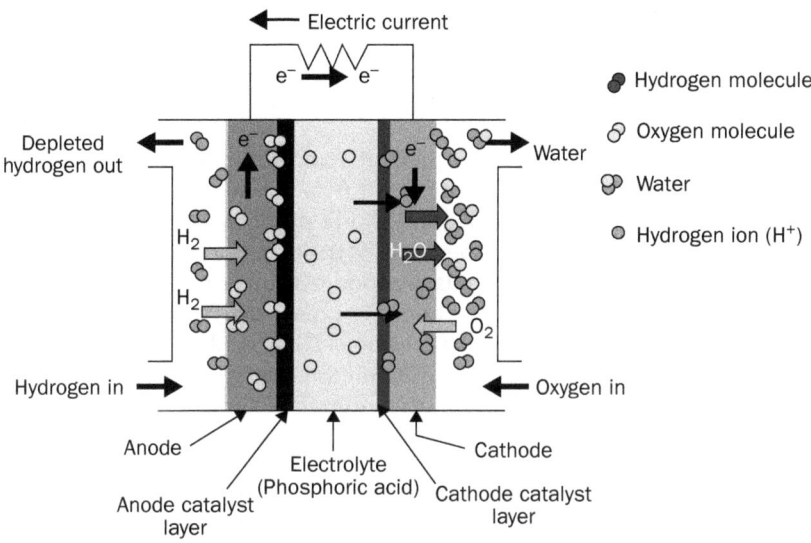

FIGURE 1.20 Phosphoric acid fuel cell.

is used for cogeneration, efficiencies of nearly 85 percent can be achieved. PAFCs use liquid phosphoric acid as the electrolyte. One of the main advantages to this type of fuel cell, besides high efficiency, is that it does not require pure hydrogen as fuel and can tolerate up to 1.5 percent CO concentration in fuel, which broadens the choice of fuels that can be used. However, any sulfur compounds present in the fuel have to be removed to a concentration of <0.1 ppmV. Temperatures of about 200°C and acid concentrations of 100 percent H_3PO_4 are commonly used, while operating pressure in excess of 8 atm has been used in an 11-MW electric utility demonstration plant.

Electrochemistry of PAFCs. The electrochemical reactions occurring in a PAFC are

At the anode:

$$H_2 \rightarrow 2H^+ + 2e^-$$

At the cathode:

$$\frac{1}{2}O_2 + 2H^+ + 2e^- \rightarrow H_2O$$

The over all reaction:

$$\frac{1}{2}O_2 + H_2 \rightarrow H_2O$$

The fuel cell operates on H_2; CO is a poison when present in a concentration greater than 0.5 percent. If a hydrocarbon such as natural gas is used as a fuel, reforming of the fuel by the reaction

$$CH_4 + H_2O \rightarrow 3H_2 + CO$$

and shifting of the reformat by the reaction

$$CO + H_2O \longrightarrow H_2 + CO_2$$

is required to generate the required fuel for the cell.

Electrolyte. The PAFC uses 100 percent concentrated phosphoric acid (H_3PO_4) as an electrolyte. The electrolyte assembly is a 0.1- to 0.2-mm-thick matrix made of silicon carbide particles held together with a small amount of PTFE. The pores of the matrix retain the electrolyte (phosphoric acid) by capillary action. At lower temperatures, H_3PO_4 is a poor ionic conductor and CO poisoning of the Pt electrocatalyst in the anode can become severe. There will be some loss of H_3PO_4 over long periods, depending on the operating conditions. Hence, as a general rule, sufficient acid reserve is kept in the matrix at the beginning.

Electrode. The PAFC (similar to a PEMFC) uses gas diffusion electrodes. Platinum or platinum alloys are used as the catalyst at both electrodes. In the mid-1960s, the conventional porous electrodes were PTFE-bonded Pt black, and the loadings of Pt were about 9 mg/cm². In recent years, Pt supported on carbon black has replaced Pt black in porous PTFE-bonded electrode structures. Pt loading has also dramatically reduced to about 0.25 mg Pt/cm² in the anode and about 0.50 mg Pt/cm² in the cathode. The porous electrodes used in a PAFC consist of a mixture of the electrocatalyst supported on carbon black and a polymeric binder to bind the carbon black particles together to form an integral structure. A porous carbon paper substrate provides structural support for the electrocatalyst layer and also acts as the current collector. The composite structure consisting of a carbon black/binder layer onto the carbon paper substrate forms a three-phase interface, with the electrolyte on one side and the reactant gases on the other side of the carbon paper. The stack consists of a repeating arrangement of a bipolar plate, the anode, electrolyte matrix, and cathode.

Hardware. A bipolar plate separates the individual cells and electrically connects them in a series in a fuel cell stack. A bipolar plate has a multifunction design; it has to separate the reactant gases in the adjacent cells in the stack, so it must be impermeable to reactant gases; it must transmit electrons to the next cell (series connection), so it has to be electrically conducting; and it must be heat conducting for proper heat transfer and thermal management of the fuel cell stack. In some designs, gas channels are also provided on the bipolar plates to feed reactant gases to the porous electrodes and to remove the reaction products. Bipolar plates should have very low porosity so as to minimize phosphoric acid absorption. These plates must be stable and corrosion-resistant in the PAFC environment. Bipolar plates are usually made of graphite–resin mixtures that are carbonized and heat treated to 2700°C to increase corrosion resistance. For 100-kW and larger power generation systems, water cooling has to be used and cooling channels are provided in the bipolar plates to cool the stack.

Temperature and humidity management.   Temperature and humidity management are essential for proper operation of a PAFC. The PAFC system has to be heated up to 130°C before the cell can start working. At lower temperatures, concentrated phosphoric acid does not get dissociated, resulting in a low availability of protons. Also, due to lower vapor pressure of the concentrated acid, the water generated will not come out with the reactant stream and the moisture retention dilutes the acid. This causes an increase in acid volume, which results in acid oozing out through the electrode. With the start of normal cell operation, its temperature increases and acid concentration gets back to its normal value that causes acid volume to shrink, resulting in drying of the electrolyte matrix pores if the acid is not replenished. Controlled stack heating at start-up is achieved by using an insertable heater system. During operation, the temperature of the stack is maintained by controlling the air flow in the oxidant channel. At high loading conditions, insertable coolers may be used to remove excess heat from the stack. Large-power PAFC systems use a water-cooling system.

Moisture generated at the cathode dilutes the acid on the cathode side of the electrolyte matrix, causing higher vapor pressure. This results in more moisture out with the oxidant stream. With the movement of protons from anode to cathode, moisture migration takes place at the cathode side also. This water evaporation results in an acid concentration gradient from anode to cathode, causing low availability of protons and a lower potential of the cell. Therefore, water management is needed to maintain humidity of the anode stream gas at a sufficient level so that the vapor pressure matches the acid concentration level at the operating temperature.

Performance.   For good performance, the normal operating temperature range of a PAFC is $180°C < T < 250°C$; below 200°C, the decrease in cell potential is significant. Although an increased temperature increases performance, higher temperatures also result in increased catalyst sintering, component corrosion, electrolyte degradation, and evaporation. PAFCs operate in the current density range of 100–400 mA/cm^2 at 600–800 mV/cell. Voltage and power limitations result from increased corrosion of platinum and carbon components at cell potentials above approximately 800 mV. Since the freezing point of phosphoric acid is 42°C, the PAFC must be kept above this temperature once commissioned to avoid the thermal stresses due to freezing and thawing. Various factors affect the PAFC life. Acid concentration management by proper humidity control is very important to prevent acid loss and performance degradation. A PAFC has a life of 10,000–50,000 h; commercially available (UTC Fuel Cells) PAFC systems operating at 207°C have shown a life of 40,000 h with reasonable performance (degradation rate $\Delta V_{\text{lifetime}}$ (mV) $= -2$ mV/1000 h) [3, 23].

1.9.10 Molten carbonate fuel cells (MCFCs)

The MCFC has evolved from work in the 1960s, aimed at producing a fuel cell that would operate directly on coal. Although direct operation on coal is no longer a goal, a remarkable feature of the MCFC is that it can directly operate on coal-derived fuel gases or natural gas and is therefore also called a direct fuel cell (DFC). MCFCs operate at high temperatures (600–650°C) compared to phosphoric acid (180–220°C) or PEM fuel cells (60–85°C). Operation at high temperatures eliminates the need for external fuel processors that the lower temperature fuel cells require to extract hydrogen from naturally available fuel. When natural gas is used as fuel, methane (the main ingredient of natural gas) and water (steam) are converted into a hydrogen-rich gas inside the MCFC stack ("internal reforming") (Fig. 1.21). High operating temperatures also result in high-temperature exhaust gas, which can be utilized for heat recovery for secondary power generation or cogeneration. MCFCs can therefore achieve a higher fuel-to-electricity and an overall energy use efficiency (>75 percent) than the low-temperature fuel cells. The MCFC is a well-developed fuel cell and is a commercially viable technology for a stationary power

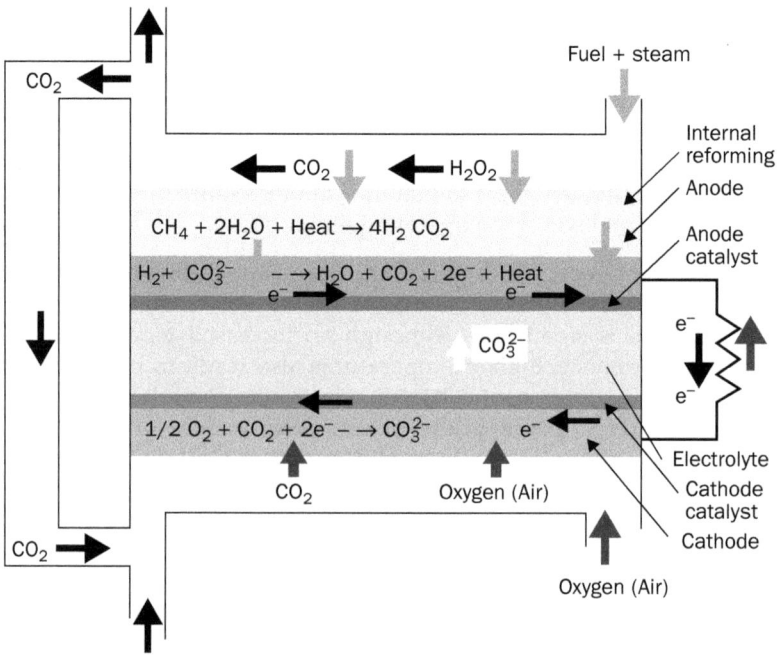

Figure 1.21 Molten carbonate fuel cell.

plant, compared to other fuel cell types. A number of MCFC proto-type units in the power range of 200 kW to 1 MW and higher are operating around the world. The cost and useful life issues are the major challenges to overcome before the MCFC can compete with the existing (thermal or other) electric power generation systems for widespread use.

Electrochemistry of MCFC.   The electrochemical reactions occurring in the cell are:

Anode half reaction.   At the anode, hydrogen reacts with carbonate ions to produce water, carbon dioxide, and electrons. The electrons travel through an external circuit—creating electricity—and return to the cathode.

$$H_2 + CO_3^{2-} \rightarrow H_2O + CO_2 + 2e^-$$

Cathode half reaction.   At the cathode, oxygen from the air and car-bon dioxide recycled from the anode react with the electrons to form carbonate ions that replenish the electrolyte and transfer the current through the fuel cell, completing the circuit.

$$\frac{1}{2}O_2 + CO_2 + 2e^- \rightarrow CO_3^-$$

The overall cell reaction is:

$$H_2 + \frac{1}{2}O_2 + CO_2(\text{cathode}) + H_2O + CO_2(\text{anode})$$

If a fuel such as natural gas is used, it has to be reformed either exter-nally or within the cell (internally) in the presence of a suitable cata-lyst to form H_2 and CO by the reaction:

$$CH_4 + H_2O \rightarrow 3H_2 + CO$$

Although CO is not directly used by the electrochemical oxidation, it produces additional H2 by the water gas shift reaction:

$$CO + H_2O \rightarrow H_2 + CO_2$$

Typically, the CO_2 generated at the anode is recycled to the cath-ode, where it is consumed. This requires additional equipment to either transfer CO_2 from the anode exit gas to the cathode inlet gas or produces CO_2 by combustion of anode exhaust gas and mix with the cathode inlet gas.

Electrolyte.   The MCFC uses a molten carbonate salt mixture as its electrolyte. At operating temperatures of about 650°C, the salt mixture is in a molten (liquid) state and is a good ionic conductor.

The composition of salts in the electrolyte may vary but usually consist of lithium/potassium carbonate (Li_2CO_3/K_2CO_3, 62–38 mol%) for operation at atmospheric pressure. For operation under pressurized conditions, lithium/sodium carbonate ($LiCO_3/NaCO_3$, 52–48 or 60–40 mol%) is used as it provides improved cathode stability and performance. This allows for the use of thicker Li/Na electrolyte for the same performance, resulting in a longer lifetime before a shorting caused by internal precipitation. The composition of the electrolyte has an effect on electrochemical activity, corrosion, and electrolyte loss rate. Li/Na offers better corrosion resistance but has greater temperature sensitivity. Additives are being developed to minimize the temperature sensitivity of the Li/Na electrolyte. The electrolyte has a low vapor pressure at the operating temperature and may evaporate very slowly; however, this does not have any serious effect on the cell life. The electrolyte is suspended in a porous, insulating, and chemically inert ceramic ($LiAlO_2$) matrix. The ceramic matrix has a significant effect on the ohmic resistance of the electrolyte. It accounts for almost 70 percent of the ohmic polarization. The electrolyte management in an MCFC ensures that the electrolyte matrix remains completely filled with the molten carbonate, while the porous electrodes are partially filled, depending on their pore size distributions.

Electrode.   The anode is made of a porous chromium-doped sintered Ni-Cr/Ni-Al alloy. Because of the high temperatures resulting in a fast anode action, a large surface area is not required on the anode as compared to the cathode. Partial flooding of the anode with molten carbonate is desirable as it acts as a reservoir that replenishes carbonate in the stack during prolonged use. The cathode is made up of porous lithiated nickel oxide. Because of the high operating temperatures, no noble catalysts are needed in the fuel cell. Nickel is used on the anode and nickel oxide on the cathode as catalysts. Bipolar plates or interconnects are made from thin stainless steel sheets with corrugated gas diffusion channels. The anode side of the plate is coated with pure nickel to protect against corrosion.

Performance.   At the high operating temperatures of an MCFC, CO is not a poison but acts as a fuel. In the MCFC, CO_2 has to be added to oxygen (air) stream at the cathode for generation of carbonate ions. The anode reaction converts these ions back to CO_2, resulting in a net transfer of two ions with every molecule of CO_2. The need for CO_2 in the oxidant stream requires that CO_2 from the spent anode gas be separated and mixed with the incoming air stream. Before this can be done, any residual hydrogen in the spent fuel stream must be burned. Systems developed in the future may incorporate membrane separators to remove the hydrogen for recirculation back to the fuel stream to increase efficiency. Internal reforming of natural gas and partially cracked hydrocarbons is possible in the inlet chamber of the MCFC, eliminating the separate fuel processing of natural gas or other hydrogen-rich fuels. The requirement for CO_2 makes the digester gas (sewage, animal waste, food processing waste, etc.) an ideal fuel for the MCFC; other fuels such as natural gas, landfill gas, propane, coal gas, and liquid fuels (diesel, methanol, ethanol, LPG, etc.) can also be used in the MCFC system. The elimination of the external fuel reformer also contributes to lower costs, and high-temperature waste heat can be utilized to make additional electricity and cogeneration. MCFCs can reach overall thermal efficiencies as high as 85 percent.

With the increase in operating temperature, the theoretical operating voltage for a fuel cell decreases but increases the rate of the electrochemical reaction, and therefore the current that can be obtained at a given voltage. This results in the MCFC having a higher operating voltage for the same current density and higher fuel efficiency than a PAFC of the same electrode area. As size and cost scale roughly with the electrode area, the MCFC is smaller and less expensive than a PAFC of comparable output. Another advantage of the MCFC is that the electrodes can be made with cheaper nickel catalysts rather than the more expensive platinum used in other low-temperature fuel cells. Endurance of the cell stack is a critical issue in commercialization, and MCFC manufacturers report an average potential degradation of -2 mV/1000 h over a cell stack lifetime of 40,000 h. The high temperature limits the use of materials in the MCFC, and safety issues prevent their application for home use. MCFC units require a few minutes of fuel burning at startup to heat up the cell to its operating temperature and therefore are not suitable for use in automobiles. However, they are very good for stationary power applications and units with up to 2 MW have been constructed, and designs for units with up to 100 MW exist.

1.9.11   Solid oxide fuel cells (SOFCs)

The SOFC has the most desirable properties for generating electricity from hydrocarbon fuels. The SOFC uses a solid electrolyte and is very efficient. It can internally reform hydrocarbon fuels and is tolerant to impurities. The SOFC operates at a high temperature (700–1000°C) and so does not require any cooling system for maintaining a fuel cell operating temperature. For small systems, insulation has to be provided to maintain the cell temperature. In large SOFC systems, the operating temperature is maintained internally by the reforming action of the fuel and by the cool outside air (oxidant) that is drawn into the fuel cell. At high operating temperatures, chemical reaction rates in the SOFC are high and air compression is not required. This results in a simpler system, quiet operation, and high efficiencies. Westinghouse has worked at developing a tubular style of the SOFC that operates at 1000°C (Fig. 1.22) for many years. These long tubes have high electrical resistance but are simple to seal. Many other

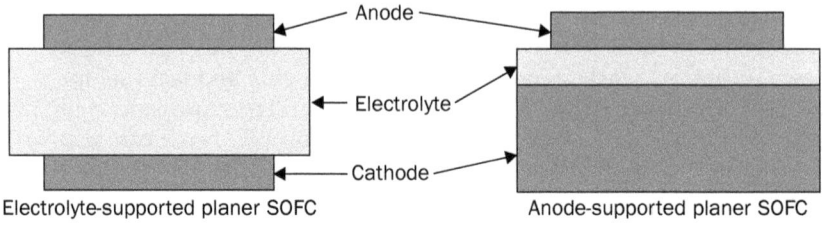

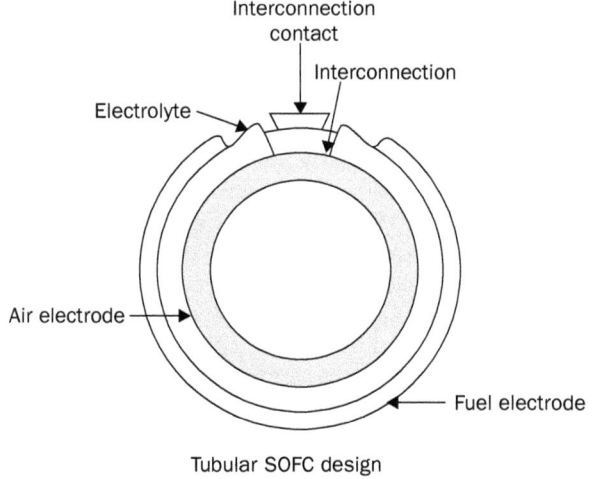

Tubular SOFC design

FIGURE 1.22 Tubular and planer solid oxide fuel cell.

manufacturers are now working on a planar SOFC composed of thin ceramic sheets that operate at 800°C or even less. Thin sheets offer low electrical resistance, and cheaper materials such as stainless steel can be used at these lower temperatures. One big advantage of the SOFC over the MCFC is that the electrolyte is a solid. Therefore, no pumps are required to circulate a hot electrolyte, and very compact, small planar SOFC systems of a few kW range could be constructed using thin sheets.

A major advantage of the SOFC is that both hydrogen and carbon monoxide are used in the cell. Therefore, in the SOFC, many common hydrocarbon fuels such as natural gas, diesel, gasoline, alcohol, and coal gas can be safely used. The SOFC can reform these fuels into hydrogen and carbon monoxide inside the cell, and the high-temperature waste thermal energy can be recycled back for fuel reforming. During operation, the SOFC is at the same time a generator and a user of heat. Heat is generated through exothermic chemical reactions and ohmic losses, while it is absorbed by the reforming reaction. It is possible to design the SOFC to be thermally balanced, thereby eliminating the requirement for external insulation and heating. Small SOFC systems are not thermally self-sustaining and may require an external heat source to start and maintain operation. In large systems, the heat generated is not fully absorbed by fuel reforming, and the excess heat can be used in gas turbines for generating electricity or for cogeneration. Another advantage of the SOFC is that expensive catalysts are not required. However, a few minutes of fuel burning is required to reach the operating temperature of the SOFC at the start. This time delay is a disadvantage for an automotive application, but for stationary electric power plants, this is not a problem as they run continuously for long periods of time.

Electrochemistry of SOFCs. Hydrogen or carbon monoxide in the fuel stream reacts with oxide ions (O^{2-}) from the electrolyte to produce water or CO_2 and to deposit electrons into the anode. The electrons pass outside the fuel cell, through the load, and back to the cathode, where oxygen from the air receives the electrons and is converted into oxide ions, which are injected into the electrolyte. In the SOFC, oxygen ions are formed at the cathode. The reaction at the cathode is

$$O_2 + 4e^- \rightarrow 2O^{2-}$$

At the operating temperature, the electrolyte offers high ionic conductivity and low electrical conductivity; therefore, oxygen ions migrate through the electrolyte to the anode. The overall reaction occurring at the anode is as follows:

The hydrogen in the fuel reacts with the oxygen ions to produce water and releases two electrons.

$$H_2 + O^{2-} \rightarrow H_2O + 2e^-$$

Carbon monoxide present in the fuel causes a shift reaction to produce additional fuel (H_2).

$$CO + H_2O \leftrightarrow H_2 + CO_2$$

The following internal reforming reaction for the hydrocarbon fuel takes place on the anode side

$$C_xH_y + xH_2O \rightarrow xCO + \left(x + \frac{y}{2}\right)H_2$$

For methane-rich fuels, this reforming reaction is

$$CH_4 + H_2O \rightarrow CO + 3H_2$$

This reaction is generally not in chemical equilibrium, and the CO shift reaction takes place to provide more hydrogen. The overall cell reaction is

$$H_2 + \frac{1}{2}O_2 \rightarrow H_2O$$

Electrolyte.   The use of a solid electrolyte in the SOFC eliminates the electrolyte management problems associated with the liquid electrolyte fuel cells and also reduces corrosion considerations to a great extent. In the SOFC, it is the migration of oxygen ions (O^{2-}) through the electrolyte that establishes the voltage difference between the anode and cathode. Therefore, the electrolyte must be a good conductor of O^{2-} ions and a bad conductor of electrons; it must also be stable at the high operating temperature. Some ceramics possess these properties and therefore are good candidates for this application. With the help of modern ceramic technology and solid-state science, many ceramics can be tailored for electrical properties unattainable in metallic or polymer materials. These tailored ceramic materials are termed electroceramics, and one group is known as *fast ion conductors* or superionic conductors. These superionic conductors when used as a solid electrolyte allow easy passage of ions from the cathode to the anode in an SOFC. The material generally used as an electrolyte in the SOFC is dense yttria-stabilized zirconia. It is an excellent conductor of negatively charged oxygen ions at high temperatures (1000°C), but its conductivity reduces drastically with the drop in temperature. Other materials such as scandia-stabilized zirconia (ScSZ), which shows good ionic conductivity at a lower temperature (800°C), are also being investigated, but the electrolyte developed with ScSZ-based materials is expensive and they degrade very quickly.

Electrode.   The anode is made of metallic Ni and Y_2O_3-stabilized ZrO_2 (YSZ). Yttria-stabilized zirconia is added in Ni to inhibit sintering of the metal particles and to provide a thermal expansion coefficient close to those of the other cell materials [26]. Nickel structure is normally obtained from NiO powders; therefore, before starting the operation for the first time, the cell is run with hydrogen in an open-circuit condition to reduce the NiO to nickel. The anode structure is fabricated with a porosity of 20 percent to 40 percent to facilitate mass transport of the reactant and product gases. The Sr-doped lanthanum manganite ($La_{1-x}Sr_xMnO_3$, $x = 0.10$–0.15; known as LSM) is most commonly used for the cathode material. LSM is a p-type semiconductor. Similar to the anode, the cathode is also a porous structure that permits rapid mass transport of the reactant and product gases.

Hardware.   In the SOFC, both CO and hydrogen are used as direct fuel. Therefore, it is important that the fuel and air streams are kept separate, and a thermal balance should be maintained to ensure that operating temperatures remain within an acceptable range. Several designs of the SOFC (tubular and planer) have been developed to accommodate these requirements. The SOFC is a solid-state device and shares certain properties and fabrication techniques with semiconductor devices.

Individual cells in the stack are connected by interconnects, which carry an electrical current between cells and can also act as a separator between the fuel and oxidant supplies. In high-temperature SOFCs, the interconnects that are used are ceramic such as lanthanum chromite, or if the temperature is limited to less than 1000°C, a refractory alloy based on Y/Cr may be used. The interconnects constitute a major proportion of the stack cost. Stack and other plant construction materials that are used also need to be refractory to withstand the high-temperature gas streams. Volatility of chromium-containing ceramics and alloys can result in contamination of the stack components, and the presence of a toxic material such as Cr^{6+} requires special disposal procedures.

The high operating temperature (1000°C) of the SOFC requires a significant startup time. The cell performance is very sensitive to operating temperatures. A 10 percent drop in temperature results in an ~12 percent drop in cell performance due to the increase in internal resistance to the flow of oxygen ions. The high temperature also demands that the system include significant thermal shielding to protect personnel and to retain heat. Also, the materials required for such high-temperature operation, particularly for interconnect and construction materials, are very expensive. Operating the SOFC at temperatures lower than 700°C would be beneficial as low-cost metallic materials, such as ferritic stainless steels, can be used as interconnect and construction materials. This will make both the stack and balance of a plant cheaper and more robust. Using ferritic materials also significantly reduces the problems associated with chromium. The other advantages of low/intermediate-temperature operation are rapid startup and shutdown and significantly reduced corrosion rates.

However, to operate at reduced temperatures, several changes are required in stack design, cell materials, reformer design and operation, and operating conditions. With the reduction in operating temperature, the ionic conductivity of the electrolyte decreases and the parasitic losses due to the conductivity of the electrodes and interconnects increase. This results in a rapid deterioration of the performance of the SOFC. This can be overcome to some extent by reducing the thickness of the electrolyte to compensate for its reduced ionic conductivity. The thickness reduction that is required to accommodate, say, a 200°C reduction in the operating temperature leads to impracticably thin membranes.

Some designs in which a thin, dense layer of the electrolyte is physically supported on one of the electrodes (electrode-supported design) are suggested. This structure of a very porous support is difficult to manufacture, and an expensive thin-film deposition technique such as chemical vapor deposition (CVD) is needed to manufacture these systems. Even then, the mechanical strength of the structure (defined by the porous electrode) is often poor, and the handling of the structure through subsequent processing and assembly is difficult. Another approach to improve SOFC performance at low operating temperatures is to use different materials for the electrolyte and the electrode. Several materials options are being investigated.

1.9.12 Biofuel cells

A biofuel cell operation is similar to a conventional fuel cell, except that it uses biocatalysts such as enzymes, or even whole organisms instead of inorganic catalysts like platinum, to catalyze the conversion of chemical energy into electricity. They can use available substrates from renewable sources and convert them into benign by-products with the generation of electricity. As mentioned earlier, in recent years, medical science is increasingly relying on implantable electronic devices for treating a number of conditions. These devices demand a reliable and maintenance-free (any maintenance that might require surgery) power source. Biofuel cells can provide solutions to most of these problems. A biofuel cell can use fuel that is readily available in the body, for example, glucose in the bloodstream, and it would ideally draw on this power for as long as the patient lives. Since they use concentrated sources of chemical energy, they can be small and light.

A biofuel cell can operate in two ways: It can utilize the chemical pathways of living cells (microbial fuel cells), or, alternatively, it can use isolated enzymes. Microbial fuel cells have high efficiency in terms of conversion of chemical energy into electrical energy; however, they suffer from the low volumetric catalytic activity of the whole organism and low power densities due to slow mass transport of the fuel across the cell wall. Isolated enzymes extracted from biological systems can be used as catalysts to oxidize fuel molecules at the anode and to enhance oxygen reduction at the cathode of the biofuel cell. Isolated enzymes are attractive catalysts for biofuel cells due to their high catalytic activity and selectivity. The theoretical value of the current that can be generated by an enzymatic catalyst with an activity of 103 U/mg is 1.6 A, a catalytic rate greater than platinum! However, practical observed currents are much lower due to the loss of catalytic activity from immobilization of the enzymes at the electrode surface and energy losses of the overall system. A major challenge in the biofuel cell design is the electrical coupling of the biological components of the system with the fuel cell electrodes. Molecules known as electron-transfer mediators are needed to provide efficient transport of electrons between the biological components (enzymes or microbial cells) and the electrodes of the biofuel cell. Integrated biocatalytic systems that include biocatalysts, electron-transfer mediators, and electrodes are under research and development. Biofuel cells have much wider fuel options; enzymatic biofuel cells can operate on a wide variety of available fuels such as ethanol, sugars, or even waste materials.

A basic microbial biofuel cell consists of two compartments, an anode compartment and a cathode compartment, separated by a PEM as shown in Fig. 1.23. Usually, Nafion-117 film (an expensive material) is used as the PEM; it allows hydrogen ions generated in the

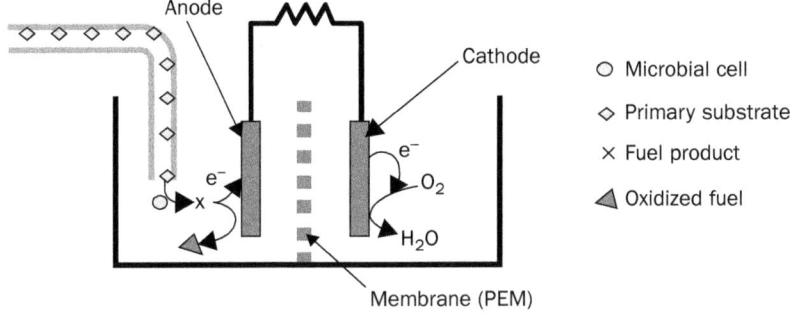

Anode

Cathode

○ Microbial cell

◇ Primary substrate

✕ Fuel product

◀ Oxidized fuel

e^-

O_2

H_2O

Membrane (PEM)

FIGURE 1.23   Biofuel cell.

anode compartment to be transferred across the membrane into the cathode compartment. Previously, graphite electrodes were used as the anode and cathode, but they are now replaced by woven graphite felt as it provides a larger surface area than a regular graphite electrode of similar dimensions.

This facilitates an increased electron transfer from the microorganisms. A microorganism (e.g., *Escherichia coli*) is used to breakdown glucose to generate adenosine triphosphate (ATP), which is utilized by cells for energy storage. Methylene blue (MB) or neutral red (NR) is used as an electron mediator to efficiently facilitate the transfer of electrons from the microorganism to the electrode. Electron mediators tap into the electron transport chain, chemically reducing nicotinamide adenine dinucleotide (NAD^+) to its protonated form NADH. The exact mechanism by which the transfer of electrons takes place through these electron mediators is not fully known; however, it is known that they insert themselves into the bacterial membrane and essentially "hijack" the electron transport process of glucose metabolism of the bio-electrodes in a biofuel cell. Their activity is dependent on pH, and a potassium phosphate buffer (pH 7.0) is used to maintain the pH value in the anode compartment. The cathode compartment contains potassium ferricyanide a potassium phosphate buffer (pH 7.0), and a woven graphite felt electrode. Potassium ferricyanide reaction helps in rapid electron uptake. Hydrogen ions (H^+) migrate across the PEM and combine with oxygen from air and the electrons to produce water at the cathode. The cathode compartment has to be oxygenated by constant bubbling with air to promote the cathode reactions. It may be worth mentioning that the electron transport chain occurs in the cell membrane of prokaryotes (a unicellular organism having cells lacking membrane-bound nuclei, such as bacteria), while this process occurs in the mitochondrial membrane of eukaryotes (animal cells). Therefore, attempts to substitute eukaryotic cells for bacterial cells in a biofuel cell may present a significant challenge.

Electrochemistry of microbial fuel cells. In a microbial fuel cell, two redox couples are required to generate a current: (a) coupling of the reduction of an electron mediator to a bacterial oxidative metabolism and (b) coupling of the oxidation of the electron mediator to the reduction of the electron acceptor on the cathode surface. The electron acceptor is subsequently regenerated by the presence of O_2 at the cathode surface. The electrochemical reactions in a biofuel cell using glucose as a fuel are

At the anode:

$$C_6H_{12}O_6 + 6H_2O \rightarrow 6CO_2 + 24e^- + 24H^+$$

At the cathode:

$$4Fe(CN)_6^{3-} + 4e^- \rightarrow 4Fe(CN)_6^{4-}$$
$$4Fe(CN)_6^{4-} + 4H^+ + O_2 \rightarrow 4Fe(CN)_6^{3-} + 2H_2O$$

Complete oxidation of glucose does not always occur. One might often get additional products besides CO_2 and water. For example, *E. coli* forms acetate, being unable to completely breakdown glucose, thereby limiting electricity production. Recently, an elegant approach to address this long-standing problem of limited enzyme stability has been reported [30]. It is suggested that the immobilization of enzymes in Nafion layers to create a bio-anode results in stable performance over months.

Another way of using a microorganism's ability to produce electrochemically active substances for energy generation is to combine a bioreactor with a biofuel cell or a hydrogen fuel cell. The fuel can be produced in a bioreactor at one place and transported to a (H_2 or bio-) fuel cell to be used as a fuel. In this case, the biocatalytic microbial reactor produces the fuel, and the biological part of the device is not directly integrated with the electrochemical part (Fig. 1.24).

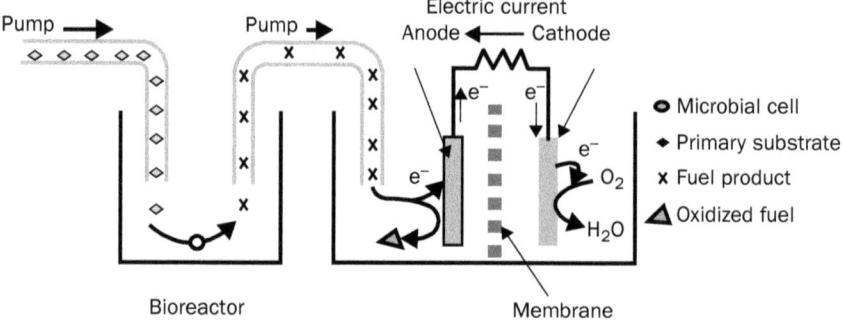

FIGURE 1.24 Bioreactor and biofuel cell combination.

The advantage of this scheme is that it allows the electrochemical part to operate under conditions that are not compatible with the biological part of the device. The two parts can even be separated in time, operating completely independently. The most widely used fuel in this scheme is hydrogen gas, allowing well-developed and highly efficient H_2/O_2 fuel cells to be conjugated with a bioreactor.

In recent years, ethanol has been developed as an alternative to the traditional methanol-powered biofuel cell due to the widespread availability of ethanol for consumer use, its nontoxicity, and increased selectivity by alcohol. Ethanol fuel cells with immobilized enzymes have provided higher power densities than the latest state-of-the-art methanol biofuel cells. Open-circuit potentials ranging from 0.61 to 0.82 V and power densities of 1.00–2.04 mW/cm^2 have been produced.

Mediatorless microbial fuel cells. Most biofuel cells need a mediator molecule to speed up the electron transfer from the enzyme to the electrode. Recently, mediatorless microbial fuel cells have been developed. These use metal-reducing bacteria, such as members of the families *Geobacteraceae* or *Shewanellaceae*, which exhibit special cytochromes bound to their membranes. These are capable of transferring electrons to the electrodes directly. *Rhodoferax ferrireducens*, an iron-reducing microorganism, has the ability to directly transfer electrons to the surface of electrodes and does not require the addition of toxic electron-shuttling mediator compounds employed in other microbial fuel cells. Also, this metal-reducing bacterium is able to oxidize glucose at 80 percent electron efficiency (other organisms, such as *Clostridium* strains, oxidize glucose at only 0.04 percent efficiency). In other fuel cells that use immobilized enzymes, glucose is oxidized to gluconic acid and generates only two electrons, whereas in microbial fuel cells (MFCs) using *R. ferrireducens*, glucose is completely oxidized to CO_2 releasing 24 electrons. These MFCs have a remarkable long-term stability, providing a steady electron flow over extended periods. Current density of 31 mA/m^2 over a period of more than 600 h has been reported [31]. MFCs using *R. ferrireducens* have the ability to be recharged and have a reasonable cycle life and low capacity loss under open-circuit conditions. They allow the harvest of electricity from many types of organic waste matter or renewable biomass. This is an advantage over other microorganisms in the family *Geobacteraceae*, which cannot metabolize sugars.

Another recent development has been the use of microfibers rather than flat electrodes and the enzyme-based electroactive coatings. The anode coating used is glucose oxidase, which is covalently bound to a reducing-potential copolymer and has osmium complexes attached to its backbone. The cathode coating contains the enzyme laccase and an oxidizing-potential copolymer. The osmium redox centers in the coatings electrically "wire" the reaction centers of the enzymes to the carbon fibers. This electrode design avoids glucose oxidation at the cathode and O_2 reduction at the anode, eliminating the need for an electrode-separating membrane. This has led to miniature "one-compartment biofuel cells" for implantable devices within humans, such as pacemakers, insulin pumps, sensors, and prosthetic units. Biofuel cells with two 7-μm-diameter, electrocatalyst-coated carbon fiber electrodes placed in 1-mm grooves machined into a polycarbonate support with a power output of 600 nW at 37°C (enough to power small silicon-based microelectronics) have been reported.

Microbial fuel cells have a long way to go before they compete with more established hydrogen fuel cells or electrical batteries. However, a number of factors provide motivation for research into microbial fuel cells for electricity production.

1. Bacteria are adapted to feeding on virtually all available carbon sources (carbohydrates or more complex organic matter present in sewage, sludge, or even marine sediments). This makes them potential catalysts for electricity generation from organic waste.

2. Bacteria are omnipresent in the environment and are self-reproducing, self-renewing catalysts; thus a simple initial inoculation of a suitable strain could be cultured continuously in an MFC for long-term operation.

3. The catalytic core of conventional fuel cells uses expensive precious metals such as platinum, and biocatalysts like bacteria may become a serious cost-reducing alternative.

Although biofuel cells are still in an early stage of development and work toward optimizing the performance of a biofuel cell system is needed, the utilization of white blood cells as a source of electrons for a biofuel cell could mark an important step in developing a perpetual power source for implantable devices. There is still a lot of work to be done as there are many unanswered questions; however, the feasibility of constructing commercially viable biofuel cell power supplies for a number of applications is promising.

1.9.13   Fuel cell system

A fuel cell power system requires the integration of many components. The fuel cell produces only dc power and utilizes only certain processed fuels. Besides the fuel cell stack, various components are incorporated in a fuel cell system. A fuel processor is required to allow operation with conventional fuels; a power conditioner is used to tie fuel cells into the ac power grid or distributed generation system; for high-temperature fuel cells, a cogeneration or bottoming cycle plant is needed to utilize rejected heat for achieving high efficiency. A schematic of a fuel cell power system with interaction among various components is shown in Fig. 1.25.

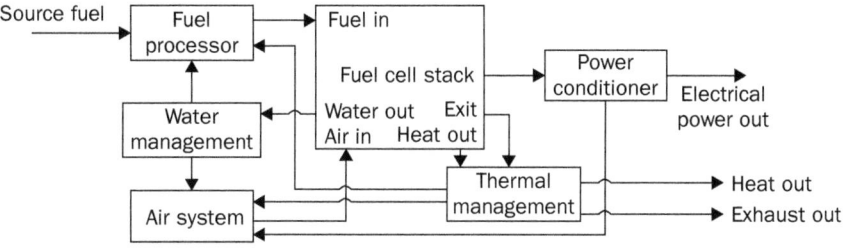

Figure 1.25 A fuel cell power system schematic.

1.9.14 Fuel processor

A fuel processor converts a commercially available fuel (gas, liquid, or solid) to a fuel gas reformate suitable for the fuel cell use. Fuel processing involves the following steps:

1. **Fuel cleaning**—It involves cleaning and removal of harmful species (sulfur, halides, and ammonia) in the fuel. This prevents fuel processor and fuel cell catalyst degradation.

2. **Fuel conversion**—In this stage, a naturally available fuel (primarily hydrocarbons such as natural gas, petrol, diesel, ethanol, methanol, biofuels [such as produced from biomass, landfill gas, biogas from anaerobic digesters, syngas from gasification of biomass and wastes], etc.) is converted to a hydrogen-rich fuel gas reformat.

3. **Downstream processing**—It involves reformate gas alteration by converting carbon monoxide (CO) and water (H_2O) in the fuel gas reformate to hydrogen (H_2) and carbon dioxide (CO_2) through the water gas shift reaction, selective oxidation to reduce CO to a few parts per million, or removal of water by condensing to increase the H_2 concentration.

A schematic showing the different stages in the fuel-processing system is presented in Fig. 1.26. Major fuel-processing techniques are steam reforming (SR), partial oxidation (POX) (catalytic and noncatalytic), and autothermal reforming (ATR). Some other techniques such as dry reforming, direct hydrocarbon oxidation, and pyrolysis are also used. Most fuel processors use the chemical and heat energy of the fuel cell effluent to provide heat for fuel processing. This enhances system efficiency.

Steam reforming is a popular method of converting light hydrocarbons to hydrogen. In SR, heated and vaporized fuel is injected with superheated steam (steam-to-carbon molar ratio of about 2.5:1) into a reaction vessel. Excess steam ensures complete reaction as well as inhibits soot formation. Although the steam reformer can operate without a catalyst, most commercial reformers use a nickel- or

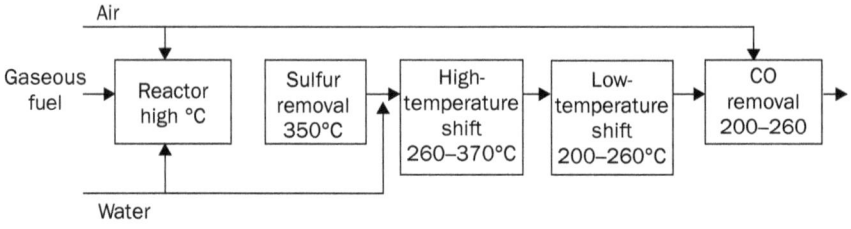

Figure 1.26 A fuel-processing system.

cobalt-based catalyst to enhance reaction rates at lower temperatures. Although the water gas shift reaction in the steam reformer reactor is exothermic, the combined SR and water gas shift reaction is endothermic. It therefore requires a high-temperature heat source (usually an adjacent high-temperature furnace that burns a small portion of the fuel or the fuel effluent from the fuel cell) to operate the reactor. SR is a slow reaction and requires a large reactor. It is suitable for pipeline gas and light distillates using a fuel cell for stationary power generation but is unsuitable for systems requiring rapid start and/or fast changes in load.

In POX, a substoichiometric amount of air or oxygen is used to partially combust the fuel. POX is highly exothermic, and the resulting high-temperature reaction products are quenched using superheated steam. This promotes the combined water gas shift and steam-reforming reactions, which cools the gas. In a well-designed POX reformer with controlled preheating of the reactants, the overall reaction is exothermic and self-sustaining. Both catalytic (870–925°C) and noncatalytic (1175–1400°C) POX reformers have been developed for hydrocarbon fuels. The advantage of POX reforming is that it does not need indirect heat transfer, resulting in a compact and lightweight reformer. Also, it is capable of higher reforming efficiencies than steam reformers.

Autothermal reforming combines SR with POX reforming in the presence of a catalyst that controls the reaction pathways and thereby determines the relative extents of the POX and SR reactions. The SR reaction absorbs part of the heat generated by the POX reaction, limiting the maximum temperature in the reactor. This results in a slightly exothermic process, which is self-sustaining, and high H_2 concentration. The ATR fuel processor operates at a lower operating cost and lower temperature than the POX reformer, and it is smaller, quicker starting, and quicker responding than the SR.

Most of the natural hydrocarbon fuels, such as natural gas and gasoline, contain some amount of sulfur, or sulfur-containing odorants are added to them for leak detection. As the fuel cells or reformer catalysts do not tolerate sulfur, it must be removed. Sulfur removal is usually achieved with the help of zinc oxide sulfur polisher, which removes the mercaptans and disulfides. A zinc oxide reactor is operated at 350–400°C to minimize bed volume. However, removing sulfur-containing odorants such as thiophane requires the addition of a hydrodesulfurizer stage before the zinc oxide polisher. Hydrogen (supplied by recycling a small amount of the natural gas-reformed product) converts thiophane into H_2S in the hydrodesulfurizer. The zinc oxide polisher easily removes H_2S.

To reduce the level of CO in the reformat gas, it must be water gas shifted. The shift conversion is often performed in two or more stages when CO levels are high. A first high-temperature stage allows high reaction rates, while a low-temperature converter allows a higher conversion. Excess steam is used to enhance the CO conversion. In a PEMFC, the reformate is passed through a preferential CO catalytic oxidizer after being shifted in a shift reactor, as a PEMFC can tolerate a CO level of only about 50 ppm.

A fuel processor is an integrated unit consisting of one or more of the above stages, as per the requirements of a particular type of fuel cell. High-temperature fuel cells such as the SOFC and MCFC are equipped with internal fuel reforming and hence do not require a high-temperature shift or low-temperature shift stage. The CO removal stage is not required for the SOFC, MCFC, PAFC, and circulating AFC. For the PEMFC, all the stages are required.

1.9.15 Air management

Besides fuel, a fuel cell also requires an oxidant (usually air). Depending on the application and design, air provided to the fuel cell cathode can be at a low pressure or a high pressure. High pressure of the air improves the reaction kinetics and increases the power density and efficiency of the stack. But increasing the air pressure reduces the water-holding capacity of the air and therefore reduces the humidification requirements of the membrane (PEMFC). It also increases the power required to compress the air to a high pressure and thereby reduces the net power available. At present, most fuel cell stacks for stationary power applications are designed for operating pressures in the range of 1–8 atm, while automotive fuel cell systems based on the PEMFC technology are designed to operate at lower pressures of 2–3 atm to increase power density and improve water management.

1.9.16 Water management

Water management is critical for fuel cell operation. Water is a product of the fuel cell reaction, and it must be removed from the exhaust gas for use in various operations such as fuel reformation and humidifying reactant gases (to avoid drying out the fuel cell membrane). For automotive applications, water condensed from the exhaust steam is recycled for reforming and reactant humidification in a closed cycle to avoid periodical recharging with water.

1.9.17 Thermal management

The reaction products of the electrochemical reaction in a fuel cell are water, electricity, and heat. The heat energy released in a fuel cell stack is approximately equal to the electrical energy generated and must be managed properly to maintain the fuel cell stack temperature at the optimal level. If this thermal energy (waste heat) is properly utilized, it will considerably increase the efficiency of a fuel cell system. In low-temperature (<200°C) fuel cells (PEMFC, AFC, and PAFC), the stack is cooled by supplying excess air in low power (<200-W) systems, whereas a liquid coolant (deionized water) is used for large-size systems. The waste heat carried out by the coolant is utilized for cogeneration (space heating, water heating, etc.). In high-temperature (<600°C) fuel cell (MCFC and SOFC) systems, all the heat of reaction is transferred to the reactants to maintain the stack temperature at the optimal level. The thermal energy of the high-temperature exhaust may be utilized to preheat the incoming air stream, or in internal or external fuel reformer. The high-temperature exhaust may also be used for cogeneration or electricity generation in a downstream gas turbine system.

1.9.18 Power-conditioning system

The power-conditioning system is an integral part of a fuel cell system. It converts the dc electric power generated by the fuel cell into regulated dc or ac for consumer use. The electrical characteristics of a fuel cell are far from that of an ideal electric power source. The dc output voltage of a fuel cell stack varies considerably with the load current (Fig. 1.27), and it has little overload capacity. It needs considerable auxiliary power for pumps, blowers, and so forth and requires considerable startup time due to heating requirements. It is slow to respond to load changes, and its performance degrades considerably with the age of the fuel cell. The various blocks of a fuel cell power-conditioning system are shown in Fig. 1.28.

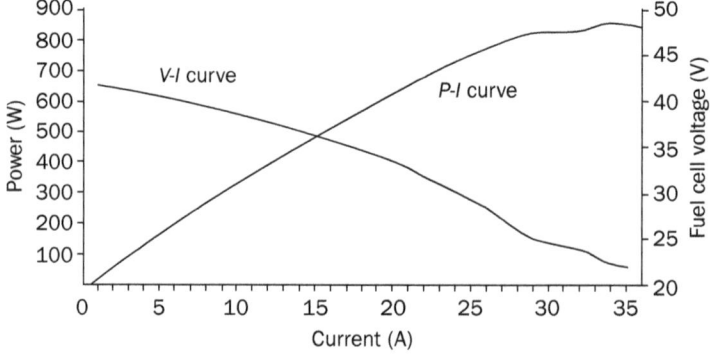

FIGURE 1.27 Voltage-current and voltage-power characteristics of a typical fuel cell.

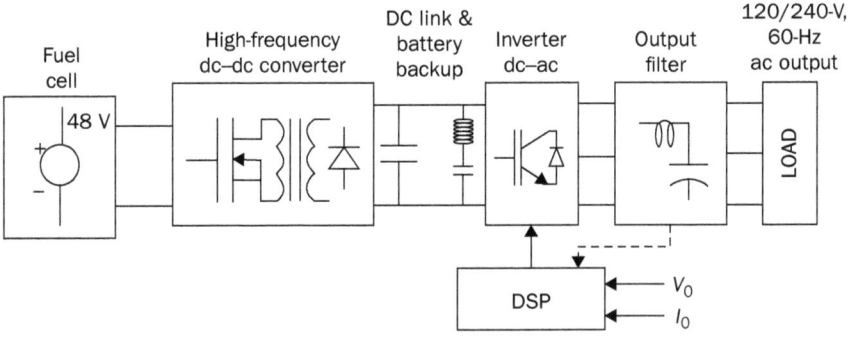

Figure 1.28 Schematic of a fuel cell power-conditioning system.

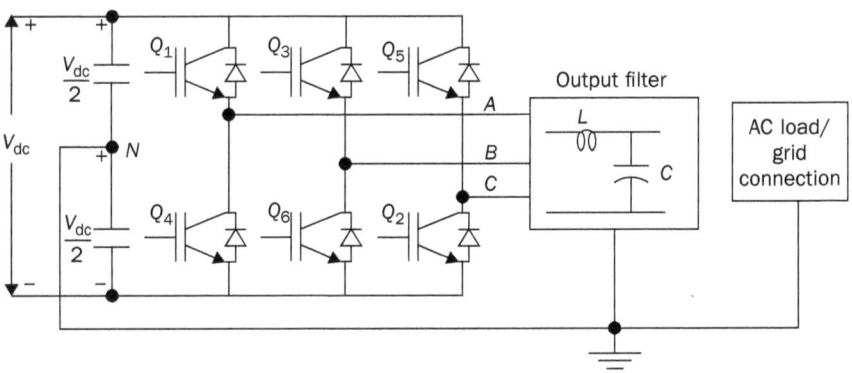

Figure 1.29 Three-phase inverter for an ac load or grid connection.

The dc voltage generated by a fuel cell stack is usually low in magnitude (<50 V for a 5- to 10-kW system, <350 V for a 300-kW system) and varies widely with the load. A dc–dc converter stage is required to regulate and step up the dc voltage to 400–600 V (typical for 120/240-V ac output). Since the dc–dc converter draws power directly from the fuel cell, it should not introduce any negative current into the fuel cell and must be designed to match the fuel cell ripple current specifications. A dc–ac conversion (inverter) stage is needed for converting the dc to ac power at 50 or 60 Hz (see Fig. 1.29). Switching frequency harmonics are filtered out using a filter connected to the output of the inverter to generate a high-quality sinusoidal ac waveform suitable for the load.

1.9.19 Fuel cell applications

The major applications for fuel cells are as stationary electric power plants (including cogeneration units), as a transportation power source for vehicles, and as portable power sources, besides an electric power source for space vehicles or other closed environments.

Stationary power applications are favorable for fuel cell systems. Stationary applications mostly require continuous operation, so startup time is not a very important constraint. Thus, high-temperature fuel cells such as the MCFC and SOFC systems are also suitable for this application in addition to the PAFC and PEMFC systems. The fuel source for stationary applications is most likely to be natural gas, which is relatively easy to reform in the internal reformer of high-temperature fuel cells or in the external reformer for low-temperature fuel cells. An advantage of using natural gas is that the distribution infrastructure for natural gas already exists. Promising applications for stationary fuel cell systems include premium power systems (high-quality uninterruptible/backup power supply systems); high-efficiency cogeneration (heat and electricity) systems for residences, commercial buildings, hospitals, and industrial facilities; and distributed power generation systems for utilities. Although some demonstration and commercial stationary fuel cell power plants in sizes from a few kilowatts to 11 MW are in operation, widespread commercialization can be expected only if their installation cost drops down from the present cost of $4000/kW to about $400–700/kW (or about $1000/kW for some premium applications).

The recent surge of interest in fuel cell technology is because of its potential use in transportation applications, including personal vehicles. This development is being sponsored by various governments in North America, Europe, and Japan, as well as by major automobile manufacturers worldwide, who have invested several billion dollars with the goal of producing a high-efficiency and low-emission fuel cell power plant at a cost that is competitive with the existing internal combustion engines. With hydrogen as the onboard fuel, such vehicles would be zero-emission vehicles. With fuels other than hydrogen, an appropriate fuel processor to convert the fuel to hydrogen will be needed. Fuel cell-powered vehicles offer the advantages of electric drive and low maintenance, because of the few critical moving parts. The major activity in transportation fuel cell development has focused on the polymer electrolyte fuel cell (PEFC), and many of the technical objectives related to the fuel cell stack have been met or are close to being met. The current development efforts are focused on decreasing cost and resolving issues related to fuel supply and system integration.

Besides exotic areas of applications such as space vehicles or submarines, another promising area of application for fuel cells is portable power systems. Portable power systems are small, lightweight systems that power portable devices (e.g., computers, laptops, cellular phones, and entertainment electronic devices), camping and recreational vehicles, military applications in the field, and so forth. These devices need power in the range of a few watts to a few hundred watts. Fuel cell systems based on DMFC or PEMFC technology are well suited for many of these applications. The convenience of transporting and storing liquid methanol makes DMFC systems attractive for this application. A small container of methanol or a cylinder of compressed hydrogen can be used as a fuel supply. When the fuel is depleted, a new fuel container may be installed in its place after removing the old one.

In recent years, there has been a lot of interest in electric power generation using renewable energy sources such as wind energy, solar energy, and tidal energy. A major problem with these energy sources is that all are intermittent in nature. Combining the renewable energy-based power generation system with a fuel cell system would solve this problem to a great extent. A hybrid wind/solar energy–fuel cell system can use wind/solar power for generating hydrogen using the electrolysis of water, and store it in cylinders at high pressure. This hydrogen can then be used as the fuel for the fuel cell stack. The stored hydrogen can also be used to fuel the fuel cell vehicles and so forth. In a grid-connected wind/solar energy–hydrogen system, wind/solar power whenever available provides electricity for hydrogen production. The grid power is used during off-peak periods for low-cost electricity and hydrogen production; whereas during peak-demand periods or no/low wind/solar energy periods, the fuel cell can generate electricity using the stored hydrogen. These hybrid systems could be configured in several ways.

1.9.20 Conclusion

Fuel cell systems are one of the most promising technologies to meet our future power generation requirements. Fuel cell systems provide a clean and efficient technology for electrical and automotive power systems. With cogeneration efficiencies higher than 80 percent, fuel cells promise to reduce primary energy use and environmental impact. Fuel cells are a good alternative for rural energy needs, especially in remote places where there are no existing power grids or power supply is unreliable. The application of fuel cells into the transportation sector will reduce greenhouse emissions considerably; if fuels from renewable energy sources are used, it would nearly eliminate greenhouse gas emissions. Utility companies are beginning to locate small, energy-saving power generators closer to loads to overcome right of way problems and transmission line costs. The modular design of fuel cells suits this distributed generation strategy very nicely as new modular units can be added when the demand increases. This reduces the financial risk for utility planners. Biofuel cells are attractive for implant devices as they can use glucose in blood to power these devices, eliminating the need for surgery for maintenance and battery replacement. Use of digester gas as a fuel in biofuel cells makes them very attractive for power generation from garbage and other organic waste. This will also help in waste disposal, a big problem in the agriculture and food industries.

All fuel cell technologies (PEMFC, DMFC, AFC, PAFC, MCFC, SOFC, and MFC) discussed in this chapter are in a very advanced stage of development and are near commercialization. Although a number of demonstration units of different types of fuel cells are operating all over the world and many PAFC and AFC units have been commercially sold and are successfully operating, fuel cells are still awaiting widespread commercialization due to their high cost and limitation in the choice of the fuel used. These barriers will be overcome in the next few years, and fuel cells will become a preferred power source with widespread applications.

References

1. L. A. Kristofferson and V. Bokalders. *Renewable Energy Technologies: Their Applications in Developing Countries*. A Study of the Beijer Institute and the Royal Swedish Academy of Sciences, Oxford, Pergamon: Sweden, 1986.
2. A. K. Raja, M. Dwivedi, and A. Prakash. *Introduction to Non-conventional Energy Resources*. New Delhi, India: Scitech Publication, 2012.
3. A. Nag and K. Vizaykumar. *Environmental Education and Solid Waste Management*. New Delhi, India: New Age Publisher, 2006.
4. N. George. "Wave power." In *Encyclopedia of Energy Technology and the Environment*. New York, NY: John Wiley and Sons, 1995.
5. *Proceedings of Bio-Energy Society* (Department of Nonconventional Energy), CGO, New Delhi, India, October 14–16, 1985.
6. R. C. Kuhad and A. Singh. Lignocellulose biotechnology: Current and future prospects, *Critical Reviews in Biotechnology* **13**, 151–172, 1993.
7. G. Hoogers (Ed.). *Fuel Cell Technology Handbook*, Boca Raton, FL: CRC Press, 2003.
8. W. Vielstich, A. Lamn, and H. A. Gasteiger (Eds.). *Handbook of Fuel Cells: Fundamentals, Technology and Applications*, Four Volumes. New York, NY: John Wiley, 2003.
9. Department of Energy. *Fuel Cell Handbook*, sixth ed., Pittsburgh, PA: National Energy Technology Laboratory, DOE, November 2002.
10. M. C. Williams. Fuel cells and the world energy future, *IEEE Power Engineering Society Summer Meeting* 1, 725, July 15–19, 2001.
11. M. A. Laughton. Fuel cells, *Engineering Science and Education Journal* **11**(1), 7–16, 2002. 12.
12. R. K. Shah. Introduction to fuel cells, In *Recent Trends in Fuel Science and Technology*, S. Basu (Ed.), New Delhi, India: Anamya Publishers, pp. 1–9, 2007.
13. R. M. Allen and H. P. Bennetto. Microbial fuel cells—Electricity production from carbohydrates, *Applied Biochemistry and Biotechnology* **39/40**, 27–40, 1993.

14. K. S. Dhathathreyan and N. Rajalakshmi. Polymer electrolyte membrane fuel cell, In *Recent Trends in Fuel Science and Technology*, S. Basu (Ed.), New Delhi, India: Anamya Publishers, pp. 41–115, 2007.

Questions

1. What is nonconventional energy, and what are its advantages and disadvantages?

2. How is solar radiation measured? What are solar panel and solar thermal collectors? Discuss the applications of solar energy.

3. What are the applications of geothermal energy? How are low and high tides formed and how have they been used for the production of tidal energy?

4. What are the different systems of wind energy? What are the advantages and disadvantages of wind energy?

5. Describe the different stages of biogas production. What are the different factors of gobar gas projects?

6. What is a fuel cell? What are the major technical challenges in fuel cell commercialization? Consider the benefits and drawbacks of the different types of fuel cells.

CHAPTER 2

Photosynthetic Plants as Renewable Energy Sources

Ahindra Nag and P. Manchikanti

2.1 Introduction

Renewable energy is an energy resource naturally regenerated over a short time scale derived from the sun (such as thermal, photochemical, and photoelectric) or from other natural environment effects (geothermal and tidal energy). It is forecasted that approximately half of the total resources in the world will be exhausted by 2025. This survey has also revealed that global warming and climate change are serious issues that need immediate action. The use of fossil fuels (coal, oil, gas, etc.) contributes significantly to global warming and climate change [1]. Worldwide there is strong support for renewable energy, as proven by a number of surveys [1, 2]. In 2003, a European Commission survey across the 15 European Union (EU) countries showed that 69 percent of the citizens supported more renewable energy-related research, compared to 13 percent for gas, 10 percent for nuclear fission, 6 percent for oil, and 5 percent for coal. Understandably, due to the inherent recycling nature as well as environmental benefits involved, renewable sources of energy are the solution for energy management. There is an increased investment globally in such technologies for not only enhancing the preservation of biological resources but also for increasing energy efficiency and pollution control [1].

Biomass is one such renewable source of energy. Out of the 1.1×10^{20} kW heat generated every second by the sun, only 47 percent

($\sim 7 \times 10^{17}$ kWh) reaches the earth's surface. Solar energy is utilized by conversion to different energy forms such as biomass, wind, or hydropower. Green plants are only able to effectively use visible light of wavelength falling between 400 and 700 nm. This photosynthetically active radiation constitutes about 43 percent of the total incident solar radiation to produce biomass. Biomass energy generally involves the utilization of energy contents of such items as agricultural residues (pulp derived from sugarcane, corn fiber, rice straw and hulls, and paper trash) and energy crops. So, *biomass* is a comprehensive term that includes essential forms of matter derived from photosynthesis or ultimately available as animal waste [2]. The production of energy from plants is not a new idea; wood burning has been in common use since ancient times. About one-seventh of the energy used around the world is derived from firewood. Biomass supplies 14 percent of the world's primary energy consumption and is considered to be one of the important renewable resources of the future. With the increase in population and the demand for resources, demand for biomass is expected to increase rapidly. On average, 38 percent of the primary energy resources in developing countries is biomass. In the United States alone, biomass sources provide about 3 percent of all the energy consumed. In terms of energy efficiency measures and stabilization of energy consumption between 2010 and 2020, the European Renewable Energy Council (EREC) survey estimates that among the various types of renewable energy resources, biomass-derived energy will be a significant portion of energy used [1]. The survey also reveals that biomass and biofuels are the top two in terms of employment that they generate. Burning new biomass does not contribute to new CO_2 into the atmosphere as replanting harvested biomass ensures that CO_2 is absorbed and returned for a cycle of new growth [2].

2.2 Mechanism and Efficiency of Photosynthesis in Plants

In photosynthesis, CO_2 from the atmosphere and water from the earth combine to produce carbohydrates, which are the components of biomass and solar energy that drive this process. When biomass is efficiently utilized, the oxygen from the atmosphere combines with the carbon in plants to produce CO_2 and water (see Fig. 2.1). Typically, photosynthesis converts less than 1 percent of the available sunlight to be stored as chemical energy.

The advantages of using plants for renewable energies (fuels and chemicals) are as follows:

- Advances in agriculture and forestry technologies have resulted in increased utilization of land resources for cultivation of energy crops.

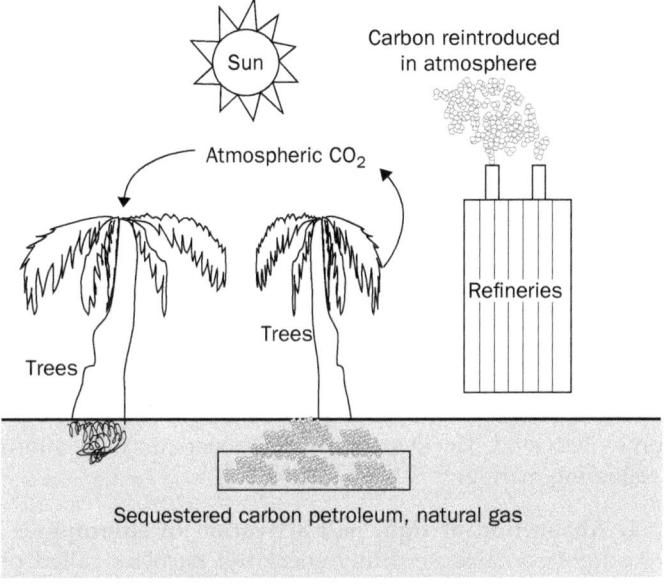

FIGURE 2.1 Simplified carbon cycle.

- By increasing harvesting of solar energy, there is effective usage of biomass-based resources.
- Multiple economic benefits can be derived—for example, sugar can be used as such for fermentation to alcohol—depending on the market.
- Biomass combustion, unlike fossil fuels, does not contribute to increased CO_2 levels in the atmosphere [2].
- There are increased employment opportunities resulting from the above.

While the advantages of using biomass-based energies are apparent, it is important to note that biomass cannot by itself provide complete replacement of fossil fuels. Hence, it is one of the solutions toward achieving energy efficiency. Further factors, such as competition for biomass between energy production and human nutritional needs, as well as the possible environmental effects, must be kept in mind. There are several factors that should be considered in using plants for the generation of energy: efficiency of solar energy absorption and conversion, quality of biomass produced, plant growth, growth under marginal conditions, soil characteristics, and cost-effectiveness of production of energy and conversion. We will focus on the utilization of terrestrial plants for production of renewable energies.

2.3   Photosynthetic Process

There are essentially two types of reactions in photosynthesis: a series of light-dependent reactions that are temperature independent (or light reaction) and a series of temperature-dependent reactions that are light independent (or dark reactions). The rate of the light reaction can be increased by increasing light intensity, and the rate of the dark reaction can be increased by increasing temperature to a certain extent (see Fig. 2.2).

2.3.1   Hill reaction (light reaction)

The process of formation of CO_2 and O_2 during photosynthesis is called the Hill reaction or photolysis of water. This primary photochemical reaction takes place in the presence of sunlight. The reaction is associated with chlorophyll, and after receiving light energy, the chlorophyll becomes activated. The steps in the Hill reaction can be summed up in the following manner:

1. **Absorption of light and activation of chlorophyll**   Radiant light contains very tiny energized particles called photons or quanta, which are absorbed by the chlorophyll and it becomes activated.

2. **Photolysis**   The dissociation of water molecules by light energy that have been absorbed by the chlorophyll is called photolysis. The reaction can be represented as

$$4H_2O \xrightarrow{\text{Light energy}} 4H^+ + 4OH^-$$

$$4H^+ \xrightarrow{\text{Chlorophyll}} 2H_2$$

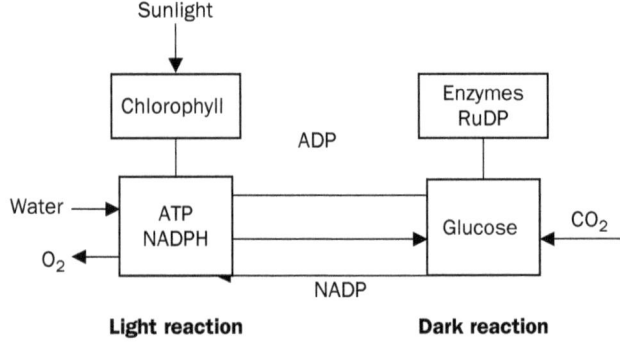

Light reaction   **Dark reaction**

FIGURE 2.2   General process of photosynthesis.

$$H_2 + NADP \rightarrow NADPH_2 \text{ (Hydrogen acceptor)}$$

$$4OH^- \xrightarrow{\text{Recombination}} 2H_2O + O_2$$

3. **Photophosphorylation**   This is the stage of formation of ATP from ADP .

$$ADP + Pi \rightarrow ATP$$

2.3.2   Blackman's reaction (dark reaction)

The dark reaction is independent of light. This reaction is purely enzymatic and is carried out in the stoma portion of the chloroplast. Ribulose-1, 5-diphosphate (RuDP), a pentose phosphate present in plant cells, acts as the initial acceptor of CO_2 and changes thereby into a very unstable C_6. The latter is converted into 3-phosphoglyceric acid (3-PGA), which is transferred to 3-phosphoglyceraldehyde. For this reaction, ATP and $NADPH_2$ (produced in the light reaction) are necessary as cofactors. Three molecules of RuDP combine with three molecules of CO_2 to give rise to six molecules of PGA. Three molecules of RuDP utilized initially as CO_2 acceptors are regenerated by five molecules of phosphoglyceraldehyde through different intermediates like xylulose-5-phosphate and ribulose-5-phosphate. The only molecule of phosphoglyceraldehyde is converted into fructose-1,6-diphosphate, which may be transformed into sucrose and starch through other reactions.

2.3.3   Efficiency of photosynthesis

While there are several factors that affect photosynthetic rate, the three main factors are light intensity, carbon dioxide level, and temperature. The net efficiency of photosynthesis is estimated by the net growth of biosynthesis and the amount used for respiration. The requirements for achieving high energy conversion are optimal temperature, light, nutrition, leaf canopy, absence of photorespiration, and so forth. Many plant species can be distinguished by the type of photosynthetic pathway they utilize. Most plants utilize the C_3 photosynthesis route. C_3 determines the mass of carbon present in the plant material. Poplar, willow, wheat, and most cereals are C_3 plant species. Plants such as perennial grass, *Miscanthus*, sweet sorghum, maize, and artichoke all use the C_4 route of photosynthesis and accumulate significantly greater dry mass of carbon than the C_3 plants. Advances in crop production, agricultural techniques, and so forth have led to potential applications in low-cost biomass production with high conversion efficiencies. Further, introduction of alternative nonfood crops on surplus land and the use of biomass as a sustainable and environmentally safe alternative make biomass an attractive renewable energy resource. The potential of biomass energy derived from forest and agricultural residues worldwide is estimated at about 30 EJ/yr. For the adoption of biomass as a renewable energy source, the cultivation of energy crops using fallow and marginal land and efficient processing methods are vital [3].

C_3 metabolism in plants and the pentose phosphate pathway

In C_3 plants, the pathway for reduction of carbon dioxide to sugar involves the reductive pentose phosphate cycle. This involves addition of CO_2 to the pentose bisphosphate, ribulose-1,5-bisphosphate (RuBP). The enzyme-bound carboxylation product is hydrolytically split, through an internal oxidation- reduction process, into two identical molecules of 3-PGA. An acyl phosphate of this acid is formed by reaction with ATP. This is further reduced with NADPH. Five molecules of the resulting triose phosphate are converted into three molecules of the pentose phosphate, ribulose 5-phosphate. Three molecules of ribulose 5-phosphate are converted with ATP to give the carbon dioxide acceptor, RuBP, thereby completing the cycle. When these three RuBP molecules are carboxylated and split into six PGA molecules and these are reduced to triose phosphate, there is a net gain of one triose phosphate molecule over the five needed to regenerate the carbon dioxide acceptor. Triose phosphate is formed in this cycle and can either be converted into starch for storage of energy inside the chloroplast, or it can serve its primary function by being transported out of the chloroplast for subsequent biosynthetic reactions. In a mature leaf, sucrose is synthesized and exported to the rest of the plant, thus providing energy and reduced carbon for growth [4]. Wheat, potato, rice, and barley are examples of C_3 plants. A representative C_3 cycle is shown in Fig. 2.3.

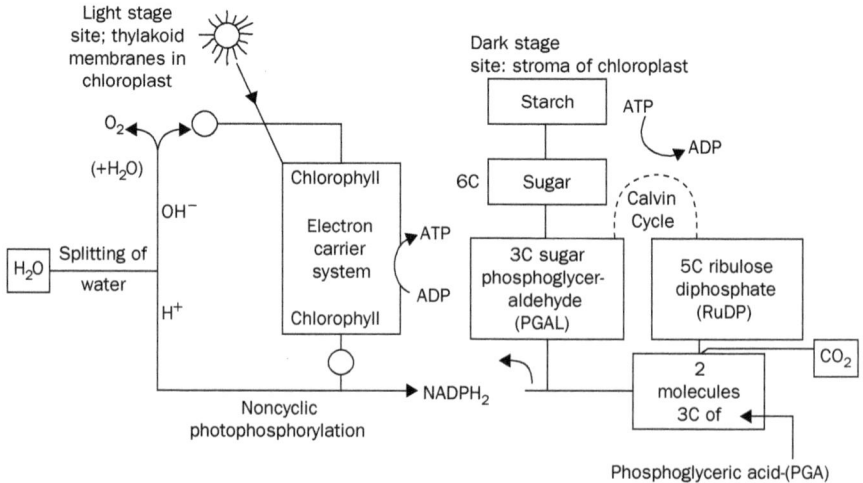

FIGURE 2.3   Representation pathways of C3 plant photosynthesis. (*With permission from Oxford University Press.*)

C_4 metabolism in plants

In air that contains low carbon dioxide in relation to oxygen, oxygen competes for the carbon dioxide binding site of the ribulose bisphosphate carboxylase. This is known to set off a process of photorespiration in plants, and it is believed that the C_4 plants have evolved from such a mechanism. Such plants possess a specialized leaf morphology called "Krantz anatomy" and a special additional CO_2 transport mechanism. This typically overcomes the problem of photorespiration. Such avoidance of photorespiration is known to result in higher growth rates. The Krantz anatomy is characterized by the fact that the vascular system of the leaves is surrounded by a vascular bundle, or bundle-sheath cells, which contain enzymes of the reductive pentose phosphate cycle. The reduction of CO_2 is similar to that of C_3 plants, except that the CO_2 for carboxylation of CO_2 is derived not from the stomata but is released in bundle-sheath cells by decarboxylation of a four-carbon acid (C_4 acid). This C_4 acid is supplied by the mesophyll cells that surround the bundle sheath cells. The C_4 pathway for the transport of CO_2 starts in a mesophyll cell with the condensation of CO_2 and phosphoenolpyruvate to form oxaloacetate, in a reaction catalyzed by phosphoenolpyruvate carboxylase (PEPCase), and the reduction of oxaloacetate to malate [5]. Figure 2.4 shows the C_4 cycle of CO_2 fixation in photosynthesis.

Due to the elimination of the photorespiration process, C_4 plants are proposed to be ideal for increased biomass production, especially in marginal conditions. Grasses are suitable for this purpose as they can be grown on a repetitive cropping mode for continuous and maximum production of biomass. Grasses such as Bermuda grass, Sudan grass, sugarcane, and sorghum are good candidates for energy generation

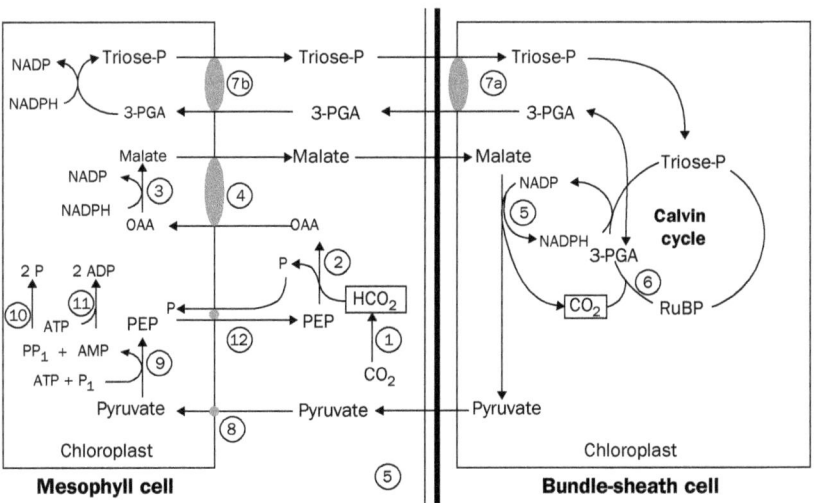

Figure 2.4   The C_4 cycle of CO_2 fixation in photosynthesis. (*Source: Häusler et al. [5]*)

Plant characteristics	C$_3$ cycle type	C$_4$ cycle type
Leaf anatomy	Mesophyll (palisade and spongy type), no chloroplasts in bundle-sheath cell	Krantz anatomy, bundle-sheath cell with chloroplasts
Chloroplasts	Single-type	Dimorphic
Carboxylase type	Primary (Rubisco)	Primary PEPCase in mesophyll, Secondary (Rubisco in bundle-sheath cell)
Primary CO$_2$ acceptor	RuBP	PEP
Primary stable product	3-phosphoglyceric acid (3-PGA)	Oxalocetate (OAA)
Ratio of CO$_2$:ATP:NADPH	1:3:2	1:5:2
Productivity (hectare/ton/yr)	~20	~30

TABLE 2.1   Differences between C$_3$ and C$_4$ plants

from biomass. A comparison of the characteristics of C$_3$ and C$_4$ plants, in terms of leaf anatomy, is shown in Table 2.1.

2.4   Plant Types and Growing Cycles

Several plants have been proposed to be good sources of energy. These include woody crops and grasses/herbaceous plants, starch and sugar crops and oilseeds, fast-growing trees such as hybrid poplars, shrubs such as willows, and so forth. Energy crops can be grown on agricultural lands not utilized for food, feed, and fiber. Farmers could plant these crops along the riverbanks, along lakeshores, between farms and natural forests, or on wetlands. These crops could be a good source of alternate income, reducing the risk of fluctuating markets and stabilizing farm income. Woody plants, herbaceous plants/grasses, and aquatic plants are different sources for biomass production. The type of biomass selected determines the form of energy conversion process. For instance, sugarcane has high moisture content, and therefore, a "wet/aqueous" bioconversion process, such as fermentation, is the predominant method of use. For a low-moisture content type such as wood, gasification, pyrolysis, or combustion are the more cost-effective ways of conversion.

Characteristics of an ideal energy crop are mentioned below:

- Low energy input to produce
- Low nutrient requirements
- Tolerance to abiotic and biotic stresses
- High yield/high conversion efficiency
- Low level of contaminants

FIGURE 2.5 Eucalyptus plantation.

Energy plantations and cropping are means of growing selected species of trees or crops that can be harvested in a shorter time for fuel, energy, and other resources. Each type of popular plant species is discussed in brief, with respect to renewable resources.

Euclayptus. It is a fast-growing plant for firewood (see Fig. 2.5). Different species such as *Eucalyptus nitens, E. fastigata,* and *E. globulus* are used in many countries such as Australia and Brazil. Eucalyptus, an exotic species from Australia, is a versatile tree which adopts itself to a variety of edaphic and climatic conditions. It comes up in different types of soils and climates varying from tropical to warm temperatures and with annual rainfall ranging from 400 to 4000 mm. It grows well in deep, fertile, and well-drained loamy soils with adequate moisture. A large eucalyptus plantation program has been successfully launched in Brazil to serve as the feedstock for its methanol plant. Amatayakul et al. suggest that if eucalyptus wood is used for electricity generation, the cost of electricity generation would be 6.2 US cents/kWh, and consequently, the cost of substituting a wood-fired plant for a coal-fired plant and a gas-fired plant would be US $107 and $196 per ton of C, respectively [6]. Eucalyptus plantations could offer economically attractive options for electricity generation and CO_2 abatement.

Casuarina. *Casuarina* is a genus of shrubs and trees of the Casurinacea family, native to Australia and islands of the Pacific. The species involve *Casuarina equisetifolia* Linn. It is a big evergreen tree with a trunk diameter of 30 cm and height 15 m, and is harvested after 5–7 years (see Fig. 2.6). The plant fixes nitrogen through symbiotic bacteria and thus adds fertility to the soil. It is very useful for afforesting sandy beaches and sand dunes. The wood is used for fuel purposes.

Mimosa. *Mimosa leucocephala* or kubabul is a fast-growing species known for energy plantation (see Fig. 2.7). It has a very high potential for nitrogen fixation and can be well adapted to poor soils, drought, and windstorms. It can fix up to 500 kg of nitrogen per hectare per annum. It coppices readily, and the sprouts, after harvesting, can grow up to 18 ft

Figure 2.6   *Casuarina* plantation.

Figure 2.7   Mimosa plant with flower (planted with gap between two plants). (*Source: Creative Commons.*)

in just 1 year. It is also called the wonder tree. Under irrigated conditions, it can give fodder yields up to 80–100 ton/(ha . yr). Three different varieties of this species (Hawaiian, Cunningham, and Brazilian) are commonly used for plantations in Hawaii, Salvador, and Peru. The Hawaii and the Cunningham varieties are used for energy plantation in India and Australia, respectively. A Hawaiian plantation of 1.27 hectares can support a 1-MW power plant. In Brazil and the Philippines, it is converted into charcoal that has 70 percent of the heating value of oil. Charcoal can be used to produce calcium carbide, acetylene, vinyl plastics, pig iron, and ferroalloys. The low silica, ash, and lignin contents and high cellulose content make this plant good for paper and pulp materials, and also for rayons and cellophanes. It not only gives a prolific fuelwood yield but is also a nutrient-rich fodder for livestock.

Sugarcane.   Sugarcane (*Saccharum officinale*) is a hardy plant that can tolerate poor drainage, can be cultivated as a rotation crop, and can be maintained for years. It is grown in fertile areas with more than 1000 mm of rain and an abundant supply of water. The ethanol yields from this are in the range of 3.8–12 kL/(ha . yr) [7].

Cassava.   Cassava (*Manihot esculenta*), like sugarcane, is grown in tropical climates with an average rainfall of 1000 mm. As it is relatively drought resistant, it can withstand lower annual rainfall. It needs to be grown annually and is difficult to mechanize, and compared to sugarcane, it is less energy efficient. Ethanol yields are estimated in the range of 0.5–4.0 kL/(ha . yr).

Sorghum.   Sorghum embraces a wide variety of plant types and, unlike sugarcane and cassava, is found in the tropical summer rainfall zones. While it can grow in as little as 200–250 mm annual rainfall, maximum yields are obtained in a minimum of 500–600 mm rainfall. Compared to other cereals, it can tolerate high temperatures. Due to its deep root system and low rate of transpiration, it is exceptionally resistant to drought. Ethanol yields of stems and grains of sorghum are in the range of 1.0–5.0 and 2.0–5.0 kL/(ha . yr), respectively.

Babassu.   Babassu (*Orbignya* sp.) is a palm popular in Brazil for the ethanol derived from it. The mesocarp of coconut is the raw material for ethanol production, with an estimate of 0.24 kL/(ha . yr).

Oil-bearing crops.   Vegetable oils are the most promising alternatives to diesel fuel. About 97 percent of all oil-bearing plants are grown in tropical and subtropical climates. There has been some research into the use of plant oils from sunflower, peanut, rapeseed, soybean, and coconut oils as biofuels in unmodified/slightly modified engines. Seed-based oils are shown to lead to slightly higher fuel consumption, probably due to their calorific value [8]. About 14 percent of the oil supplied in the world market is palm oil, yielding an average 3.4 ton/(ha . yr) of oil [9]. Individual palm seeds, however, are capable of producing much higher yields. The extraction of palm kernel oil increases fuel oil yields by 10 percent. Current cultivation is mostly in lowland humid tropics such as Malaysia, West Africa, and Indonesia. While the conditions to grow coconut palms are similar to oil palms, the yield potential of coconut palms has not yet been developed to that potential. Soybeans and peanuts are annual leguminous crops that are used as sources of both oil and protein. Soybeans thrive best in subtropical climates. The individual varieties differ greatly in terms of their reaction to the length of a day and normally can be grown in a limited geographical area. Peanut cultivation requires an ambient temperature for growth, as less than optimal temperatures are known to result in poor yields. Due to its deep root system, it is relatively resistant to drought. It is also a suitable crop for mixed cultivation along with oil palms and corn. In terms of calorific value of seed, oil plants such as *Simmondsia chinensis, Pittosporium resinifreum, Ricinus communis, Jatropha curcas,* and *Cucurbita foetidissima* are found to be ideal. Buffalo gourd (*Cucurbita foetidissima*), a desert-adapted plant, produces high-quality oil and fermented starch. The oil has a high ratio of unsaturated to saturated fatty acids. Crude protein and fat content in the whole seeds are 32.9 percent and 33 percent, respectively [8]. With a seed yield of 3000 kg/ha and estimated 16 percent hydrocarbon, about 35 barrels of crude oil could be produced per hectare, in addition to carbohydrate from roots, forage from vines, and protein-rich oil cakes. Jojoba (*Simmondsia chinensis*) is a shrub that grows naturally in the United States and Mexico. Its seeds contain about 50 percent of oil by weight and does not decrease with long-term storage. The oil is remarkably resistant to degradation by bacteria, probably because it cannot cleave and metabolize the long-chain esters it contains (mostly hydrocarbons containing 38–44 carbon atoms). Jojoba oil has potential uses as a fuel and chemical feedstock, and can also be used as a replacement for vegetable oils in foods, hair oils, and cosmetics since it does not become rancid.

Additionally, it can be used as a source of long-chain alcohols for antifoaming agents and lubricants. The hydrogenated oil is a white, hard crystalline wax and has potential uses in preparation of floor and automobile waxes, waxing fruit, impregnating paper containers, and manufacturing of carbon paper and candles. Physic nut (*Jatropha curcas*), a tropical American species, is a large shrub, or a small tree. The seeds yield 46–58 percent oil of kernel weight and 30–40 percent of seed weight. In trade, this oil is called curcas oil. All parts of the plant exude sticky, opalescent, acidic, and astringent latex, containing resinous substances. The bark of this plant is a rich source of tannin (31 percent) and also yields a dark-blue dye. Now *Jatropha* oil, a semidrying oil, is in high demand for use as biodiesel in Asian countries. It is employed in preparation of soaps and candles and used as an illuminant and lubricant. In China, a varnish is prepared by boiling the oil with iron oxide, and in England, it is used in wool spinning. The oil is used for medicinal purposes for skin diseases, for rheumatism, as an abortifacient, and it is also effective in dropsy, sciatica, and paralysis.

Miscanthus.   *Miscanthus,* a thin-stemmed grass, has been identified as an ideal fuel crop as it gives a high dry-matter yield (see Fig. 2.8). Under

Figure 2.8 *Miscanthus. (Source: www.bluestem.ca/ miscanthusgracillimus.htm. Used with permission.)*

adequate rainfall conditions, light-arable soils give good yield. It has been found that dark-colored soils produce better yield than light-colored soils. It has been evaluated as a bioenergy crop in Europe for over 10 years and is grown in several European countries. Annual harvesting ability, low mineral content, and good energy yield per hectare are desirable characteristics. It is propagated as rhizomes planted in double rows about 75 cm apart, with 175-cm gaps between the rows. While disease control is not a significant issue, weed control measures are important. In Germany and Denmark, yields are 13–30 ton/ha for 3- to 10-year-old plantation [10].

Panicum. *Panicum virgatum* or switchgrass (see Fig. 2.9) is another thin-stemmed herb that has been used as a model plant [10]. It is a C_4 species, and though it has lower moisture content than wood, it has similar calorific value. It has been found suitable for the development of ethanol for petrol replacement. The low ash and alkali content makes it a suitable fuel for combustion.

Switchgrass has been identified to be a good model bioenergy species, due to its high yield, high nutrient-use efficiency, and broad geographical distribution. Further, it also has good attributes in terms of soil quality and stability, cover value for wildlife, and low inputs of energy, water, and agrochemicals. Evaluation of the use of switchgrass with coal in existing coal-fired boilers and the handling, operation, combustion, and emission characteristics of the co-firing process have been studied. Switchgrass has supplied up to 10 percent of the fuel energy input. In comparison to the use of corn for the source of bioethanol, switchgrass has been found to generate 15 times more efficiency of

FIGURE 2.9 *Panicum. (Source: www.biology.missouristate. edu/ Herbarium/Plants. Used with permission.)*

FIGURE 2.10 Hemp. (*Source: www.greenspirit.com. Used with permission.*)

energy production, and it is predicted that switchgrass may entail more profits than conventional crops for a specific area [10].

Hemp. Hemp is a member of the mulberry family that includes mulberry, paper mulberry, and the hop plant (see Fig. 2.10). It has a cellulose content of about 80 percent and has been grown for the production of medicinal, nutritional, and chemical production. Hemp is the earliest recorded plant cultivated for production of textile fiber. It has a low-moisture content for biomass feedstock [11].

Artocarpus hirsute and Ficus elastica.   Stem and leaf samples of *A. hirsute* and *F. elastica* have been evaluated for their potential as a renewable energy source. Stem and leaf samples of *F. elastica* and *A. hirsute* were evaluated for polyphenol, oil, and hydrocarbon contents. *F. elastica* shows the maximum accumulation of protein (24.5 percent), polyphenol (4.2 percent), oil (6.1 percent), and hydrocarbon (2.0 percent) contents. The leaf of *F. elastica* has been identified to be a good renewable energy source [12].

Calotropis procera.   Latex obtained from *C. procera* could be hydro-cracked to obtain hydrocarbons under severe thermochemical conditions. Instead, biodegradation is a less energy-intensive technique for latex degradation. Enhancements in the heptane level have been found in *C. procera* latex that was subjected to different fungal and bacterial treatments, compared to those of untreated ones. Nuclear magnetic resonance (NMR) and fourier transform infrared spectroscopy (FTIR) analyses reveals that the latex has undergone demethylation, dehydrogenation, carboxylation, and aromatization during microbial treatment. Petroleum obtained by hydrotreatment of the biotransformed latex is proposed to be used as fuel [13]. Some of the important latex-bearing plants are *Hevea brasiliensis, Euphorbia* sp., *Parthenium agentatum, Pedilanthus macrocarpus, F. elastica,* and *Manihot glaziorii.* Several resin-rich plants such as *Cappaifera multijuga* (diesel tree), *Copaifera langsdorffi, Pinus, Dipterocarpus, Shorea* sp., and *Pithosporum resiniferum* produce prolific terpene and oleoresins, and are as such very desirable fuel crops. Woody and herbaceous plants have specific growth conditions, depending on the soil type, soil moisture, nutrient content, and sunlight. These factors determine their suitability and growth rates for specific geographical locations. Cereals such as wheat and maize and perennial grasses such as sugarcane have varied yields with respect to the climatic conditions. Depending on the habitat, plants differ in their characteristic makeup. Their cell walls have varying amounts of cellulose, hemicellulose, lignin, and other minor components. The relative proportion of cellulose and lignin is one of the selection criteria in identifying the suitability of a given plant species as an energy crop. Herbaceous plants are usually perennial, having a lower proportion of lignin that binds together with cellulose fibers. Woody plants characterized by slow growth are composed of tightly bound fibers resulting in their hard external surface. Generally, cellulose is the largest component, representing about 40–50 percent of the biomass by weight; the hemicellulose portion represents 20–40 percent of the material by weight. Cellulose is a straight-chain polysaccharide composed of D-glucose units. These units are joined by β-glycosidic linkage between C-1 of one glucose unit and C-4 of the next glucose unit. The number of D-glucose units in cellulose ranges from 300 to 2500. Hemicellulose is a mixture of polysaccharides, composed almost entirely of sugars—such as glucose, mannose, xylose, and arabinose—and methylglucuronic and galacturonic acids, with an average molecular weight of <30,000 g. Cellulose is crystalline, strong, and resistant to hydrolysis, whereas hemicellulose has a random, amorphous structure with little strength. It is easily hydrolyzed by dilute acid or base.

A complete structure of lignin is not well defined because the lignin structure itself differs between plant species. Generally, lignin consists of a group of amorphous, high-molecular-weight, chemically related compounds. Phenylpropanes, three carbon chains attached to rings of six carbon atoms, are the building blocks of lignin. These might have one or two methoxyl groups attached to the rings. Sugar/starch feedstocks, such as cereals, have been traditionally used in biochemical conversion of biomass to liquids such as ethanol. High-cellulose content of biomass is generally more efficient and therefore preferred over the lignin-rich biomass for conversion of glucose to ethanol. Depending on the end use and type of bioconversion preferred, the choice of the plant species varies. In northern Europe, the C_3 woody species especially grown on short rotation coppice, such as willow and poplar, and forestry residues, are used [14]. In Europe, there is wide interest in the use of oilseed rape for producing biofuel [15]. Brazil was one of the first countries to begin large-scale fuel alcohol production from sugarcane.

2.5 Harvesting Plants for Bioenergy

Biomass can be converted into different types of products, including:

1. Electrical/heat energy

2. Transport fuel

3. Chemical feedstock

Woody and herbaceous species are the ones used most often by biomass researchers and industry. Several parameters are important in the biomass conversion process. The principal considerations in terms of the material type are moisture content, calorific value, fixed carbon and volatile proportion, ash/residue content, alkali metal content, and cellulose–lignin ratio. In a wet-biomass conversion process, the moisture content and cellulose–lignin ratio is of prime concern, while in a dry-biomass conversion process, it is the alkali metal content and cellulose–lignin ratio. The Laticiferous plant species of Apocyanaceae, Asclepiadaceae, Convolvulaceae, and Euphorbiaceae have been analyzed for use as renewable energy sources. Analysis of oil and hydrocarbon contents of 15 different plant species tested has revealed that *Carissa carandas* L., *Ceropegia juncea* Roxb., *Hemidesmus indicus* R. Br., and *Sarcostemma brunourianum* W. A. are the most suitable species [16]. In another study, five different plant species *Plumeria alba*, *C. procera*, *Euphorbia nerifolia*, *Nerium indicum*, and *Mimusops elengi* have been evaluated as potential renewable energy sources. Whole plants and plant parts (leaf, stem, and bark) have been analyzed for oil, polyphenol, hydrocarbons, crude protein, α-cellulose, lignin, ash, and mineral content. The barks of these plants were identified to have greater hydrocarbon content than the leaves. Based on the dry-biomass yields, hydrocarbon content, and other properties, these plant species most suitable for renewable energy sources have been identified [17]. In a study conducted on 51 plant species in Tennessee, in the United States, an examination of the oil, polyphenol, hydrocarbon, protein, and ash content reveals that *Lapsana communis* yields the maximum oil (6.1 percent dry, ash-free plant sample basis). *Chrysopsis graminifolia, Solidago erecta,* and *Verbesina alternifolia* have been identified as rubber-producing species with 0.4–0.7 percent hydrocarbon [18].

2.6   Products

Several processes similar to petroleum refining are involved in the conversion of biomass into different products. Biorefineries convert biomass into different products in different stages. The different stages involved in the conversion of biomass to products are depicted in Fig. 2.11.

There has been a tremendous increase in biobased products such as ethanol, high-fructose syrups, citric acid, monosodium glutamate, lysine, enzymes, and specialty chemicals worldwide. It is estimated that in 2000–2006 in the United States alone, there will be an increase in the use of liquid fuels, organic chemicals, and biopolymers from the current level of ~2 percent, 10 percent, and 90 percent each to 10 percent, 25 percent, and 95 percent, respectively [19].

2.6.1   Gaseous products

In Chap. 1, gasification (pyrolysis) of biomass, biogas, gobargas, hydrogen, and biohydrogen were discussed in detail.

2.6.2   Liquid products

An important renewable energy resource for transportation purposes is liquid fuel based on plant oils. However, pure plant oils are generally not suitable for use in modern diesel engines. This can be overcome by

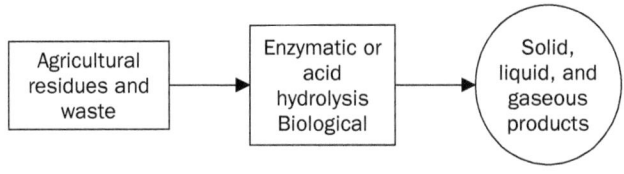

FIGURE 2.11   Different products from biomass.

the process of transesterification. The resultant fatty-acid methyl esters have properties similar to those of diesel and are commonly called biodiesel. Biodiesel presents several advantages, such as better CO_2 balance than diesel, low soot content, reduced hydrocarbon emissions, and low carcinogenic potential [20]. The specification standards for the European Union (EU) and the United States are EN14214 and ASTM D6751, respectively. The EU directive established a minimum content of 2 percent and 5.75 percent biodiesel for all petrol and diesel used in transport by December 31, 2005, and December 31, 2010, respectively. Biodiesel refers to the pure oil before blending with diesel fuel. Biodiesel blends are represented as "BXX," with "XX" representing the percentage of biodiesel component in the blend (National Biodiesel Board, 2005) [21]. In the biomass-to-liquid conversion processes, biomass is broken down into a gaseous constituent and a solid constituent by low-temperature gasification. The next step involves production of synthetic gas, which is converted into fuel (termed SunFuel) by the Fischer-Tropsch synthesis process, with downstream fuel optimization by hydrogen after treatment [22]. Ethanol has already been introduced in countries such as Brazil, the United States, and some European countries. In Brazil, it is currently produced from sugar and, in the United States, from starch at competitive prices. Ethanol is currently produced from sugarcane and starch-containing materials, where the conversion of starch to ethanol includes a liquefaction step (to make the starch soluble) and a hydrolysis step (to produce glucose). There are generally two types of processes for production of bioethanol: the lignocellulosic process and the starch process. Unlike the starch-based process, the lignocellulosic process has not been as widely adopted due to techno-economic reasons. High ethanol yield requires complete hydrolysis of both cellulosic and hemicellulose with a minimum of sugar dehydration, followed by efficient fermentation of all sugars in the biomass. Certain advantages of using lignocellulose-based liquid biofuels are that they are evenly distributed across the globe and hence are readily available, less expensive compared to agricultural feedstock, produced at a lower cost, and have low net greenhouse gas emissions. Enzymatic processes (essentially using bacteria, yeasts, or filamentous fungi) have been considered for lignocellulosic processes. The enzymatic process when coupled with the fermentation process is known as simultaneous saccharification and fermentation. This has proved to be efficient in the fermentation of hexose and pentose sugars [23]. Genencor International (www.genencor.com/) and Novozymes, Inc. (www.novozymes.com) have been awarded $17 million each by the U.S. Department of Energy with a goal to reduce the enzyme cost tenfold (www.eere.energy.gov/). The Iogen Corp. (www.iogen.ca/) demo-plant is the only one that produces bioethanol from lignocellulose, using the enzymatic hydrolysis process. This plant is known to handle about 40 ton/day of wheat, oat, barley, and straw and is designed to produce up to 3 ML/yr of cellulose ethanol. Refer to Chap. 3 for bioethanol preparation, Chap. 6 for boidiesel processing, and Chap. 7 for ethanol and methanol used in engines.

2.6.3 Solid products

Refer to Sec. 1.14, Chap. 1, for more details on biomass. Solid products fall under the following categories:

1. Direct outcome of photosynthesis: Products from forest, shrubs, agricultures, and aquacultures.
2. Nonphotosynthesis: Mushrooms, animal biomass, indirect from photofixation.
3. Wastes: Forests and agricultural products.
4. Municipal solid wastes: Not all solid biomass may be suitable for different end uses, i.e., energy production or energy recovery. For example, mushrooms are notably useful as food, feed, or fodder, not otherwise. Biomass properties are guidelines to further and more fruitful end uses. The properties depend on the following:
   a. Water or moisture content (aqueous/dry)
   b. Calorific or combustion value
   c. Dry residues/ash content/silicates, and so forth
   d. Alkali metal/oxides in the ash
   e. Ratio of cellulose/liquid/oils/fats/of other carbonaceous matters
   f. Ratio of solid/liquid/volatiles

Direct combustion of biomass for heat generation is the most inefficient technique in energy economy, heat being the most inefficient of all forms of energy. The best way to utilize biomass is to recycle biomass for production of other or further biomass, namely, agriculture, horticulture, aquaculture, poultry, animal farming, and so forth. Randomness is reduced (low entropy change), and environmental chaos is lessened. Properties (a), (c), and (d) are significant for farming; (b) and (f) are important for hydrolytic processes; and (e) is important for biofuels and biodiesel. All the points are important for fermentations and in biorefineries. Biorefinery has become a new science and technology harmony for a promising future, which takes care of different aspects of biosafety, minimizes waste, and maximizes energy efficiency. It is a field of engineering and technology for the future. Biorefinery is a system similar to that of petroleum in its requirements for producing fuels and chemicals from biomass. A biorefinery is a capital-intensive project and is based on a conversion technology process of biomass. Hence, several technologies—thermochemical, chemical, biochemical, and so

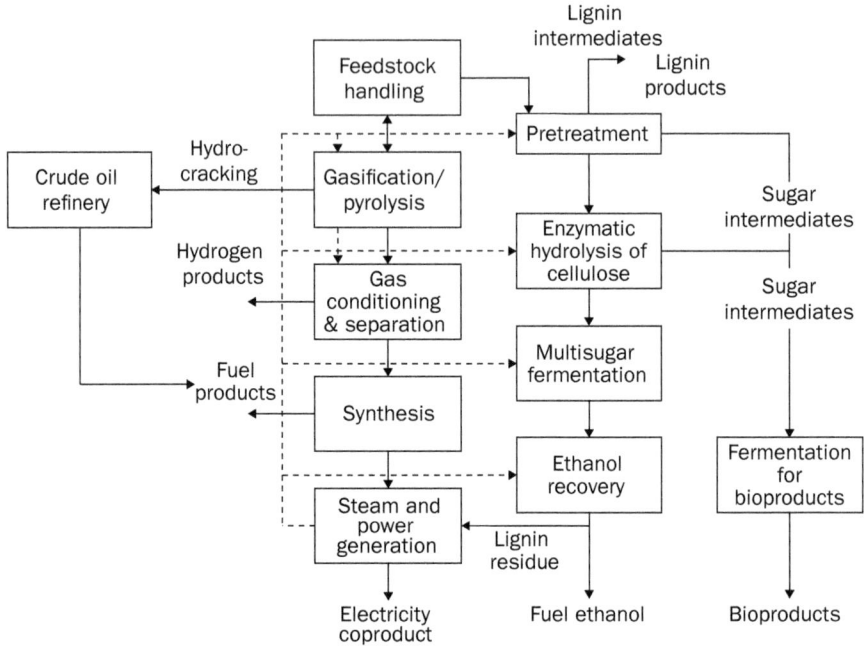

FIGURE 2.12 An integrated biorefinery process. (*Permission from S. Fernando, Associate Editor, FPEI—American Society of Agricultural and Biological Engineers [ASABE], Mississippi State University, USA.*)

forth—are combined to reduce the overall cost. Fernando et al. suggest an integrated biorefinery process from bio-oil produced from pyrolysis of biomas (see Fig. 2.12), which will not only produce sugar but also different by-products and electricity [24]. The process can produce its own power.

Fermentation is equally important. Anaerobic and restricted aerobic digestion with selected algae species allow us to harvest hydrogen and clean fuels, without much loss of biomass and with the least amount of waste products. In an aerobic process, the process is carried out by oxidizing the volatile matter into biodegradable organic fractions of solid waste. Air acts as a source of oxygen, and aerobic bacteria act as a catalyst. The change occurring during the process may be represented as

$$\text{Biomass} + O_2 \text{ (Aerobic bacteriad)} \rightarrow CO_2 + H_2O + \text{Organic manure}$$

Anaerobic digestion is carried out by segregating the nonbiodegradables and the biodegradables at the same time. This may be done manually or mechanically. The smaller pieces of inorganic materials like clay and sand may be removed by washing the biomass with water. The washed material is then shredded into a size that will not interfere with mixing and may be more amenable to bacterial action. The shredded biomass is then mixed with sufficient quantity of water, and slurry is fed into a digester system. If necessary, nutrients like nitrogen, phosphorus, and potassium have to be added to the digester. The process involves four groups of bacteria in the digested slurry as follows:

1. Hydrolytic bacteria catabolize carbohydrates, proteins, lipids, and so forth contained in the biomass to fatty acids, H_2, and CO_2.

2. Hydrogen-producing acetogenic bacteria catabolize certain fatty acids and some neutral end products to acetate, CO_2, and H_2.

3. Homoacetogenic bacteria synthesize acetate, using H_2, CO_2, and formate.

4. In the final phase, called the methanogenic phase, methanogenic bacteria cleave acetate to methane and CO_2.

Water acts as a catalytic agent in methane formation. Thus, water is acted upon by enzymes, itself breaking down to hydrogen and oxygen. Hydrogen is used by microorganisms to reduce CO_2 to CH_4, while oxygen oxidizes carbon dioxide, i.e., makes it acidic (H_2CO_3). In simple terms, acetate (in presence of CoI) is simultaneously oxidized to CO_2 and reduced to CH_4. For details, refer to Chap. 1, methanation, and Baker's and Ganzalus pathway. Thus, methane-forming bacteria play an important role in the circulation of substances and energy turnover in nature. They absorb CO, CO_2, and H_2 to give hydrocarbon and methane and help synthesis of their own cell substances. During anaerobic digestion, gas containing mainly CH_4 and CO_2 is produced. The gas is known as biogas, which is used for the generation of electricity or fuel. The residual biomass comes out of the digester in the form of a slurry, which is separated into a sludge, which is used as fertilizer and a stream of waste water. Research is ongoing to produce renewable energies from different plant sources, which will necessarily dominate the world's energy supply in the long-term. Using renewable-energy system technologies will create employment at much higher rates than any other technologies would [1]. There are economic opportunities for industries and craft jobs through production, installation, and maintenance of renewable energy systems.

References

1. European Renewable Energy Council (EREC). *Integration of Renewable Energy Sources: Targets and Benefits of Large-Scale Deployment of Renewable Energy Sources*, Workshop—Renewable Energy Market Development, Riga, Latvia, May 2004.

2. J. A. Bassham. Increasing crop production through more controlled photosynthesis, *Science 197*, 630–638, 1977.

3. P. McKendry. Energy production from biomass (Part I): Overview of biomass, *Bioresource Technology 83*, 37, 2002.

4. A. Nag. *Analytical Techniques in Agriculture, Biotechnology and Environmental Engineering*, New Delhi: Prentice-Hall, 2006.

5. R. E Häusler, H.-J. Hirsch, F. Kreuzaler, and C. Peterhänsel. Overexpression of C_4-cycle enzymes, *Journal of Experimental Botany 53*(369), 591–607, 2002.

6. W. Amatayakul and C. Azaul. Eucalyptus for fossil fuel substitution and carbon sequestration: The costs of carbon dioxide abatement in Thailand, *International Journal of Sustainable Development 6*(3), 359–377, 2003.

7. C. L. Schulze, E. Schnepf, and K. Motbes. Uber die Localisation der Kautschukpartikel in verschiedenen Typen von Milchröhren. Flora, *Abstracts* **158**, 458–460, 1967.

8. Energy Information Administration. Forecast and analysis of energy data, *International Energy Outlook 2005*, Report #: DOE/EIA-0484 (2005): www.eia.doe.gov/oiaf/ieo/oil.html.

9. E. Chlorent and R. P. Overend. Liquid fuels from lignocellulosics, In: *Biomass Regenerable Energy*, Hall, D. O. and Overend, R. P. (Eds.), Rochester, UK: John Wiley & Sons, pp. 257–269, 1987.

10. S. B. McLaughlin, R. Samson, D. Bransby, and A. Wiselogel. Evaluating physical, chemical, and energetic properties of perennial grasses as biofuels, In: *Proceedings of the Seventh National Bioenergy Conference—Bioenergy '96*, Nashville, TN, September 15–20, 1996.

11. L. Dewey. *Hemp Hurds as Papermaking Material*, USDA Bulletin No. 404, US Government Printing Office, Washington, DC, October 14, 1916.

12. R. Palaniraj and S. C. Sati. Evaluation of *Artocarpus hirsute* and *Ficus elastica* as renewable source of energy, *Indian Journal of Agricultural Chemistry 36*(1), 23, 2003.

13. B. K. Behera, M. Arora, and D. K. Sharma. Studies on biotransformation of *Calotropis procera* latex–A renewable source of petroleum, value-added chemicals, and products, *Energy Sources 22*(9), 781, 2000.

14. Ove Arup and Partners. *Monitoring of a Commercial Demonstration of Harvesting and Combustion of Forestry Wastes*, ETSU B/1171-P1, London, UK, 1989.

15. F. Culshaw and C. Butler. *A Review of the Potential of Bio-Diesel as a Transport Fuel*, ETSU-R-71, The Stationary Office, London, UK, 1992.

16. T. Sekar and K. Francis. Some plant species screened for energy, hydrocarbons and phytochemicals, *Bioresource Technology 65*(3), 257–259, 1998.

17. D. Kalita and C. N. Saikia. Chemical constituents and energy content of some latex bearing plants, *Bioresource Technology 92*(3), 219–227, 2004.

18. M. E. Carr and M. O. Bagby. Tennessee plant species screened for renewable energy sources, *Economic Botany 41*(1), 78–85, 1987.
19. S. Fernando, C. Hall, and S. Jha. NO_x reduction from biodiesel fuels, *Energy Fuels 20*, 376–382, 2006.
20. G. Vicente, M. Martinez, and J. Aracil. Kinetics of *Brassica carinata* oil methanolysis, *Energy Fuels 20*, 1722–1726, 2006.
21. W. Steiger. Biomass-to-liquid fuels: Energy for future, In: *Proceedings of World Renewable Energy Congress VIII: Linking the World with Renewable Energy*, Denver, co, August 29–September, 2004, 32–33.
22. B. Hahn-Hagerdal, S. Galbe, G. Gorwa-Grauslund, and Z. G. Liden. The fuel of tomorrow from the residues of today, *Trends in Biotechnology 24*(12), 549–556, 2006.
23. National Biodiesel Board. Biodiesel 101: www.biodiesel.org/resources/biodiesel_basics/ default.shtm.
24. S. Fernando, S. Adhikari, C. Chandrapal, and N. Murali. Biorefineries: Current status, challenges and future direction, *Energy and Fuels 20*, 1727–1737, 2006.

Questions

1. Describe the photosynthetic process. What is efficiency of photosynthesis?

2. Discuss C_3 metabolism in plants. What are differences between C_3 and C_4 metabolism?

3. What are the characteristics of an ideal energy crop? What are the conversion products from biomass?

4. What are the different categories of solid products?

5. Discuss the integrated biorefinery process.

CHAPTER 3

Microbial and Algae Oil Biodiesel Production and Engine Performance

M. P. Dorado and M. Sedano

3.1 Introduction

Climate change is forcing leading industrialized countries to protect the fragile environment by preventing or, at least, reducing the use of contaminant fuels and technologies. Among the most pollutant fuels are coal and petroleum-derived fuels, i.e., diesel fuel and gasoline. In fact, the Paris Agreement goals are inconsistent with emissions derived from the use of coal. Although in 2016 there were 7273 coal-fired thermal power plants in the world, providing 1964 GW (plus 1082 planned stations, that aimed to provide 596 GW), their future is uncertain [1]. As an example, EU Industrial Emissions Directive (IED) has forced European large combustion plant operators to meet "Best Available Techniques Reference" (BREF) requirements by 2021. These techniques aim to reduce exhaust emissions, mainly SO_x and NO_x, while improving plant efficiency. Energy plants that do not comply with these requirements will have to close by 2021 [2].

The use of diesel fuel and gasoline in transport is also controversial. Recently, many industrialized European countries have proposed a set of actions against the use of compression ignition (diesel) engine–based cars. In fact, diesel engine–based cars (followed by spark ignition engine–based cars) will be banned in the EU in the near future, ranging from 2025 to 2050. However, diesel engines will continue to be required, mainly due to two reasons. On the one hand, most countries are not embracing such regulations. While petroleum and coal-based technologies are usually affordable by developing countries, renewable energy–based systems require significant investments including facilities, skilled workforce, and appropriate climatic/geographic conditions. On the other hand, there is no reliable alternative to fuel stationary diesel power generators (i.e., as an emergency energy supplier if the grid fails or in nuclear plants), low-price domestic heating, and, most important, maritime and air transport are exempted from restrictions concerning the use of fossil fuels. It is important to mention that diesel engines present the highest thermal efficiency compared to any combustion engine. It would be disgraceful if research stopped seeking alternative less-pollutant fuels to reduce emissions.

In sum, it is noteworthy that diesel engines will remain the main driving force of economy sectors such as aviation, ocean freight, and long-haul trucking. And if only their emissions are considered, they already exceed the 2°C target adopted at the Conference of the Parties to the UN Framework Convention on Climate Change (COP21). For this reason, and forced by the new scenario of limiting air pollution legislation, development and promotion of new environmentally friendly biofuels is a must. Biodiesel may definitely assist climate change mitigation [3].

Vegetable oil–derived biodiesel is in the doldrums. While no NGO is concerned about the use of land to grow products that are far from essential goods, such as tobacco, spirits (rum, wine, beer, etc.), coffee, tea, cut flowers, hashish, Christmas trees, etc., many NGOs have adopted a highly conflictual approach with the use of land to produce energy-based commodities. Moreover, governments have been persuaded to promote laws against the use of so-called first-generation biodiesel, that is, biodiesel from vegetable oils. In this context, oleaginous microorganisms, including microalgae, to provide next-generation biodiesel have clearly moved up the political agenda.

3.2 Microalgae Oil Biodiesel

As mentioned above, oils (usual biodiesel raw material) have traditionally been produced from crops, although they can also be provided by several oleaginous microorganisms, such as many microalgae, fungi, and bacteria [4]. Oleaginous microorganisms are those that are able to accumulate above 20 percent lipids of their total dry weight.

Microalgae are aquatic microscopic organisms that have been identified as a promising raw material for biodiesel production, due to their ability to synthesize lipids. As they do not need agricultural land to grow (they are mainly produced in oceans) research has been promoted by governments. However, technological and economic issues need to be faced before transfer to an industrial scale.

Several steps are required after cultivation to produce microalgae oil biodiesel. The most significant are mentioned below (cultivation, recovery or harvesting, lipid extraction, and biodiesel production through transesterification).

Among other classifications, the most frequent production techniques are mentioned below (Fig. 3.1).

3.2.1 Open pond system

This system can be performed in natural water (i.e., lakes) or in artificial containers. Microalgae may be grown in tanks, circular ponds, or shallow ponds, but raceway ponds are the preferred method. In this last method, water is continuously flowing along the track, so microalgae, carbon dioxide, and nutrients are circulated. This method provides many benefits, including simple, cheap, easy construction and

FIGURE 3.1   (a) Raceway pond, (b) flat-plate photobioreactor, (c) column photobioreactor, and (d) tubular photobioreactor. Reprinted from *Computers and Electronics in Agriculture*, 76(2), Bitog, J. P., Lee, I. B., Lee, C. G., Kim, K.-S., Hwang, H.-S., Hong, S.-W., Seo, I.-H., Kwon, H.-S., Mostafa, E., Application of computational fluid dynamics for modeling and designing photobioreactors for microalgae production: A review, 131–147, Copyright (2011), with permission from Elsevier and authors.

plant operation. Water evaporation and CO_2 losses, poor light use by cells, and the needs of large extensions of land are major constraints. Another significant drawback is due to its low productivity, when the strain growth is limited by temperature [5] (Fig. 3.2(a) and 3.2(b)).

3.2.2   Desert cultivation

This system is characterized by the culture of robust microalgae strains that can grow under extreme conditions (high levels of salinity and poor-quality land) (Fig. 3.3).

Figure 3.2 (a)   Microalgae grown in open ponds. (*Courtesy of Pacific Northwest National Laboratory, PNNL.*)

Figure 3.2 (b)   Algae "farm" developed as part of research by the University of Queensland. (*Courtesy of beefcentral.com.*)

FIGURE 3.3 Microalgae cultured in Atacama desert (Chile). (*Photo courtesy of Fitoplancton Marino, S.L.*)

3.2.3   Closed ponds or closed reactors

This system consists of covered ponds, i.e., with a greenhouse, although photobioreactor is the preferred technology. It allows controlling culture abiotic conditions (pollution, evaporation, temperature, CO_2), which is its main advantage compared to open ponds system [5]. For this reason, it allows more species to be cultured. Some of the most representative types of closed ponds are listed below.

3.2.3.1   *Horizontal photobioreactors*

This system consists of a network of tubes laying on the ground. At a certain frequency, a pump raises vertically the culture into the reactor, helping to mix it (Fig. 3.4(a) and 3.4(b)).

3.2.3.2   *Vertical systems*

Microalgae are grown in a vertical framed structure (Fig. 3.5). The main drawback of this system is its low productivity. It is usually associated with poor mixing and deoxygenation. However, continuous bubbling air may improve it [5].

3.2.3.2.1 Air-lift method   The system is made of tubes interconnecting two areas, a gassed area where a gas is bubbled (riser tube) and an ungassed area (downcomer tube). This situation, caused by the irregular presence of gas, generates different fluid bulk densities, leading to fluid circulation (Fig. 3.6). In this technology, air circulates through microalgae to make sure that water reaches the culture during growth. Bubbling the gas through the sparger in the riser tube ensures mixing [6]. It includes medium frequency light/dark cycles, which are responsible for productivity and biomass yield [7].

FIGURE **3.4 (a)**   Pilot plant facility at Coyote Gulch outside Durango, Colorado (US). (*Courtesy of Dr. Jason Quinn, Dr. Jeff Moody, and Dr. Chris McGinty from Utah State University, and BGG-Solix Combination, former Solix BioSystems.*)

FIGURE **3.4 (b)**   Horizontal tubular microalgae reactor (*Photograph courtesy of Prof. Dr. Rene Wijffels and Dr. Iago Dominguez Teles, Bioprocess engineering, Wageningen University & Research.*)

3.2.3.3   Flat plate reactors

The system presents narrow panels, with the mission of increasing the surface area to maximize sunlight. It also forces circulation within microalgae culture, thanks to a gas unit working across panels [8] (Fig. 3.7).

FIGURE 3.5 30 m^3 Phyco-Flow™ serpentine photobioreactor, manufactured by Varicon Aqua, UK, 2017. (*Courtesy of Varicon Aqua.*)

FIGURE 3.6 100 L Phyco-Lift airlift photobioreactor, manufactured by Varicon Aqua, UK, 2017. (*Courtesy of Varicon Aqua.*)

Figure 3.7 Flat plate reactor. (*Copyright Subitec GmbH.*)

3.2.3.4 Fermenter-type reactors

In these bioreactors, fermentation takes place. The use of this system is not highly extended due to low efficiency. Recent developments suggest that light control may enhance productivity, making fermenter-type reactors an interesting alternative to produce algal biomass [8] (Fig. 3.8).

3.3 Recovery of Microalgae Biomass: Harvesting

Microalgae biomass appears diluted in water, so drainage methods for biomass recovery must be carried out. The efficiency of selected

Figure 3.8 Fermenter-type reactors. (*Courtesy of BIONET and author Martín García Pérez.*)

harvesting method will depend on microalgae species and will influence fuel economic viability and sustainability. Among the recovery methods, thickening methods (sedimentation through gravity settling, coagulation/flocculation, flotation or electroflotation) and biomass dewatering methods (filtration, including micro/ultra filtration through membrane separation, centrifugation) are the most extended. Once recovered, to improve downstream process, biomass needs to be dried. Drying methods include solar or greenhouse drying, lyophilization, or spray drying [9].

With the purpose of lipid extraction, the usual biomass recovery method is centrifugation [10, 11]. Although it has not been clearly stated whether recovery method selection may interfere in lipid content, the fatty acid profile remains unaffected [9].

3.4 Lipid Extraction Processes

Microalgae lipid content may reach up to 75 percent w/w of biomass, depending on culture conditions, selected microalgae strain, lipid extracting method, and, with the certain previously mentioned doubt, chosen biomass harvesting method [12]. Among more than 30 tested oleaginous strains, considering both lipid content and growth speed, *Chlorella sp.* (20 percent to 30 percent lipid content), *Dunaliella sp.* (17.5 percent to 67 percent), *Nannochloropsis sp.* and *Scenedesmus sp.* (11 percent to 55 percent) are promising candidates for biodiesel production [13]. In terms of fatty acid profile, *Chlorella sp.*, *Saccharomyces cerevisiae*, *Picochlorum sp.*, *Botryococcus sp.*, *Scenedesmus sp.*, and *Nannochlopsis oculata* are preferred for biodiesel production due to their high content of oleic acid [13–15].

There are a number of lipid extraction methods, mainly physical and chemical. The best fitting lipid extraction method is the more efficient one, depending on microalgae strain chemical composition.

Considering lipid extraction for biodiesel production, the most extended approach includes solvent-based extraction. Lipid extraction is usually carried out with a solvent-assisted process using either vortex and centrifugation, shaking at 700 rpm, or ultrasound-aided at least at 40 kHz, as the most used. Preferred solvents are chloroform, hexane, ethanol, and methanol, among many others [16].

3.5   Microalgae Oil Biodiesel Production

Some microalgae strains potentially produce up to 141,000 L/ha/yr of oil, which is around 200 times above the highest vegetable oil productivity. However, there is a significant drawback, as culture energy requirements outweigh microalgae supplied energy [17, 18]. After extraction, lipids may be transformed into biodiesel trough chemical or biochemical paths, namely transesterification or interesterification, respectively. Transesterification is the chemical process that transforms a fatty acid ester with high viscosity (inferred by the presence of glycerol) into another ester of the same fatty acid with lower viscosity. This reaction is carried out, for microalgae lipids, in the presence of a lower alcohol (mainly methanol, which will help in removing glycerol from triacylglycerol molecules), a catalyst (mainly acid catalysts) under heating and vigorous stirring for up to one hour of reaction time [19]. Alkaline catalyzed transesterification is discarded, as high free fatty acid content of lipids will lead to saponification reactions.

3.5.1   Direct or *in-situ* transesterification

In this process, previous oil extraction and further purification are not requested, as transesterification and oil extraction occur in the same step. Thus, it helps reduce time and final costs. Biomass may be dried or wet. Efficiency of direct transesterification and further biodiesel yield depend on biomass water content (it may reach up to 55 percent of final cost), reaction time, and temperature [20].

After transesterification, biomass residues are removed from biodiesel by centrifugation or filtration. This method does not distinguish between lipid classifications; if it is needed, solvent extraction has to be applied. Direct transesterification provides similar or even higher biodiesel yield than conventional (with a previous oil extraction step) transesterification [21, 22].

3.5.2   Interesterification

This is an enzyme-catalyzed (i.e., high activity enzyme *Candida antarctica*) process to produce fatty acid esters. This process is gaining interest among researchers, as it provides a final product with low purification requirements (as a biocatalyst substitutes acid/basic catalyst), although the high cost of enzymes is a significant drawback that prevents its extensive application [23].

Best biodiesel properties are linked to several factors, i.e., high fatty acid methyl ester chain length, low unsaturation degree, microalgae strain, culture conditions, biomass harvesting, and lipid extraction method, if applied [4, 24].

3.6 Microbial Oil Biodiesel

The main advantage of heterotrophic microorganisms (fungi, bacteria, and heterotrophic microalgae) is that they can grow on biomass or organic waste (i.e., agricultural and food industry residues) as a carbon source [25]. Although microalgae are also microorganisms, this section is devoted to the use of bacteria, fungi, and yeasts to produce biodiesel. Also, oleaginous yeast presents fast growth and high lipid accumulation, compared to microalgae [26]. Provided that bacteria can only synthesize and accumulate specific lipids (mainly, phospholipid and glycolipid, and only a few are able to accumulate triglycerides, i.e., *Gordonia sp.* DG, *Nocardia globerula,* and *Rhodococcus opacus* [27–29]) and polyunsaturated fatty acids, the preferred lipid producers, apart from microalgae, are yeast and fungi [30]. Moreover, bacteria that may accumulate triglycerides are neither working under an economically sound process nor providing significant oil yields [31].

Selecting the appropriate oleaginous microorganism to successfully implement oleochemical production from single cell oil (SCO or microbial oil, MO) through a fermentation process based on high carbon sources that lead to high productivity and lipid content in cellular biomass is key. Most bacterial strains accumulate SCO in amounts up to 40 percent of total cellular dry weight [32]. But more impressive is the fact that some yeast strains (e.g., *Rhodosporidium sp., Rhodotorula sp., Lipomyces sp.*) may accumulate intracellularly about 70 percent w/w of SCO [33]. Moreover, SCO accumulated by many yeasts depicts a fatty acid composition similar to that of most common vegetable oils [32]. In this sense, considering palm oil among the preferred oils for biodiesel production, *C. curvatus* provides SCO with similar composition to palm oil [34].

There is sufficient knowledge concerning basic physiology of oleaginous microorganisms leading to lipid accumulation. The carbon-to-nitrogen ratio (C/N) in the fermentation medium is considered the most decisive factor for efficient lipid accumulation. Other parameters influencing lipid accumulation are aeration, minerals, and nitrogen source. Lipid accumulation is initiated when nitrogen or another nutrient source is depleted from the fermentation broth and an excess of carbon source is still present. Nitrogen is needed for protein and nucleic acid synthesis during growth. Carbon is needed for energy-consuming and anabolic processes, involving lipid, carbohydrate, protein, and nucleic acid synthesis. When nitrogen source is depleted, protein and nucleic acid synthesis stops and lipid accumulation is enhanced [35]. Therefore, a fed-batch fermentation process, where all these parameters are easily controlled, should result in high lipid production. Microbial growth takes place in the first stage by supplying a nutrient-complete fermentation medium, while lipid accumulation is achieved in the second stage, after nitrogen exhaustion, by supplying a medium rich in a carbon source.

A major challenge for microbial oil is lipid production increase, which has been studied under advanced microbiological approaches, i.e., cell fusion, mutation breeding, and genetic modifications [30]. However, both microbial oil and biodiesel production costs must be reduced by 50 percent to make microbial oil biodiesel competitive with that provided by vegetable oil [36]. For this reason, the use of organic waste as low-cost substrate is becoming an attractive culture alternative.

3.6.1 Yeasts

Although preferred yeasts to produce microbial oil are *Rhodosporidium toruloides Y4* [37], *Lipomyces starkeyi* [38], *Cryptococcus curvatus* [39], and *Yarrowia lipolytica* [40], other yeasts with significant lipid accumulation capacity include *Rhodotorula glutinis*, *Rhodotorula graminis*, *Rhodotorula mucilaginosa*, *Pichia guilliermondii*, *Pichia kudriavzevii*, and *Candida tropicalis*, among many others [30, 31]. It is remarkable that only 5 percent of yeasts may accumulate lipids above 25 percent [41]. Moreover, fatty acid composition depends on both species and substrate [31]. Oleic acid, which is one of the preferred fatty acids in the biodiesel industry, besides palmitic acid, are the most common fatty acids accumulated by yeast [31]. However, most substrates tend to increase saturated fatty acids accumulation, providing similar fatty acid composition to palm oil, i.e., *C. curvatus*; *C. curvatus*, however, grown in cheese, provides a fatty acid composition similar to rapeseed oil [42]. Saturated fatty acids lead to improve cetane number and oxidation stability, while properties under cold weather and viscosity get worsen. As long as polyunsaturated fatty acids are not present, microbial oil will be suitable for biodiesel production [24]. As previously mentioned, an interesting fact related to oleaginous yeasts is that they may grow on agro-industrial waste.

The main challenge is derived from oil extraction, as cell walls are resistant to the use of many solvents. An extended alternative, but far from green chemistry, is provided by the use of a mixture of chloroform and methanol. Other researchers have found efficient alternatives based on the use of β-1,3-glucomannanase plMAN5C enzyme, assisted by microwave pretreatment. After cell disruption, above 96 percent of oil may be extracted with the use of ethyl acetate [43]. Extracted oil shows a total lipids composition that includes glycolipids, waxes, phospholipids, etc. that cannot be transesterified; thus, lipid fraction pretreatment is needed [44]. In sum, yeast is considered among the preferred oleaginous microorganisms to produce biodiesel, as it exhibits rapid growth, it is easy to scale-up, provides high lipid production, and has appropriate fatty acid composition to produce biodiesel [42].

3.6.2   Filamentous fungi

These microbes are able to accumulate lipids, but to a lesser extent compared to yeast. They may grow in submerged cultures, where they show the capacity of building pellets due to filamentous growth during fermentation. This leads to several advantages, namely, reduced broth viscosity, which enhances mass transfer and mixing, and easy harvesting by simple low-cost cell filtration [45].

Oil extraction may be carried out through alternate low-cost methanolysis applied to fungal biomass. Among the preferred fungi that are able to grow in organic waste culture, as a carbon source, are Zygomycetes, i.e., *Thamnidium elegans*, *Mortierella isabelline*, *Aspergillus terreus*, *Cunninghamella echinulate*, *Mortierella ramanniana*, *Zygorhynchus moelleri*, etc. [31, 42, 46, 47].

3.7   Engine Performance and Exhaust Emissions

To date, the majority of diesel engine studies to test feasibility of bio-diesel have been carried out under steady-state conditions. However, engines should be tested under transient cycles (Worldwide harmonized Light vehicles Test Cycle, WLTC), provided that engine working parameters (turbo lag, EGR response, injection timing, etc.) significantly differ from those considered under steady-state cycles [48]. Moreover, changes in engine operating parameters will also affect exhaust and noise emissions, including particulate matter (PM). In this context, the use of oxygenated fuels, i.e., biodiesel, may assist combustion, providing excess of oxygen, thus reducing exhaust emissions [49].

As oxide nitrogen emissions may be reduced by the use of urea-based catalysts, a major focus is placed on PM emissions. Diesel fuel PM is responsible for respiratory and cardiovascular deaths. Depending on particle size, PM may be distributed in different regions of the respiratory system. In this sense, adverse health effects seem to be associated with both ultrafine and fine particles (10–100 nm). In fact, after exposure to ultrafine particles, human mechanisms are activated providing different biological responses, including oxidative stress, endothelial activation, and systemic inflammatory cytokines release, linked to allergy processes [50]. PM accumulate in the lung pulmonary area, which contains alveolar sacs with easy access to the bloodstream. There, they may have longer lung residence time than particles deposited in the extra-thoracic regions. Moreover, compared to PM of bigger size, fine and ultrafine PM contain toxic materials (heavy metals and organic compounds) and could promote carcinogenic processes. It also worsens asthma [50, 51].

To reduce PM, transport sector needs to be upgraded to the latest regulations in terms of exhaust emissions (in Europe, EURO 6C). In fact, recent studies show lack of efficiency of PM filters when old EURO 3 diesel engine–based cars are tested, as fine and ultrafine PM are not trapped by filters, although they reduce total PM and toxicity [52]. In any case, current legislation that applies to new European cars is EURO 6, so new studies about filter efficiency should be carried out for these new vehicles.

As microalgae and microbial oil biodiesel are bringing a recent alternative to diesel fuel, there is an important lack of studies concerning performance and exhaust emissions analysis of a diesel engine fueled with these new fuels. For this reason, the next section will cover the expected behavior of a diesel engine running on the studied biofuels, based on fuel properties. And finally, a section will include the few engine tests carried out using the proposed alternative fuels.

3.7.1   Fuel properties of microalgae and microbial oil biodiesel

When no engine tests are available, statistical models may help predict both biodiesel properties and engine performance, based on oil fatty acid composition and properties [24, 53].

Wahlen et al. [54] analyzed fuel properties of biodiesel from microalgae (*Chaetoceros gracilis*), yeast (*Cryptococcus curvatus*), and bacteria (*Rhodococcus opacus*) oils and results were compared with soybean oil biodiesel properties. Although oils depicted different fatty acid composition, most significant biodiesel properties, i.e., heating value, viscosity, density, and cetane index, were similar to those of soybean biodiesel. Moreover, the use of glucose as substrate to produce *C. curvatus* oil provided a desirable fatty acid composition for biodiesel production (up to 60 percent mono-unsaturated fatty acids with a chain length similar to common vegetable oils). Moreover, its biodiesel depicted the highest cetane index among tested biodiesel from different origin [77].

Concerning yeast oil biodiesel, fuel properties have been predicted by Leiva-Candia et al. [42]. Authors included the majority of known oleaginous yeast strains grown in substrates from different residues, i.e., *Cryptococcus curvatus*, *Rhodotorula mucilaginosa*, *Lipomyces starkeyi*, *Yarrowia lipolytica*, *Candida curvata*, *Candida oleophila*, *Zygosaccharomyces rouxii*, *Pichia membranifaciens*, *R. toruloides*, and *Cryptococcus albidus*. According to the authors, most samples showed a cetane number higher than that of rapeseed oil biodiesel, and slightly closer to that of palm oil biodiesel. All samples met European biodiesel standard EN 14214. Other predicted properties, such as cold filter plugging point (CFPP), significantly differed between strains and compared to conventional vegetable oil–based biodiesel. Considering predicted kinematic viscosity, only a few samples showed values in agreement with EN 14214. However, it is worth mentioning that extensively used palm oil biodiesel also has problems meeting EN 14214 kinematic viscosity threshold. In general terms, the authors concluded that biodiesel from selected yeast oils present appropriate fuel properties to fuel diesel engines, but only when blended with diesel fuel. This will enhance cold weather behavior, among other properties. Patel et al. assessed yeast oil biodiesel properties of 32 strains, grown on different substrates, based on oil fatty acid composition [55].

Considering PM emissions, higher density, viscosity, boiling point, surface tension, and lower cetane number of algae biodiesel blends were considered responsible for PM reduction [56]. The authors also found that algae oil biodiesel blends exhibit higher carbon chain length and unsaturation degree compared to conventional biodiesel. This was considered the main reason for PM emissions to increase at certain engine loads.

3.7.2 Exhaust emissions of microalgae and microbial oil biodiesel

Yadav et al. tested up to 50 percent microalgae oil biodiesel blended with diesel fuel in a single cylinder diesel engine and found out that 30 percent biodiesel blend showed the minimum unburnt HC and CO emissions. NO_x and CO_2 emissions increased due to unsaturation moiety and oxygen content, respectively [57]. Rajak et al. tested different blends of *Spirulina platensis* microalgae oil biodiesel with diesel fuel, including straight biodiesel, and found out that best option was provided by 20 percent biodiesel blend. Results provided a reduction of CO, NO_x, and smoke emissions, although brake-specific fuel consumption and CO_2 emissions slightly increased compared to the use of straight diesel fuel [58]. The use of 50 percent dinoagellate *Crypthecodinium cohnii* (fire algae) oil biodiesel blend to fuel a diesel engine was also assessed. The authors found a significant exhaust emissions improvement [59]. Fisher et al. concluded that for a content above 20 percent of algae oil biodiesel in the blend, NO_x emissions increased [60]. And more recently, other authors proposed the addition of butanol to algae oil biodiesel to reduce NO_x emissions [61]. Another research group reported a decrease in PM, but an increase in particulate number when algae oil biodiesel was used [56]. It may be explained by the high content of C22:5 and C22:6 of microalgae oil biodiesel. Although it seems that PM reduction is linked to NO_x increase, some authors have suggested the use of emulsions to reduce both emissions [62].

Considering microbial oil biodiesel, *Metschnikowia pulcherrima* oil biodiesel was tested as a fuel for diesel engines. As it contains fewer carbon atoms than diesel fuel, CO and unburnt HC emissions were lower, while CO_2 emissions were higher due to better combustion. The higher the presence of biodiesel in the blend, the higher the NO_x emissions, as expected. However, NO_x emissions decreased proportionally to engine load increase. Smoke was also reduced. These findings are attributed to 20 percent blend [63].

Soccol et al. worked a pilot plant for biodiesel production from *Rhodosporidium toruloides* DEBB 5533 using sugarcane juice. Further diesel engine operation showed lower pollutant emissions (CO_2, CO, NO_x, and unburnt HC) compared to those of soybean oil biodiesel [64].

Some authors have compared engine behavior with different microalgae and microbial oil biodiesel. Engine performance and exhaust emissions of a two-cylinder diesel engine fueled with biodiesel from microalgae (*Chaetoceros gracilis*), yeast (*Cryptococcus curvatus*), and bacteria (*Rhodococcus opacus*) oils have been tested and results compared with both no. 2 diesel fuel and soybean oil biodiesel. Results considering engine power, torque, and brake-specific fuel consumption were similar to those provided by the use of soybean oil biodiesel. In general terms, exhaust emissions were similar among biofuels. However, biofuels produced significantly less CO and unburnt HC emissions than diesel fuel. Particularly and unexpectedly, microalgae oil biodiesel reduced NO_x emissions below to values provided by the use of diesel fuel, while yeast oil biodiesel increased them [54].

Aviation is becoming an interesting market for these new biofuels. In September 2008 the U.S. company Solazyme launched the first algae-based jet fuel. It meets ASTM D1655 specification for aviation turbine fuel [65].

3.8 Future Lines

Today, microalgae/microorganism oil biodiesel is not providing an economically sound alternative to diesel fuel. Improving technologies mentioned above are key to extend its use, and this includes increasing the knowledge about both microorganism growing conditions and potential substrates. Genetic engineering may provide a significant tool to succeed with this challenge, as oil composition may be genetically modified to meet biodiesel standards. Moreover, this tool may also provide the best substrate. Research in most oleaginous microalgae and microorganisms previously mentioned is insufficient. To determine the viability of their use as a source to produce biodiesel, and to optimize both transesterification as well as engine performance/exhaust emissions more research is needed.

Acknowledgments

My sincere thanks to the following people, organizations, and companies for their generosity of letting me use their photos: Pacific

Northwest National Laboratory (PNNL), SUBITEC GmbH (special thanks to Gabriele Fees for her kindness and great support), Jessie Pascual P. Bitog and coauthors, beefcentral.com (special thanks to Jon Condon for great support), BGG-Solix Combination, Jason Quinn, Jeff Moody and Chris McGinty from Utah State University, Hans Reith and Iago Dominguez Teles (Wageningen University, Agrotechnology & Food Sciences), Fitoplancton Marino (special thanks to Lalia Mantecon for great support), Varicon Aqua (special thanks to Joe McDonald and Marco Lizzul for great support), BIONET (Carmen García Álvarez y Raquel Riquelme are specially acknowledged) and photographer Martín García Pérez, for an excellent photo. I also would like to thank those people who helped me find the permissions or the owners of the photographs: Timo Enderle, from Cofactor consulting, Mary-Ann Muffoletto (College of Science's public relations specialist of Utah State University), and Submariner network (Angela Schultz-Zehden and Irina Zimmermann).

References

1. Anonymous, *Implications of the Paris Agreement of Coal Use in the Power Sector*. 2016, Climate Analytics.
2. G. Wynn and P. Coghe. *Europe's Coal-Fired Power Plants: Rough Times Ahead. Analysis of the Impact of a New Round of Pollution Controls*. 2017, Institute for Energy Economics and Financial Analysis.
3. L. Lynd. The grand challenge of cellulosic biofuels. *Nature Biotechnology* **35**(10), 912–915, 2017.
4. S. Pinzi et al. Latest trends in feedstocks for biodiesel production. *Biofuels, Bioproducts & Biorefining* **8**, 126–143, 2014.
5. M. Tredici and R. Materassi. From open ponds to vertical alveolar panels: the Italian experience in the development of reactors for the mass cultivation of phototrophic microorganisms. *Journal of Applied Phycology* **4**(3), 221–231, 1992.
6. M. Chisti. *Airlift Bioreactors*. New York, NY: Elsevier Applied Science Ltd., 1991.
7. S. Ammar. Cultivation of microalgae Chlorella vulgaris in airlift photobioreactor for biomass prodution using commercial NPK nutrients. *Al-Khwarizmi Engineering Journal* **12**(1), 90–99, 2016.
8. A. Carvalho, L. Meireles, and F. Malcata. Microalgal reactors: A review of enclosed system designs and performances. *Biotechnology Progress*, **22**(6), 1490–1506, 2006.
9. M. Menegazzo and G. Fonseca. Biomass recovery and lipid extraction processes for microalgae biofuels production: A review. *Renewable & Sustainable Energy Reviews* **107**, 87–107, 2019.
10. T. Doan, B. Sivaloganathan, and J. Obbard. Screening of marine microalgae for biodiesel feedstok. *Biomass & Bioenergy* **35**, 2534–2544, 2011.

11. E. Sydney et al. Microalgae as raw material for biofuels production. *Journal of Industrial* **88**, 3291–3294, 2011.

12. C.-Y. Chen et al. Cultivation, photobioreactor design and harvesting of microalgae for biodiesel productoin: A critical review. *Bioresource Technology* **102**, 71–81, 2011.

13. J. Milano et al. Microalgae biofuels as an alternative to fossil fuel for power generation. *Renewable & Sustainable Energy Reviews* **58**, 180–197, 2016.

14. I. Nascimento et al. Screening microalgae strains for biodiesel production: lipid productivity and estimation of fuel quality based on fatty acids profiles as selective criteria. *Bioenergy Research* **6**, 1–13, 2013.

15. E. B. Sydney, A. Tokarski, A. C. Novak, J. C. de Carvalho, A. L. Woiciecohwski, and S. C. Larroche. Screening of microalgae with potential for biodiesel production and nutrient removal from treated domestic sewage. *Applied Energy* **88**, 3291–3294, 2011.

16. L. Brennan and P. Owende. Biofuels from microalgae—a review of technologies for production, processing, and extractions of biofuels and co-products. *Renewable & Sustainable Energy Reviews* **14**(2), 557–577, 2010.

17. A. Demirbas and M. Demirbas. Importance of algae oil as a source of biodiesel. *Energy Conversion & Management* **52**, 163–170, 2011.

18. Y. Chisti. Constraints to commercialization of algal fuels. *Journal of Biotechnology* **167**, 201–214, 2013.

19. I. Rawat et al. Biodiesel from microalgae: a critical evaluation from laboratory to large scale production. *Applied Energy* **103**, 444–467, 2013.

20. V. Skorupskaite, V. Makareviciene, and M. Gumbyte. Opportunities for simultaneous oil extraction and transesterification during biodiesel fuel production from microalgae: a review. *Fuel Processing Technology* **150**, 78–87, 2016.

21. M. Griffiths, R. Van Hille, and S. Harrison. Selection of direct transesterification as the preferred method for assay of fatty acid content of microalgae. *Lipids* **45**, 1053–1060, 2010.

22. R. Menezes et al. Evaluation of the potentiality of freshwater microalgae as a source of raw material for biodiesel production. *Quimica Nova* **36**, 10–15, 2013.

23. S. Razzak et al. Integrated CO_2 capture, wastewater treatment and biofuel production by microalgae culturing—a review. *Renewable & Sustainable Energy Reviews* **27**, 622–653, 2013.

24. S. Pinzi et al. The ideal vegetable oil-based biodiesel composition: a review of social, economical and technical implications. *Energy & Fuels* **23**, 2325–2341, 2009.

25. J. Ling et al. Lipid production by a mixed culture of oleaginous yeast and microalga from distillery and domestic mixed wastewater. *Bioresource Technology* **173**, 132–139, 2014.

26. Q. Li, W. Du, and D. Liu. Perspectives of microbial oils for biodiesel production. *Applied Microbiology Biotechnology* **80**, 749–756, 2008.
27. H. M. Alvarez et al. Biosynthesis of fatty acids and triacylglycerols by 2, 6, 10, 14-tetramethyl pentadecane-grown cells of Nocardia globerula 432. *FEMS Microbiology Letters* **200**, 195–200, 2001.
28. M. K. Gouda, S. H. Omar, and L. M. Aouad. Single cell oil production by Gordonia sp DG using agro-industrial wastes. *World Journal of Microbiology & Biotechnology* **24**(9), 1703–1711, 2008.
29. S. Hetzler and A. Steinbüchel. Establishment of cellobiose utilization for lipid production in Rhodococcus opacus. *Applied and Environmental Microbiology* **79**, 3122–3125, 2013.
30. Y. Ma et al. Biodiesels from microbial oils: Opportunity and challenges. *Bioresource Technology* **263**, 631–641, 2018.
31. D. E. Leiva-Candia and M. P. Dorado. New frontiers in the production of biodiesel: biodiesel derived from macro and microorganisms. In *Liquid Biofuels: Emergence, Development and Prospects*, M.S. Antônio Domingos Padula, Omar Santos y Denis Borenstein, Editor. London, UK: Springer-Verlag, 2009, pp. 205–225.
32. X. Meng et al. Biodiesel production from oleaginous microorganisms. *Renewable Energy* **34**, 1–5, 2009.
33. Y. Li, Z. Zhao, and F. Bai. High density cultivation of oleaginous yeast Rhodosporidium toruloides Y4 in fed-batch culture. *Enzyme and Microbial Technology* **41**, 312–317, 2007.
34. R. Davies. Yeast oil from cheese whey: process development. In *Single Cell Oil*, M. RS, Editor. London, UK: Longman, 1988. pp. 99–145.
35. A. Amaretti et al. Single cell oils of the cold-adapted oleaginous yeast Rhodotorula glacialis DBVPG 4785. *Microbial Cell Factories* **9**(1), 73, 2010.
36. A. A. Koutinas et al. Design and techno-economic evaluation of microbial oil production as a renewable resource for biodiesel and oleochemical production. *Fuel* **116**, 566–577, 2014.
37. X. Zhao et al. Lipid production by Rhodosporidium toruloides Y4 using different substrate feeding strategies. *Journal of Industrial Microbiology and Biotechnology* **38**(5), 627–632, 2014.
38. C. H. Calvey et al. Nitrogen limitation, oxygen limitation, and lipid accumulation in Lipomyces starkeyi. *Bioresource Technology* **200**, 780–788, 2016.
39. B. G. Ryu et al. High-cell-density cultivation of oleaginous yeast Cryptococcus curvatus for biodiesel production using organic waste from the brewery industry. *Bioresource Technology* **135**, 357–364, 2013.
40. K. Mathiazhakan, D. Ayed, and R. D. Tyagi. Kinetics of lipid production at lab scale fermenters by a new isolate of Yarrowia lipolytica SKYJ. *Bioresource Technology* **221**, 234–240, 2016.

41. A. Beopoulos, J. M. Nicaud, and C. Gaillardin. An overview of lipid metabolism in yeasts and its impact on biotechnological processes.*Applied Microbiology Biotechnology* **90**(4), 1193–1206, 2011.
42. D. E. Leiva-Candia et al. The potential for agro-industrial waste utilization using oleaginous yeast for the production of biodiesel. *Fuel* **123**, 33–42, 2014.
43. L. Zheng et al. Exploring the potential of grease from yellow mealworm beetle (Tenebrio molitor) as a novel biodiesel feedstock. *Applied Energy* **101**, 618–621, 2013.
44. A. Chatzifragkou et al. Biotechnological conversions of biodiesel derived waste glyceroll by yeast and fungal species.*Energy*, 2011. **36**: p. 1097–1108.
45. C. J. Xia et al. A new cultivation method for microbial oil production: cell pelletization and lipid accumulation by Mucor circinelloides. *Biotechnology & Biofuels* **4**, 15, 2011.
46. Z. Ruan et al. Evaluation of lipid accumulation from lignocellulosic sugars by Mortierella isabellina for biodiesel production. *Bioresource Technology* **110**, 198–205, 2012.
47. E. Zikou et al. Evaluating glucose and xylose as cosubstrates for lipid accumulation and γ-linolenic acid biosynthesis of Thanmnidium elegans. *Journal of Applied Microbiology* **114**(4), 1020–1032, 2013.
48. M. D. Cardenas et al. Performance and pollutant emissions from transient operation of a common rail diesel engine fueled with different biodiesel fuels. *Fuel* **185**, 743–762, 2016.
49. O. Armas, M. D. Cardenas, and C. Mata. Smoke opacity and NOx emissions from a bioethanol/diesel blend during engine transient operatijon. SAE paper, 2007.
50. R. Bengalli et al. In vitro pulmonary and vascular effects induced by different diesel exhaust particles. *Toxicology Letters* **306**, 13–24, 2019.
51. A. F. Behndig et al. Proinflammatory doses of diesel exhaust in healthy subjects fail to elicit equivalent or augmented airway inflammation in subjects with asthma. *Thorax* **66**(1), 12–19, 2011.
52. I. M. Kooter et al. Toxicological characterization of diesel engine emissions using biodiesel and a closed soot filter. *Atmospheric Environment* **45**(8), 1574–1580, 2011.
53. S. Pinzi et al. The effect of biodiesel fatty acid composition on combustion and diesel engine exhaust emissions. *Fuel* **104**, 170–182, 2013.
54. B. D. Wahlen et al. Biodiesel from microalgae, yeast, and bacteria: engine performance and exhaust emissions. *Energy & Fuels* **27**(1), 220–228, 2013.
55. A. Patel et al. Assessment of fuel properties on the basis of fatty acid profiles of oleaginous yeast for potential biodiesel production. *Renewable & Sustainable Energy Reviews* **77**, 604–616, 2017.

56. M. M. Rahman et al. Particle emissions from microalgae biodiesel combustion and their relative oxidative potential. *Environmental Science: Processes & Impacts* **17**, 1601–1610, 2015.

57. M. Yadav et al. Experimental study on emissions of algal biodiesel and its blends on a diesel engine. *Journal of the Taiwan Institute of Chemical Engineers* **96**, 160–168, 2019.

58. U. Rajak, P. Nashine, and T.N. Verma. Assessment of diesel engine performance using spirulina microalgae biodiesel. *Energy* **166**, 1025–1036, 2019.

59. M. A. Islam et al. Combustion analysis of microalgae methyl ester in a common rail direct injection diesel engine. *Fuel* **143**, 351–360, 2015.

60. B. C. Fisher et al. Measurement of gaseous and particulate emissions from algae-based fatty acid methyl esters. *SAE International Journal of Fuels and Lubricants* **3**(2), 292–321, 2010.

61. V. Kumar et al. Production of biodiesel and bioethanol using algal biomass harvested from fresh water river. *Renewable Energy* **116**, 606–612, 2018.

62. J. K. Mwangi et al. Emission reductions of nitrogen oxides, particulate matter and polycyclic aromatic hydrsocarbons by using microalgae biodiesel, butanol and water in diesel engine. *Aerosol and Air Quality Research* **15**, 901–914, 2015.

63. A. Tamilalagan, J. Singaram, and S. Rajamohan. Generation of biodiesel from industrial wastewater using oleaginous yeast: performance and emission characteristics of microbial biodiesel and its blends on a compression injection diesel engine. *Environmental Science and Pollution Research* **26**(11), 11371–11386, 2019.

64. C. R. Soccol et al. Pilot scale biodiesel production from microbial oil of Rhodosporidium toruloides DEBB 5533 using sugarcane juice: performance in diesel engine and preliminary economic study. *Bioresource Technology* **223**, 259–268, 2017.

65. O. M. Adeniyi, U. Azimov, and A. Burluka. Algae biofuel: current status and future applications. *Renewable & Sustainable Energy Reviews* **90**, 316–335, 2018.

Questions

1. Which species are able to produce microbial oil?

2. Which steps are needed to produce microalgae oil?

3. What fatty acid composition is preferred to produce biodiesel from oil?

4. Are there any differences concerning engine behavior between the studied biodiesels discussed in this chapter?

5. How can exhaust emissions be improved when using the studied biodiesel?

CHAPTER 4

Nanotechnology in Biofuel Production: Applications, Challenges, and Future Perspectives

Hajra Javed, Muhammad Naveed Anwar, Abdul Sattar Nizami, and Mujtaba Baqar

4.1   Introduction

There has been a significant increase in the demand for petroleum-based fuels due to rapidly growing industrial and transportation sectors along with raised living standards. Currently, fossil fuels are meeting 80 percent of the worldwide energy demands with the transportation sector consuming approximately 60 percent. Overexploitation of fossil resources is resulting in high oil prices and the emission of greenhouse gases (GHG), causing vast impacts on the environment and human health. Therefore, it is becoming vital to shift toward alternative energy resources that are renewable, sustainable, and economically viable. Biofuels have emerged as a sustainable and renewable source of energy across the globe [1]. Among new developments in science and research, nanotechnology is emerged as a promising field and is being successfully applied in different disciplines. Nanotechnology is providing sustainable solutions in the fields of biofuels and bioenergy to meet the increased energy demands. Specifically, it is helping to modify feedstocks and develop catalysts that are more efficient to increase the yield of biofuels [2].

The term *Nano* originates from a Greek word meaning *dwarf* [3]. Nanoscale dimensions are typically between 1 to 100 nm or 0.2 to 100 nm. This scale enables the material to possess unique physical, chemical, and biological properties as compared to a largerscale

111

material. A single nanometer (nm) is represented as a billionth part of a meter [4]. The concept of nanotechnology was introduced initially by Richard P. Feynman about 50 years ago when he delivered a lecture titled "There Is Plenty of Room at the Bottom" on December 26, 1959, to the American Physical Society, where he explained the benefits of manufacturing materials to a small scale, thus presenting the concept of nanotechnology [3]. Nanotechnology is defined as to understand and control materials at a nanoscale level, where a unique phenomenon enables novel applications. However, nanotechnology has some limitations regarding size, and this results in the exclusion of various materials and devices, especially in the agriculture and pharmaceutical fields [5].

A nanomaterial is any material that has one or more than one dimensions in the nanoscale. Nanomaterials include natural or synthetically produced materials consisting of particles that are not bound, aggregated, or agglomerated with one another and where usually half or more than half of the total particles have dimensions between 1 to 100 nm. Nanomaterials can be classified as one-dimensional layered structures like graphene, thin films, or coatings for surfaces, or two-dimensional such as nanowires and nanotubes. They can also be three-dimensional particles such as precipitates, dendrimers, colloids or semiconductor materials, and tiny particles like quantum dots [4].

Nanomaterials are manufactured using several methods that can be classified as [6] (1) addition of acid or base to suspension to change its pH value, (2) addition of surface-active agents and/or dispersants for the dispersal of particles into the fluid, and (3) use of ultrasonic vibration. Their features vary in terms of their conductance capacity of electricity or heat, reflection of light, reactivity rates, and strength [7]. Example includes silver that shows enhanced antimicrobial activities, inert materials such as platinum and gold that possess catalytic activities, and stable materials like aluminum [6]. An engineered or synthetic nanomaterial is the one that has been produced on a nanoscale so that it has specific properties or composition. Such nanoparticles possess unique features that are different from their conventional counterparts [4].

Biofuel is defined as any solid, liquid, or gaseous fuel derived from biomass in the form of bioethanol, bio-methanol, biodiesel, bio-oil Fischer Tropsch (FT) diesel, hydrogen, and methane. Biofuel is considered a sustainable, ecofriendly source of energy [1]. Biofuels are classified into three categories: first-, second-, and third-generation biofuels. First-generation biofuels are produced from sources such as starch, sugars, fats, and oils. First-generation biofuels have three main types: biodiesel (bio-esters), bioethanol, and biogas [8]. Second-generation biofuels are produced using cellulosic biomass such as cellulose, hemicellulose, and lignin. Second-generation biofuels are considered as carbon neutral or carbon negative based on their impacts on the concentration of carbon dioxide. Examples include bioethanol, biodiesel, and dimethyl ether [9]. Third-generation biofuels are derived from algae, having greater quality and yield as compared to fuels of the first and second generations. They can produce a wide variety of fuels and value-added chemicals that include bio-methane, bio-ethanol, gasoline, and jet fuel [10].

This chapter aims to review the possibilities of using nanotechnology as an effective tool to solve the problems in the sector of bioenergy and the production of different types of biofuels. The chapter discusses several conventional methods of biofuel production along with their technical limitations. It also includes the current challenges and limitations regarding the use of nanotechnology in biofuel production systems. Furthermore, this chapter includes a section highlighting the prospects and opportunities regarding the development of environment-friendly and economical nanotechnology-based biofuels.

4.2   Methods and Limitations of Biofuel Production

4.2.1   Conventional methods of biofuel production

4.2.1.1   Production of biodiesel

A potential substitute for conventional petroleum diesel is biodiesel, which is a fatty acid methyl ester (FAME) derived from vegetable oil. Biological renewable resources such as oils, edible, nonedible, and animal fats can be used to obtain biodiesel by chemically reacting these oils with alcohols such as ethanol or methanol, and catalyzed by either acids or bases. Such reaction results in a product that is a combination of fatty acid alkyl ester (FAAE) and a co-product such as glycerol that has a high value. Catalysts can be acidic or basic, or homogeneous or heterogeneous, for the production of biodiesel. Transesterification is homogenous, reactants are mixed with a catalyst that can be a liquid acid or base, and this process is a reversible reaction [1, 11] (Fig. 4.1).

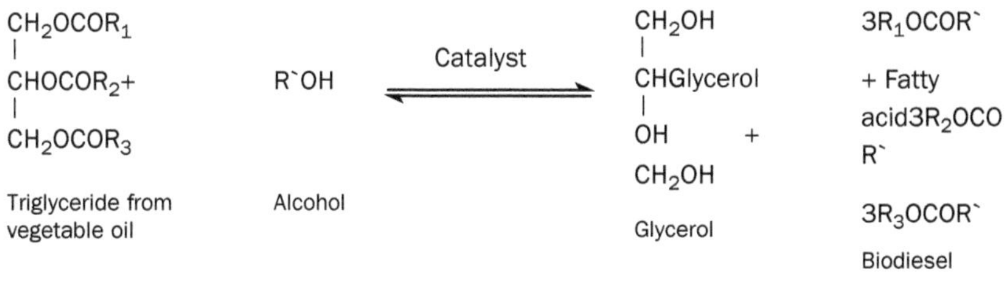

$$CH_2OCOR_1$$
$$|$$
$$CHOCOR_2 +$$
$$|$$
$$CH_2OCOR_3$$

Triglyceride from vegetable oil

$$R`OH$$

Alcohol

Catalyst

$$CH_2OH$$
$$|$$
$$CHGlycerol$$
$$|$$
$$OH   +$$
$$CH_2OH$$

Glycerol

$$3R_1OCOR`$$

+ Fatty acid $3R_2OCO$ R`

$$3R_3OCOR`$$

Biodiesel

Figure 4.1   Transesterification of vegetable oils [1].

For the homogenous transesterification, liquid bases are preferred over liquid acids as a catalyst due to the latter's corrosive nature. Although homogenous catalysis has a high rate of reaction under mild reaction conditions, still they present some drawbacks such as being corrosive, are not recycled and environment-friendly, and produce biodiesel and glycerol contaminated with sodium or potassium ions. These catalysts are also less attractive due to the production of wastewater in large amounts [1]. Heterogeneous transesterification is a process that uses solid acid or base as catalysts for the conversion of reactants to biodiesel, providing better yields as compared to the homogenous catalysts. Heterogeneous catalysts provide easy recovery and produce a product that is free from the catalyst and does not require neutralizing the product or purification. These catalysts help to economize the production of biodiesel since they are consumed in smaller amounts and can be reused. Although both solid acids and bases can be used to produce biodiesel, solid acid catalysts are reported to be more practical for the transesterification process of the fatty acid content present in the vegetable oils. Solid base catalysts are less preferred since they lead the process of saponification in oils that have a high content of free fatty acids [12] (Fig. 4.2).

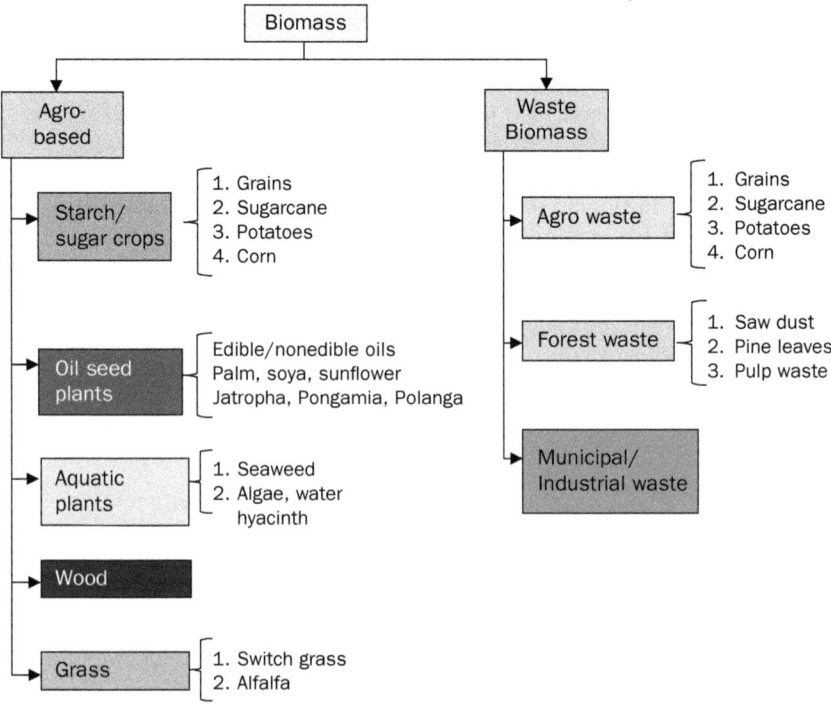

Figure 4.2 Sources of biomass feedstocks for biofuel production [1].

4.2.1.2 Process for bioethanol conversion

Feed stocks that contain carbohydrates such as crops containing sugar (sugarcane, fruits, and beetroot), starch (all kinds of grains rich in carbohydrates), and cellulosic biomass (wood, agricultural residue, and wood waste) have been used to produce ethanol. Ethanol can be produced through fermentation or different bio-chemical processes. Ethanol produced from food crops is known as grain-ethanol. Ethanol produced using biomass as feedstock such as lignocellulosic biomass or agricultural waste is known as bioethanol. Since carbohydrates are polymers of glucose, they cannot be converted to ethanol directly through a conventional fermentation process or biochemical methods. Therefore, this polymeric structure needs to be broken down into smaller units, finally converting glucose to ethanol and other valuable chemicals [13].

4.2.1.3 Process for biogas production

Biogas is a mixture of different gases, such as methane, carbon dioxide, carbon monoxide, hydrogen sulfide, and water, but its main constituents are methane and carbon dioxide. Biodegradable organic materials produce biogas under anaerobic (without oxygen) conditions. Therefore, this process of conversion is called anaerobic digestion (AD). It is a natural process of decomposition of feedstock by microorganisms in a digester tank/vessel that has no oxygen resulting in the production of biogas and a digestate. AD has four different stages such as hydrolysis, acidogenesis, acetogenesis, and methanogenesis leading to the formation of biogas that contains about 60 percent to 70 percent methane and 30 percent to 40 percent carbon dioxide [14]. In the present era where the demand and per capita cost of energy is increasing exponentially, AD of biomass to produce biofuel and biofertilizers is an attractive, environment-friendly option. For the management of solid waste in landfill sites that are a major source of bio-methane production, AD can be a promising choice. This process results in the production of landfill gas (LFG), a natural gas of low quality that is a mixture containing methane, oxygen, nitrogen, carbon dioxide, and trace amounts of organic compounds. It is essential that carbon dioxide and volatile organic contaminants are removed to improve LFG's commercial value [15]. Conventionally, LFG is used for energy generation, using internal combustion engines for power generation, and using it directly in boilers, turbines, micro-turbines, and for co-generation. Since these conventional methods have higher costs to produce, purify, separate, and collect methane, it is being emphasized to convert LFG into liquid fuel rather than a gaseous fuel through AD. Different catalytic routes can be used to convert methane to methanol [16].

4.2.1.4 Conversion processes of second-generation biofuel

Currently, conversion processes through biochemical and thermochemical routes are producing second-generation biofuels.

Thermochemical conversion processes include gasification, direct combustion, liquefaction, and pyrolysis. Heating biomass under limited oxygen supply results in the production of syngas, which is primarily composed of hydrogen and carbon monoxide. Syngas can be used directly for burning or can be upgraded to other fuels, which are in the gaseous or liquid states. The most conventional route to produce energy is the combustion of biomass, i.e., burning biomass directly in aerobic conditions (oxygen or air is available) [8].

Biomass can be liquefied (liquefaction) directly or through solution of alkalis, glycerin, or alcohols (propanol/butanol). Specific reducing gases such as hydrogen or carbon monoxide, solvents, as well as some catalysts are required by the process to produce a viscous oil that is water insoluble. In addition, syngas and solid catalysts can be used to convert lignocellulosic biomass to liquid fuel that is like heavy oils. Sometimes the handling of highly viscous oil becomes difficult and thus requires organic solvents to reduce their viscosity. Cellulose and hemicellulose present in lignocellulosic biomass can be degraded into smaller fractions by the process of hydrolysis (alkaline). These smaller fractions act as precursors for different value-added products such as gasoline and fuel additives [8, 17]. Liquefaction of biomass, dried in air at higher pressure or by using the path of fast pyrolysis, produces another product known as bio-oil. A multitudinous combination consists of organic acids of volatile nature, hydroxyl aldehydes, and ketones, ethers, esters, alcohols, phenolics, and sugars. Catalytic upgradation of bio-oil can produce organic distillate chemical products that have high quantities of hydrocarbons [18].

For the last 40 years, a well-known process called gasification has been investigated that is the conversion of biomass to viable fuels. In this process, biomass reacts with air/oxygen/steam resulting in the production of a mixture of hydrogen, methane, carbon dioxide, carbon monoxide, and nitrogen gas called producer gas or synthesis gas (syngas). Syngas can be used to produce a variety of fuels and intermediates of chemicals; however, producer gas proves useful only for stationary power generation [19]. Degrading biomass thermally under anaerobic conditions is called pyrolysis and produces bio-oil, a gaseous mixture, and solid charcoal called biochar. Pyrolysis can be categorized as conventional, fast, and flash pyrolysis by different operating conditions [20]. Conventional pyrolysis is a process that has a slow rate of heating with high retention time. In a pre-pyrolysis stage, biomass is decomposed at a temperature range of 276.85°C to 676.85°C. Initial molecular changes occur during this stage, including removal of water, cleavage of bonds, and generation of free radicals. In the second stage, solid decomposition of biomass takes place, producing the pyrolysis products. In the third stage, the solid formation of carbon-rich occurs by slowly decomposing char [21].

Fast pyrolysis is mostly used for producing products that are liquid or gaseous and takes place at higher temperature, i.e., 576.85°C to 976.85°C. It has short residence time and higher rate of heating using feedstock in the form of fine particles of biomass. In this process, biomass is decomposed to generate vapors, aerosol, and char. Fast pyrolysis results in a product that has 60 percent to 75 percent bio-oil, 10 percent to 20 percent noncondensed gases, and 15 percent to 25 percent char [22]. Flash pyrolysis is not used often due to its operating conditions and occurs in temperatures that are very high, i.e., 826.85°C to 1126.85°C. The heating speed of the process is extremely quick (greater than 726.85°C/s) and a very short residence (0.5 s). It uses fine particles of biomass that are less than 0.2 nm in size as feedstock [23].

There is another greener form of biodiesel called green diesel that is produced through catalyzed hydro-treatment of vegetable oils. Renewable feedstocks such as vegetable oils are extensively used to produce green diesel. This existing technology depends on the process of transesterification of edible or nonedible oils. The production of biodiesel has some technical issues; therefore, widespread use of biofuels in the future greatly depends on the development of new and advanced technologies for producing high-quality fuel using biological sources as feedstock [1, 24]. Using crops such as corn and sugarcane to produce ethanol is being carried out commercially through biochemical routes. Different techniques of operation such as pretreatment, hydrolysis, fermentation, production of enzymes, and product recovery need to be understood to produce ethanol from biological resources (Fig. 4.3).

Currently, technologies being used to produce ethanol are not very economical, with the focus of research being on the reduction of overall cost of production. Improved and advanced techniques are required to convert biomass (cellulosic and hemicellulosic) to sugars, with lower consumption of energy for pretreatment. Separation technologies with greater efficiency are needed so that value could be added to lignin. Lignocellulosic biomass can also be produced from agricultural and forestry wastes, and processing of industrial crops postharvest, in addition to the sources mentioned above. This waste presents a promising source to produce bioethanol, along with other valuable chemicals and fuel additives. Oxygenated fuels derived from biomass and other fuel additives can be used to produce fuels, which have fewer GHG emissions after blending with gasoline. Ethanol blending with gasoline in ratio of about 5 percent to 25 percent is being carried out, without the need to modify engine systems or their settings [25].

4.2.1.5 Processes of conversion for third-generation biofuel

Generally, biofuels derived from algae are called third-generation biofuels. To produce such biofuels, conversion technology depends

118

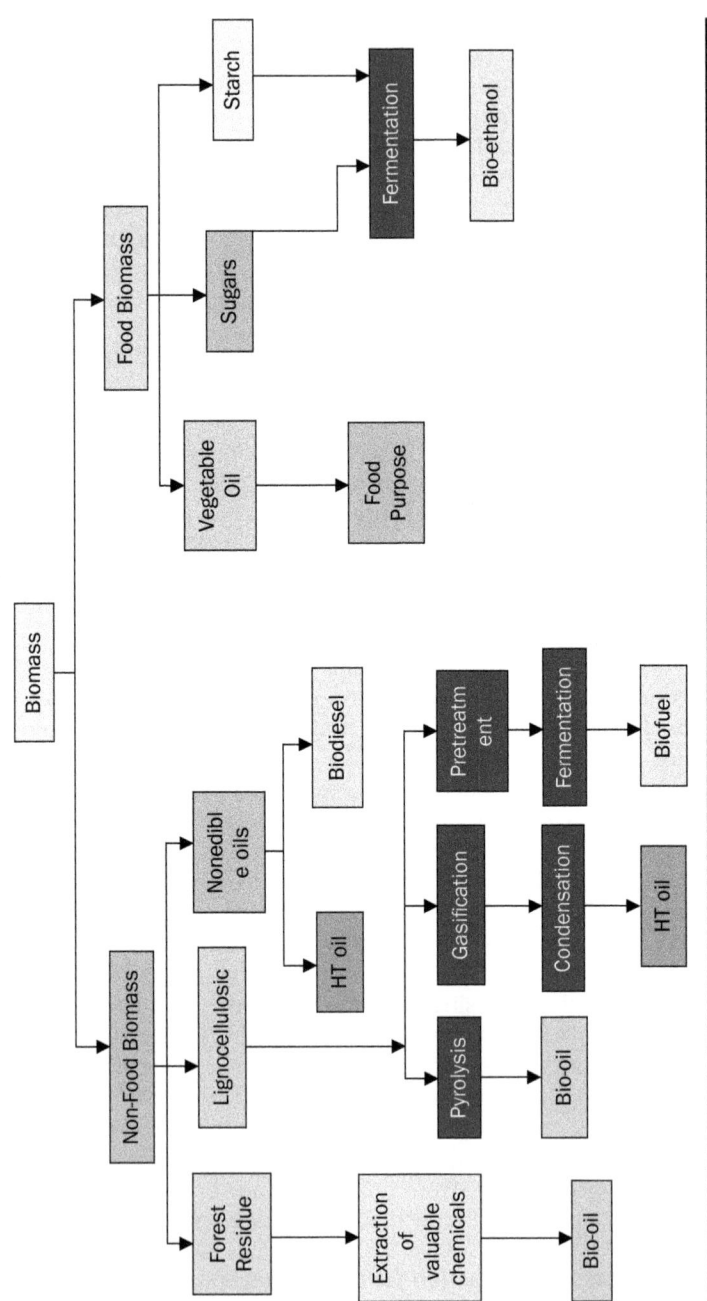

Figure 4.3 Lignocellulosic biofuel conversion process [1].

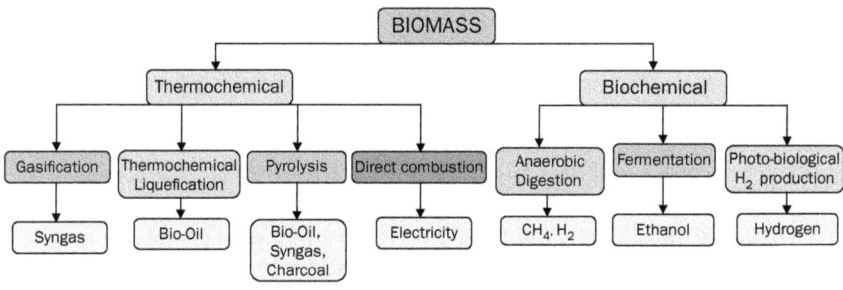

Figure 4.4 Algal biofuel conversion processes [1].

on various factors such as type of feedstock and its quantity, end-product, cost of production, the kind of energy required, and the form of energy desired. Like second-generation biofuels, technologies used for conversion of third-generation biofuels fall under two categories of thermochemical and biochemical processes along with the production of biodiesel [26]. Thermochemical conversion is a process in which thermal decomposition of organic matter (biomass) occurs. A variety of co-processes are also included such as gasification, thermochemical liquefaction, combustion, and pyrolysis that have been discussed earlier in section 1.3. Using biological routes to harvest energy from biomass is known as biochemical conversion. Commonly used techniques are AD, fermentation, and photocatalytic hydrogen production [27] (Fig. 4.4).

AD of algae includes the conversion of organic waste into biogas, a mixture of methane and carbon dioxide with trace amounts of hydrogen sulfide. This process is suitable for organic wastes that have a high moisture content, greater than 75 percent. There are three sequential stages of AD, hydrolysis, fermentation, and methanogenesis. The waste contains complex compounds that get broken down to simple sugars during the process of hydrolysis. During fermentation, conversion of simple sugar takes place to alcohols, acetic acid, volatile fatty acids, and the mixture of hydrogen and carbon dioxide. During methanogenesis, metabolites of the second step are converted to methane (approximately 70 percent) and carbon dioxide (about 30 percent). AD also produces waste (digestate) that is rich in nutrients and can be recycled to be used as a medium for new algal growth [28]. In the process of fermentation, cellulosic biomass is degraded to lower sugars and then to ethanol. The first crushing of biomass takes place and the hydrolysis to starch molecules, followed by the conversion of these molecules to sugars. Then fermentation of these sugars takes place using yeast. For ethanol production, the algae selection depends on the starch content present in algae. The greater the content of starch in algae, the greater the yield of ethanol. Therefore, *Chlorella vulgaris* acts as an excellent feedstock since its starch content is about 37 percent of its dry weight [29].

Hydrogen acts as a carrier of energy as it is ultraclean, efficient, and occurs naturally. Algae possess the genetic metabolites and enzymes required to photo produce hydrogen. Hydrogen gas is produced in the absence of oxygen, either through the electron donor pathway during fixation of carbon dioxide or through the involvement of both light and darkness. During photosynthesis, microalgae result in the conversion of water into hydrogen ions and oxygen. Hydrogenase then catalyzes the conversion of hydrogen ions into hydrogen gas in the absence of oxygen. This is a reversible process. Therefore, hydrogen either is produced or consumed. There are two approaches to produce hydrogen gas [30], (1) a two-stage process that produces hydrogen as well as oxygen, however, both are produced separately; and (2) simultaneous production of hydrogen and oxygen via photosynthesis. Theoretically, the most preferred process is the second one, but the first process is beneficial because of its green and ecofriendly nature. The conversion processes are shown in Figure 4.5.

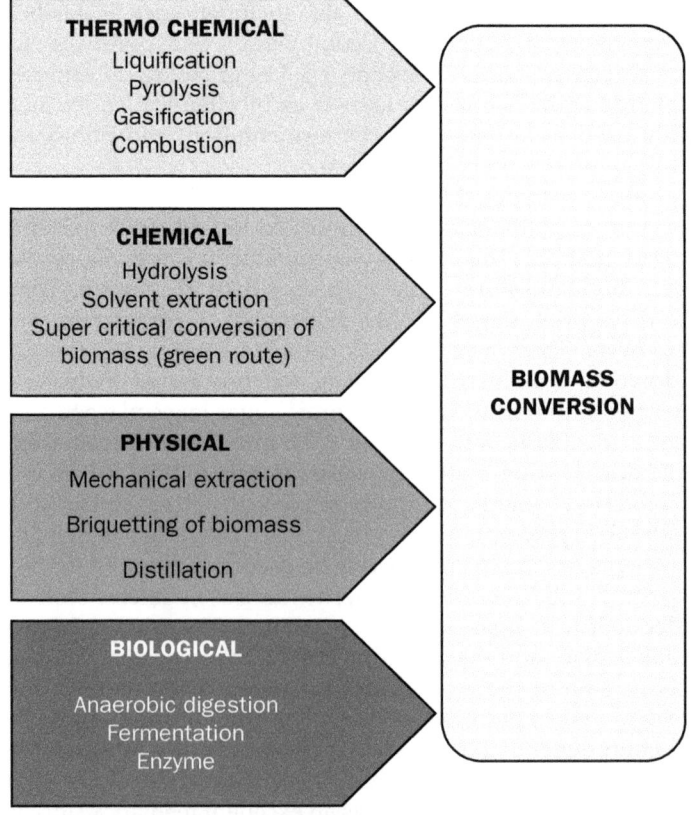

FIGURE 4.5   Biomass conversion processes [8].

4.3   Limitations of Conventional Methods

The biofuel production at commercial scale faces technical, environmental, and economic constraints as discussed below [1].

4.3.1   Availability of feedstock, food security changes in land use, and water source

The success and usefulness of biofuel and its technologies depend greatly on biomass feedstock, and it should be sufficiently available in a way that is sustainable considering its impacts on environment issues such as food versus fuel. All around the world, researchers are stressing that non-fodder feedstocks that are carbon neutral should be promoted for producing biofuels since it will not affect the biodiversity of the native food crops, which are essential for food production [31]. The food versus fuel debate has also been addressed in many studies. Since the world's population is growing day by day, the demand for food and water is increasing worldwide. Biofuel and food prices did not correlate previously, but food crops are being used to produce biofuels, and this has resulted in the strengthening of the correlation. The production and demand of biofuel are also increasing continuously along with the competitiveness of the sector, resulting in the increased costs of source materials. These trends, therefore, suggest that in the future there will be a strong linkage between food and markets of biofuel, which will affect the prices of food. This triggers the shift toward developing biofuel using non-fodder feedstocks. Therefore, biofuel production is considered to have an impact on food security as well as on changes in land use and water resource [32]. In developing countries, farmers have started growing nonfood crops so that their income and employment can be boosted. This resulted in a decrease in food availability and security and ultimately increase in prices. There are other barriers such as low energy and density of biofuels along with high collection and transportation cost of biomass. Crop production for biofuel can put significant pressure on the water resources if the water's quality and availability are not taken into consideration. This agricultural shift could result in the change of availability of water that is clean and potable. Consequently, this is putting significant pressure on the local resources of water. Biofuels are also associated with poor economics due to intensive requirements of inputs such as land and land use, water, energy, and feed. Therefore, it is recognized as a constraint in the sustainable development of biofuels [33].

4.4   Environmental Constraints

Although biofuels are termed as environment-friendly, their extensive production is found to be associated with some environmental concerns. Being a product that has been derived from biological resources, they are potentially considered as carbon-neutral over their whole life cycle. When biofuels are combusted, they release the same amount of carbon dioxide that was sequestered by the feedstock crops during photosynthesis. Therefore they contribute to the mitigation of GHG emissions. Apart from the advantages offered by biofuels as compared to petroleum-based fuels, the biofuel industry's extensive development is causing adverse environmental effects either directly or indirectly. There is an additional land requirement to grow oil crops to meet the demand, resulting in extensive deforestation. Deforestation of large scale can have many negative impacts such as erosion of soil, habitat loss, loss of biodiversity, local climate change, and flooding. Serious issues related to emissions produced from the burning of fuel that has been blended with ethanol have also been reported. Since alcohols have high oxidizing nature, they result in the production of toxic aldehydes such as formaldehyde and acetaldehyde. A primary challenge associated with third-generation biofuels is microalgae's cultivation at a larger scale to meet the industry demand. This process requires a lot of water and nutrients to aid growth such as phosphorus and nitrogen. This requires use and production of fertilizers that also requires energy, resulting in the release of GHG emissions, a major contributor to global warming and climate change. Excessive enrichment of nutrients in water bodies due to agricultural runoff can result in eutrophication [34].

4.5 Cost Competitiveness

Biofuels have advantages over conventional fuels, but other challenges such as their economic viability as compared to traditional fuels need to be sorted. Currently, research is focusing on the improvement of the economics of biofuel development so that it can be sustainable. Biofuel production has several constraints such as geographical, societal acceptance, environmental and technical, and the high cost of production, which make biofuels less preferable over conventional fuels [35]. In the transport sector, blends of gasoline and ethanol are being used, but the route of production of bioethanol is still fermentation that has a higher initial cost of pretreatment and requires expensive enzymes [25]. Biodiesel has also gained much attention as an alternative fuel since it has good compatibility and is being used either directly or as a blend in internal combustion engines. However, its complete commercialization still needs attainment. Presently, the most commonly used source of biodiesel production is a vegetable oil, i.e., greater than 90 percent worldwide. Limited availability of feedstock and efficient production technology greatly affects the economics of producing biofuel [36].

4.6   Applications of Nanotechnology in Biofuel Production

4.6.1   Various types of biofuels produced using nanotechnology

Nanotechnology is emerging as a new branch of science that aims to offer valuable solutions such as a change in feed material characteristics, and biocatalysts, which are being used to produce biofuels. Nanotechnology in an essential field of research in modern science allowing scientists and engineers to manipulate materials at molecular and cellular levels. Nanomaterials of different types like carbon nanotubes, magnetic nanoparticles, nanoparticles of metal oxides, and many others have proven to be advantageous to produce sustainable bioenergy [2].

4.6.2   Nanotechnology in production of biofuel

First-generation biofuel that is produced from food feedstocks such as starch, sugarcane, and vegetable oils presents concerns. Therefore, second-generation biofuels gained attention since they are produced from nonfood crops such as lignocellulosic materials. Despite their benefits, they do present some drawbacks such as high production cost and problems about infrastructure and technology. Keeping these realities in mind, alternative proficient technologies need to be determined to increase mass processing and biofuel productivity. Nanotechnology can play a significant role to overcome the difficulties mentioned above by modifying the feedstock characteristics used for biofuel production [37, 38]. Nanoparticles possess physiochemical properties that are exceptionally enabling them to be used strategically in biofuel production. Many nanomaterials having unique properties have been used in biofuel production such as titanium oxide, ferrous oxide, zinc oxide, carbon, and graphene. Furthermore, magnetic nanoparticles are also being used in biofuel production since they have a high surface to volume ratio, quantum and immobilizing properties. Besides these properties, the nanoparticles possess easy recovery from the reaction medium, i.e., application of a suitable magnetic field [39].

4.6.3   Use of nanotechnology in biodiesel production

Vegetable oils have properties such as density, air to fuel ratio, and heat of vaporization that are like petroleum diesel and known to reduce emissions of carbon oxide, oxides of sulfur, and smoke. The conventional method of biofuel production derived from vegetable oils, transesterification of oils, is done using methanol and sodium hydroxide (NaOH) as a catalyst. However, this process has some drawbacks such as saponification, catalyst deactivation, and slow reaction rate. The problems mentioned above can be solved by using nanotechnology such as titanium-incorporated SBA-15 mesoporous silica. It has proven to have high efficiency and recyclability as a solid catalyst to produce vegetable oil derived biofuels [40]. Bio-oil that is a mixture of both volatile and nonvolatile organic components is produced through liquefication at high pressure of air-dried wood. A study illustrated that biodiesel could be produced from used cooking oil through four different ways: blending with mineral diesel directly, micro-emulsion method, thermal cracking, and transesterification. As highlighted before, the process of transesterification has some problems regarding the production of biodiesel such as free fatty acids being present, the moisture content of oils/fats, the temperature of reaction time, and glycerides to alcohol molar ratio. These shortcomings can be effectively overcome using nanotechnology. Biodiesel can help in the substantial reduction of unburnt hydrocarbon and carbon monoxide emissions. By adding nanoparticles of cerium oxide to biodiesel, complete combustion gets promoted, and hydrocarbon oxidation is enhanced, thereby resulting in reduced hydrocarbons as well as nitrous oxide emissions [41]. Different types of nanocatalysts were explored to enhance the process of transesterification under various operating conditions. Table 4.1 lists different nanocatalysts and feedstock used for biodiesel production.

Bioreactors of carbonanotube-enzyme were produced through the incorporation of nanoparticles of iron oxide into carbon nanotubes with a single wall, resulting in magnetic carbon nanotubes with a singlewall. This bioreactor enables recycling of immobilized enzyme, resulting in improved efficiency of bioreactor and reduction in capital costs [57]. Nanomaterials in comparison to other conventional bulky materials have been proven to be good support for enzyme immobilization since they have a large ratio of surface area to volume, leading to a higher loading of enzyme and enhanced biocatalytic potential [58].

Biodiesel production has also been enhanced by using microbial enzymes bounded to nanomaterials. Use of lipase enzyme from *Pseudomonas cepacia* has seen to affect the soybean oil's transesterification. The effectiveness of this enzyme has also been explored by different studies when the enzymeis bounded to different types of nanomaterials such as nanoporous gold, nanoparticles of Fe_3O_4, and PAN nanofiber. By binding the same enzyme to polyacryonitrile nanofiber, transesterification of rapeseed oil gets affected. Biodiesel derived from soybean oil has been produced using lipase from *Thermomyces lanuginosa*, which has been bounded covalently to nanoparticles of Fe_3O_4. Lipase extracted from *Rhizopus miehei* encapsulated in a nanoparticle of silica for triolein-derived biodiesel.

Nanocatalyst	Feedstock	Yield (%)	Reference
MgO	Sunflower oil and rapeseed Oil	98.0	Verziu et al. (2008) [42]
$K_2O/\gamma-Al_2O_3$	Rapeseed oil	94.0	Heyou and Yanping (2009) [43]
KF/CaO	Chinese tallow seed oil	96.8	Wen et al. (2010) [44]
Lithium-impregnated calcium oxide (Li-Cao)	Karanja oil and Jatropha oil	99.0	Kaur and Ali (2011) [45]
ZrO_2 loaded with $C_4H_4O_6$HK	Soybean oil	98.03	Qiu et al. (2011) [46]
Hydrotalcite-derived particles with Mg/Al oxides	Jatropha oil	95.2	Deng et al. (2011) [47]
ZnO nanorods	Olive oil	94.8	Molina (2013) [48]
MgO nanoparticles on TiO_2 support	Soybean oil	95.0	Mguni et al. (2012) [49]
$Cs/Al/Fe_3O_4$	Sunflower oil	94.8	Feyzi et al. (2013) [50]
TiO_2-ZnO	Palm oil	92.2	Madhuvilakku and Piraman (2013) [51]
Iron/cadmium and iron/tin oxide nanoparticles	Soybean oil	84.0	Alves et al. (2014) [52]
$KF/\gamma-Al_2O_3$/honeycomb ceramic (HC) monolithic catalyst	Palm oil	96.0	Gao et al. (2015) [53]
Sulfamic and sulfonic acid-functionalized silica-coated crystalline Fe/Fe_3O_4 core/shell magnetic nanoparticles	Glyceryl trioleate	95.0	Wang et al. (2015) [54]
$Ca/Fe_3O_4@SiO_2$	Sunflower oil	97.0	Feyzi and Norouzi (2016) [55]
CaO	Jatropha oil	98.54	Reddy et al. (2016) [56]

TABLE 4.1 Different nanocatalysts and feedstock used for biodiesel production

Transesterification of olive oil was effectively improved by adsorbing lipase extracted from *Burkholderia sp.* magnetic nanoparticles and nanocomposite of ferric-silica [7].

4.6.4   Nanotechnology in production of bioethanol

Bioethanol is derived from carbon sources such as sugarcane and grains; however, bioethanol can also be produced from fermentation of sugars, which have been released from lignocellulosic materials. These materials are primarily composed of cellulose, hemicellulose, and lignin. Certain types of pretreatments are required to use the sugars present in cellulosic and hemicellulosic fractions of plant biomass. These pretreatments help break down the recalcitrance of the biomass by disrupting the polymeric fraction present in fermentable monomers. Usually, after carrying out initial pretreatment, enzymatic hydrolysis of the cellulosic fractions occurs. Using this method, production of monomeric glucose can be completed under mild operating conditions such as low temperature, with no pressure requirement. This process also does not produce undesirable inhibitor compounds when compared with other chemical processes [59].

For example, using cellulases for hydrolyzing lignocellulosic biomass accounts for 18 percent of the total cost required by the process that produces bioethanol. Developing advanced strategies that can provide enzyme recovery and recycling will help in reducing the cost of production. Hence, nanotechnology is one solution to this problem that helps immobilize various enzymes involved in the process of bioethanol production such as cellulases and hemicelluloses on different nanomaterials. For example, enzymesimmobilized on magnetic nanomaterials is a promising method providing easy recovery of enzymes through the application of suitable magnetic field, allowing enzymes to be reused and recover for several cycles [60, 2].

Various studies have been carried out that used magnetic nanoparticles to immobilize the enzymes for producing bioethanol. Generally, enzymes are immobilized on nanoparticles either through covalent binding or by physical adsorption. However, the most suitable method is covalent binding since protein desorption is reduced as covalent bonds are formed between the enzymes and nanoparticles. Modification of compounds is required to make sure that enzyme immobilization is stable. Moreover, the enzymes can be coated with the chemically active polymer so that enzyme linkage could be enhanced through the provision of the functional group [61]. Table 4.2 presents the literature review of various studies that have been carried out employing covalent bonding and entrapment as methods of immobilization.

Method of Immobilization	Nanomaterials Used	Key Findings	Reference
	Magnetic nanofibers	β-glucosidase entrapment on magnetic nanofibers makes the enzyme stable, making its recyclability possible and easy recovery through the application of a magnetic field.	Lee et al. (2010) [62]
	Magnetic nanofibers	β-glucosidase enzyme-binding efficiency of 93% was recorded, retaining its initial activity by 50% even in the 16th cycle	Verma et al. (2013b) [63]
	Carbon nanotubes	Rh particles were entrapped on carbon nanotubes, which have free pores that help incorporate materials of our interest.	Pan et al. (2007) [64]
Entrapment	Silica nanoparticles	Cellulose was immobilized on nanoparticles of silica, demonstrating greater efficacy of immobilized enzymes during hydrolyzation of cellulose to glucose. Results showed that glucose yield increased when immobilized enzymes were used, in comparison to free enzymes verifying that these enzymes can be used in the process of saccharification and fermentation simultaneously.	Lupoi and Smith (2011) [65]
	Carbon nanotubes having a single wall, incorporated with magnetic iron oxide nanoparticles	Enzyme immobilization in carbon nanotubes having a single wall, in which nanoparticles of magnetic iron oxide were already incorporated to provide magnetic properties was studied. Concentration alteration of the nanoparticles of iron oxide in the nanotubes helped control the performance of the immobilized enzymes. Therefore, these enzymes could be stored for a longer time at 4 degrees centigrade in acetate buffer.	Goh et al. (2012) [66]
Covalent bonding	Ferrous oxide nanoparticles	Recyclability was studied in microcrystalline cellulose hydrolyzation in which carbodiimide was used as a linking polymer to immobilize the enzymes of nanoparticles of ferrous oxide. Since the nanoparticles were magnetic, recovery of enzymes was easy and was recycled six times.	Jordan et al. (2011) [67]
	Manganese dioxide nanoparticles	Results showed that immobilized enzyme has better thermostability up to 70 degrees Celsius, as compared to the free enzymes. Hydrolysis mediated by immobilized cellulase followed by using yeast led to bioethanol production from the agricultural waste of 21.96 g/L. The enzyme still showed 60% of its activity even after repeated use.	Cherian et al. (2015) [68]

TABLE 4.2 Use of nanomaterials for enzyme mobilization by methods of entrapment and covalent bonding

Other methods such as physical adsorption and co-precipitation have also been studied for enzyme immobilization. For example, lipase was immobilized on magnetic chitosan microspheres using a method of chemical co-precipitation. The linking molecules used were glutaraldehyde to bind the enzyme with magnetic chitosan microspheres covalently [69]. Moreover, titanium oxide nanoparticles were used successfully to immobilize the enzymes by adsorption methods for hydrolyzing the lignocellulosic materials to produce bioethanol. Nanomaterials nanoparticles of silica and titanium oxide, polymeric nanoparticles, and materials of carbon such as graphene, fullerene, and carbon nanotubes have been successfully used for enzyme immobilization to enhance bioethanol production [70].

Immobilization of the completely microbial cell can be done on nanoparticles and can be used in the fermentation step to produce bioethanol. For example, an approach was developed in which *Saccharomyces cerevisiae* was immobilized on magnetic nanoparticles. The results showed that when fermentation was continued, immobilized cells of *Saccharomyces cerevisiae* demonstrated the enhanced capability to produce bioethanol. Whether immobilized cells or enzymes reused on different types of nanoparticles, these techniques promote safe and economic bioethanol production by using cheap lignocellulosic materials [71].

4.6.5   Use of nanotechnology in biogas production

Different forms of nanoparticles can be used for the enhancement of bioenergy since nanoparticles have properties that differentiate them for physical, chemical, and electrical bulk materials. They can serve as a prominent solution and help increase the production of energy. Nanotechnology is being used extensively to enhance the production of renewable energies such as biogas [6]. Biogas, flammable gas is a gaseous mixture mainly of methane and carbon dioxide produced through digesting biomass and sewage sludge anaerobically, known as AD. Biogas can be produced through sewage treatment plants, landfill sites, and digestion of organic wastes, under thermophilic and mesophilic conditions. In the process of biogas production, the AD has four major steps: hydrolysis, acidogenesis, acetogenesis, and methanogenesis. Several nanoparticles have been used as additives to improve the yield of biogas and can be found in various studies upon review of literature [6]. Table 4.3 lists some of the studies that used nanoparticles to enhance biogas yield.

Future research can focus on the use of bioactive nano-metal oxides and nano-zero valence metals in varying proportions. Another area that can be explored is the use of nano-metal oxides that are photoactive so that the amount of hydrogen can be increased eventually, leading to increased production of methane.

Nanoparticles	Size range/ Concentration	Yield	Reference
Titanium oxide (TiO_2)	7.5 nm size and 1120 mg/l concentration	10% increase in biogas yield from wastewater sludge	García et al. (2012) [72]
Cerium oxide (CeO_2)	Size range 192 nm and 10 mg/l concentration	11% increase in the production of biogas from UASB sludge	Sajith et al. (2010) [41]
Ferrous oxide (Fe_3O_4)	Size 7 nm at 10 ppm concentration	180% increase in biogas production increased by and 234% increase in methane production	Casals et al. (2014) [73]
Nano zero valence iron (NZVI) particles	Size 20 nm	30.4% increase of biogas production and 40.4% methane production	Su et al. (2013) [74]
NZVI particles	Size 50 nm	Methane production increased from 58 mmol to 275 mmol	Yang et al. (2013) [75]
Metal nano substances encapsulated in porous Silicon dioxide SiO_2	Not available	Methane production increased 70%, 48%, 7%, and 6% with nanoparticles of Nickle, Cobalt, Iron, and Platinum, respectively	Al-Ahmad et al. (2013) [76]
Micro/nano fly (MNFA) and	Not available	Biogas production was increased by 2.9 times	Lo et al. (2012) [77]
Micro/nano bottom (MNBA) ash	Not available	Biogas production was increased 3.5 times by using MNBA	Lo et al. (2012) [77]

TABLE 4.3   Literature review of use of nanoparticles for the enhancement of biogas yield

Currently all around the world, nonrenewable energy resources are being used to generate energy. However, the resulting consequences of overutilization of nonrenewable resources on the environment and our future generation's needs are being ignored. Biofuels are emerging as a more sustainable and renewable option of energy. Rapid growth is being observed in the biofuel industry and has significant potential to tackle the problems of environmental pollution and climate change along with producing renewable energy. The different components of the biofuel industry are being improved continuously by scientists and engineers such as pretreatment of biomass/feedstock, process parameters, the design of reactors, quality and yield of product, optimization of the process, along with costs, social acceptance, and availability of biofuels in the market [78].

Nanotechnology and nanomaterials are emerging as a promising tool in the biofuel industry to provide for methods that are cost and process effective so that the production of biofuels can be improved. The essential principle in the exploitation of nanotechnology in the biofuel industry is the application of two solutions together, scientific and engineering, in the pursuit of environment-friendly energy resources. Research is being conducted in designing and fabrication of nanomaterials for application in the biofuel and energy sector. Due to the high surface areas and unique characteristics, nanomaterials act as exceptional candidates in biofuel systems. Characteristics include a greater degree of crystallinity, catalysis, stability, capacity to adsorb, durability, and storage efficiency that can help in the optimization of the overall process [79].

They also possess higher recovery, reuse, and recycle potential. Successful incorporation of nanotechnology in the technology of biofuel production can be shown through examples such as transesterification, AD, pyrolysis, gasification, and hydrogeneration to produce fatty esters and biogas. Combination of these technologies with nanotechnology has shown greater efficiency, economic value, and maturity; however, these are still being operated at laboratory and pilot scales. The development on a commercial scale will act as a promising way forward for the replacement of conventional systems [55].

For instance, the production of biodiesel can be optimized through nanotechnology using catalysts based on nanomaterials. Nanocatalysts are being developed to achieve greater quality and yield of product, which assure higher efficiency, durability, and stability along with economic value. Nanocatalysts of metal oxides have been developed that have higher catalytic activity for producing biodiesel such as titanium dioxide, oxides of calcium, magnesium, and strontium [80]. Magnetic catalysts based on nanoparticles have proven to be advantageous for producing biodiesel on industrial scale mainly due to their easy recovery from reaction media, and reusability leading to production that is economical. Mesoporous nanocatalysts have also been used successfully for producing biodiesel since they show enhanced catalytic activities due to their excellent structural properties. Nanoparticles based on carbon are also being used such as nanotubes and nanofibers of carbon, graphene oxide, and biochar utilizing a variety of feedstocks especially the ones that are nonfood based [81].

Nanotechnology can also be applied in the biofuel industry as immobilized enzymes (biocatalysts) during the processes of production of lipase-catalyzed biodiesel and cellulosic ethanol. Nanostructures offer benefits such as larger surface area for higher loading of the enzyme, increased stability of enzymes, and reusability of enzymes that will help reduce the cost of operation of biofuel production on a larger scale. Examples of techniques that use nanotechnology to develop immobilized enzymes include nano-encapsulation, using silaffin for self-entrapment, and adsorption. Nanoparticles are also being used for the extraction of oils from algae without causing harm to cells. Nanomaterials are being used during the different stages of lipid accumulation, extraction, and transesterification such as silica, nanoparticles of metal oxides, carbon nanotubes having a single wall, and nano-clay to produce algal fuels. The cost of production of algal biofuels can be reduced using these techniques [82].

Another application of nanotechnology is using nanoparticles as fuel additives to boost the performance of fuel blends. Using nanoparticles such as alumina and hallownanotubes of carbon have shown enhanced characteristics of combustion of engines operated on biodiesel that result in less harmful emissions. Methane production in AD has been affected and increased positively by using nano-iron oxide, nano zero-valence iron, nano fly and bottom ash, and nano-metal oxides that are bioactive [83].

Finally, nanotechnology can act as a breakthrough in different areas of biofuel industry such as advanced fermentation, fast catalytic pyrolysis, a biofuel cell, capturing and storing of carbon, gasification, volatile products, conversion, and upgrading of ethanol and tars. In energy systems based on biomass, using farming techniques that are nano-based could result in the optimization of crop production for higher yield of biofuels. Furthermore, future research should not only concentrate on the sources of biofuels and production of energy but also efficiency and storage of energy, transport, transformation, and use of end-products of biofuels. The scientific and engineering limitations about these areas can be addressed by using nanotechnology [84].

4.7 Current Challenges and Future Perspective of Nanotechnology in Biofuel Production

4.7.1 Limitations and challenges of nanotechnology applications in biofuel production

Application of nanotechnology in large-scale processes such as producing biofuel has certain limitations.

4.7.2 Temperature control

An essential parameter to consider in the synthesis of nanoparticles is temperature. For metallic nanoparticles, the temperature for calcination varies between 100°C to 700°C, depending on the method used to synthesize the nanoparticles. Methods such as physical and chemical require high temperatures, i.e., greater than 300°C moderate or even ambient temperature is required by biological methods such as less than 100°C. The overall morphology of the nanoparticles such as their size of pores, shape, and stability is affected by temperature [85, 86].

4.7.3 Catalyst coating

Nanotechnology is being widely used to immobilize catalyst for biofuel production. The process of catalyst coating on different nanomaterials has some limitations. For example, sometimes the catalyst's size is either bigger or smaller as compared to the pore size of the nanomaterial, and results in blockage of the channels causing hinderance in diffusion. This coating cannot be controlled; therefore, the catalysts are left on the surface and not embedded in the channels. Another problem associated with the nano-catalysts is its selectivity, although the target reaction is enhanced, but it also accelerates unwanted reactions [87, 88].

4.7.4 High chemical activity

Despite widespread applications and benefits, nano-biofuels are not flawless. A major reason for this is that as microparticles are transformed into nanoparticles, the size range starts to decrease, resulting in an increased number of surface atoms. Due to this increase in the surface area of the nanoparticles, augmented chemical reactivity leads to ambiguity about the behavior of these particles when exposed to different operating conditions [72, 71].

4.7.5 Environmental impacts

Use of nanoparticles is also attributing environmental toxicity, causing toxic effects to ecosystems. It is being observed that nanoparticles that are being used in a variety of consumer products are released into the aquatic environment such as silver nanoparticles. This silver acts as a source of dissolved Ag, exerting adverse effects on the aquatic life such as bacteria, algae, and fishes. In hard water and seawater, aggregation of nanoparticles occurs, and the type of organic matter or natural particles present in water greatly influence these nanoparticles. The state of dispersion alters the ecotoxicity; however, it is influenced by different abiotic factors such as pH, salinity, and organic matter and is being investigated in various ecotoxicological studies.

A major concern regarding the use of nanoparticles is that they tend to accumulate in organisms and, thus, the food chains. On interaction with biomolecules such as proteins, lipids, and DNA, due to their small size, these nanoparticles cause severe cellular damage. Animal models are being used to study the ecotoxicity of nanoparticles further so that a better understanding could be generated as to what damage can be caused by these particles and what is the real danger when these particles accumulate [89]. These studies will play a defining role to prevent and reduce the toxic effects of these particles to both humans and the ecosystem.

Not only the benefits should be considered of these useful nanoparticles, but also their adverse effects should be taken into consideration primarily when these particles are being applied in large-scale processes such as that of biofuel production. By getting to know these toxic effects better, researchers can help propose ways to reduce these effects and help maintain a balance between their application and the environment's safety being exposed to such particles. Biofuels contain nanoparticles that are nonbiodegradable, resulting in pollution of land, water, or air. The prediction of the effects of the nanoparticles on their surroundings is difficult. If these nanoparticles are successful in entering the bionetwork via plants, their elimination would be a highly demanding process [90].

4.7.6 Economic burden and uncertainty

Nanotechnology is an expensive field, and its employment in the field of bioenergy necessitates a lot of money for the generation and assembling of particles requiring many different technologies. Huge initial investment must be made by manufacturers to start a nanotech energy plant, and if the product fails to meet the demands of the consumers, the manufacturers face a huge loss. Furthermore, recovery of the original product is also not feasible since it will require double the cost [75, 91].

4.7.7 Safety issues

Limited research has been completed regarding the safety assessment of nanomaterials and nanoparticles used for producing biofuels and bioenergy. Nanoparticles are released in the environment when they are synthesized and applied to pose a threat to both humans and the environment. Nanoparticles can enter the human body, affecting the most sensitive areas, through ingestion, inhalation, and absorption through skin that can be either intact or fractured. Since nanomaterials have a small size, they have easy penetration into the cells of humans and animals, affecting the cells' normal functioning. For example, metallic nanocatalysts and nanomaterials that are used to support catalysts such as carbon nanotubes and fibers of carbon, as well as nanoparticles based on zirconia, induce toxicity [92].

Biofuel derived from nanoparticles causes harmful effects when they are released through vehicular and industrial emissions. These particles can be deposited in the tissues of the lungs via inhalation, causing abnormality. Various respiratory ailments are caused such as asthma and bronchitis [93]. Moreover, nanoparticles also cause occupational health and safety issues for the workers involved in the manufacturing of nanoparticles who get exposed either through absorption via skin or inhalation. Nanoparticles of platinum have been studied for their effects on exposure and have been found to lower the heart rate, delay the process of hatching, and affect the touch response [94].

Although the mechanism of toxicity is not fully understood, it is dependent on the size of the nanoparticles, their shape, dose, composition, and structure. At cellular levels, nanoparticles can interact with the cell's lipid bilayer, thereby disrupting the normal function of the cell membrane. They cause pit formation in the cell membrane, resulting in leakage of the cell contents. Nanoparticles upon exposure can also interact with the mitochondria present in the biological systems, disrupting the potential of the mitochondrial membrane [95]. The energy balance of the cell is disrupted when the mitochondria are damaged, consequently disrupting the metabolism of the cell. These nanoparticles also induce reactive species of oxygen such as oxygen ions. These reactive particles accumulate and react with the enzymatic machinery of the cell. Nanoparticles have also been reported to cause damage to the DNA due to the induced interaction of these particles [96, 97].

4.8 Techno-Economic Analysis of Nanotechnology Role in Biofuel Production

Based on the increase in the use of nanoparticles for producing biofuels, it is obvious that they will have effects on humans and the environment. It is essential to assess the safety of such particles. Currently, different approaches are being used to assess the toxicity of nanoparticles; most common is in-vitro evaluation. However, extensive research is required that will focus on the in-vivo interaction of the nanoparticles used to produce biofuels and bioenergy. The demand for energy is expected to increase by more than 30 percent by the year 2035; therefore, affordable and safe ways need to be developed to consider such fluctuations and market uncertainties, also aiming to reduce adverse effects on the environment such as carbon emissions. These challenges have prompted policymakers and researchers to invest in green nanotechnology options that play a pivotal role in shaping the future road map of the larger field of nanotechnology [98].

There are several ways to assess the economic and societal impact of nanotechnologies used for large-scale processes such as the production of biofuels and bioenergy. Various economic indicators are considered to assess the market value of green nanotechnologies. However, investment in the form of research and technological development, production, the setting of standards and regulations, adoption by users, and needs for monitoring must be determined before market realization. The costs required, i.e., private and public capital by green nanotechnologies, need to be balanced against the benefits achieved, taking into consideration the factors involved: time, cost-benefit distribution, interest rate, and opportunity costs. Also, the advantages being offered by these technologies as compared to conventional technologies must be considered. In the process of production and usage of nanotechnology-based products, a series of indirect effects are likely such as supply chains, third parties involved, and the environment. However, it is not possible to measure all the costs and benefits, where some impacts such as that on human health and the environment will not be apparent in the beginning but will show after years of use [99].

The economic contribution by nanotechnology-based products can be assessed using simple analysis. These analysis would consider the net costs used for developing the technology and market entry of the products relative to the value of outputs gained, by considering time and perspective. The inputs include public R&D costs, testing of products, commercialization, product start-up, and costs of markets. These expenditures have certain outcomes having immediate economic value, such as contribution to science and knowledge, technological development both generic and specific, development of standards, and intellectual property. However, these outcomes will present the greatest economic potential through profitable sales of products, increased productivity, improvement of processes, cost savings, generation of wage and employment, and consumer benefits. These outcomes can lead to benefits to public and development, contributing to the gross domestic product (GDP), improving the competitiveness and balance of trade, and other benefits to society and the environment. Using nanomaterials can also have strategic benefits, e.g., reducing reliance on rare and precious metals as well as materials [84].

Policymakers are interested in the effects of economic development of new technologies and product, including impacts on wages and jobs. Using nanotechnology in the bioenergy sector for producing biofuels will generate employment. Employment will be generated through research, manufacturing, use, and maintenance of the products and the processes, as well as the industries and sectors that will be associated. However, employment prediction is difficult. It is a possibility that workers will shift to nanotechnology-based industries due to the replacement of conventional technology. Furthermore, it must be considered that not all jobs of nanotechnology-based industries will be well paid, and some employment offset is likely due to the replacement of conventional products and skills [84].

Another reason why employment prediction is difficult is due to economic development's lengthy scale. The timescale for costs and benefits varies for such technologies. For example, there are certain costs that occur at the early stage of a product development cycle such as cost of research and development (R&D). There are certain immediate outcomes such as R&D partnering and patents that are visible in the short term, whereas outcomes such as the development of new product and processes are visible in the medium term. However, some outcomes are only visible in the long run such as benefits to industry, economy, and society. These terms differ concerning the acceptability of technology and market [84].

A well-known example of assessment of the economic value of nanotechnologies is of a comparative method used by Walsh et al. (2010). It was a study conducted for the Department for Environment, Food and Rural Affairs (DEFRA), UK. It uses a methodology to estimate the value addition done by nanotechnologies, defining the difference between value added by nano-based product and conventional product. The net value addition is comprised of three elements: producer surplus, consumer surplus, and other externalities. A multistep process is used to determine value addition. Furthermore, identification of market scenarios is done, also considering geographic allocation by employing an S-curve of both producer surplus and externalities. This model has been successfully employed to various nanotechnology products such as nano-based food packaging, fuel catalysts, nano-enabled anti-fouling paints, and many more. This model can also be used for nano-based biofuels as well. This is an instructive approach to assess economic value; however, it does not include indirect effects. Effects on the environment can also be included in the form of externalities but only if they can be foreseen and monetized, such as valuation of carbon emission reduction based on carbon prices traded [100].

From an economic and social perspective, as well as viewing these technologies as sustainable innovations, it is essential to take into consideration the concept of life cycle assessment (LCA), i.e., considering the environmental impacts of the full life cycle of the product. There are a variety of approaches regarding LCA that take into consideration the outputs, environmental impacts, and energy requirements both direct and indirect. From the perspective of LCA, certain concerns have been raised regarding green nanotechnologies. These technologies save energy and cost and also cause carbon emission reduction; however, energy in significant amounts is required to synthesize nanomaterials and nanoparticles. It is likely that these costs may reduce in the long run due to improvement and emergence of new materials, but it is essential to take into consideration the energy costs and costs required to extract resources for nanomaterial production. Another concern is the implications nanotechnology-based products have on environment, health, and safety (EHS). Laboratory and factory workers are exposed to potential risks when they are manufacturing different nanomaterials and nanostructures. There are concerns related to the use and disposal of nanomaterials as well [84].

Green nanotechnologies have certain potential risks, and a comparison should be drawn with current technologies as well as with human and environmental costs that are not addressing global challenges effectively. However, labeling a nanotech-based product as sustainable or free of risk is not appropriate, and neither does it mean that these products do not have any economic costs or benefits. A point reinforced here is that indicators of economic assessment should be considered highlighting economic, environmental, and social implications. From a broader perspective, LCA needs to play a central role to help govern socially responsible innovation. This approach will effectively consider both social and environmental benefits [101].

4.9 Policy Analysis of Nanotechnology Role in Biofuel Production

The development and production of nanotechnology-based products are geographically widespread and being incorporated in the decision-making process in the form of research and innovation programs in 60 different countries. Currently, the regulation and oversight of nanotechnology are national. However, efforts are being made to exchange information, promote harmonization, and set standards on international levels. Some of the best examples of policy implementation regarding nanotechnology are emerging with slight differences among countries regarding governance and regulation. Several countries, OECD and non-OECD, are providing their goals for investment for nanotechnology through policymaking and program introductions [102]. Table 4.4 indicates the total investment being made in the sector of nanotechnology and nanoscience in various OECD and non-OECD countries. This shows that huge investments are being dedicated globally in both these sectors.

4.10 Future Outlook

The modern advancements in the nanotechnological sector are of importance for biofuel and bioenergy generation. As the products and employed technologies are developing, nanoparticles are foreseen as an integral component of the sustainable energy sector in the future. A wide variety of applications of different types of nanoparticles and nanocatalysts in modern biofuels have been discussed, but there are certain gaps that need to be bridged through an interdisciplinary approach based in further research in this field.

Country	Funding Programs	Nano-specific	Period	Value
Brazil	Ministry for Science & Technology	No	Annual estimate	R$11.87 million (€4.9 million)
China	Medium- & Long-Term Development Plan	Yes	2006–2008	US$38.2 million (€29.1 million)
European Union	Framework Programme 7	No	2007–2013	€3.5 billion
France	Nano 2012 Programme	Yes	2008–2012	€500 million
Germany	Nano Initiative—Action Plan 2010	Yes	2008–2013	€370 million
India	Nano Mission	Yes	2007–2012	Rs. 1000 crore (€144.8 million)
Japan	MEXT	No	Annual estimate	€470 million
Russia	Development of nanotechnology infrastructure in the Russian Federation for 2008–2011	Yes	2008–2011	€693.3 million
UK	Research Councils UK/ Technology Strategy Board	No	Annual estimate	€256 million
USA	National Nanotechnology Initiative	Yes	2012	$2.1 billion (€1.6 billion)

TABLE 4.4 Country funding schemes indicating the main nanotechnology funding programs and their value provided in currency of source and € equivalent in parentheses [108]

4.10.1 Nanocatalysis for feedstock processing

There are certain challenges associated with the catalysts involved in the processing of feedstock before biofuel generation. These include enzyme inactivation, reduced susceptibility, difficult recovery, and limited multiple uses, which have been addressed to a major extent by the incorporation of nanotechnology-based enzyme immobilization procedures. The use of nanoparticles for catalyst immobilization has resulted in the development of enzymes that have improved stability, environmental resistance, and decreased susceptibility to denaturation. However, the immobilized enzymes have less specific activity and enzyme loading potential than free soluble catalysts. The reason behind the limited enzyme loading can be linked with their incapacity of surpassing single layer adsorption on nano-matrices [103].

Studies are carried out to overcome these obstacles and different potential measures such as the concept of applying aggregate coatings on enzymes through glutaraldehyde pretreatment have been introduced. The resultant enzyme designs obtained have presented better specific activity and improved loading capacity but are limited by the economic burdens associated with such approaches. Lack of knowledge regarding their long-term effectiveness is also a major problem. Recently, the use of functionalized nanoparticles for immobilizing enzymes and the co-immobilization of diverse catalysts have raised novel ideas that need to be thoroughly evaluated and researched. This is not possible without analyzing the structural conformation of local proteins. A few new technologies such as Rosetta have presented highly accurate predictions of protein structure. These approaches have proved valuable in research and development. However, structural and functional comprehension of properties of enzymes, and their interactions at the "nanoscale," are integral for promoting the use of nanomaterials in the bioenergy sector.

Thus, further study and elaborated research are crucial to analyze and overcome the highlighted techno-economical drawbacks. Development of alternatives is essential for capital and resource intensive procedures to improve the specific activity and enzyme loading capacity of immobilized enzymes. There is a dire need to develop sustainable nanotechnology-based solutions that are introduced after considering the toxicity and safety concerns involved with the synthesis and utilization of nanomaterials [104].

4.10.2 Engineering robust biofuel producers

As the activity of enzymes is controlled by the abundance of desired cofactors and their potential of regeneration, the idea of utilizing living cell biocatalysts that have the genetic mechanism required for the synthesis of essential enzymes and cofactor is gaining importance rapidly. Encapsulating the biocatalyst by incorporating the applications of a diverse variety of nanomaterials including silica glasses and silica gels has shown certain limitations that are addressed by the development of nano-porous latexes. In the bioenergy sector, predominately in microbial fuel cells and photosynthesis-based hydrogen generation processes, these latexes are encapsulating the cells of natural as well as recombinants organisms. This process has limited application because there is a limited number of naturally occurring biocatalysts. However, with the development of innovative ideas and novel products, robust producers of biofuels can be engineered by employing nanotechnology [105].

This can be done by either enhancing the naturally occurring biofuel producers or by developing engineered nonnative and synthetic biocatalysts. New and different phenotypical characteristics can be introduced by involving certain modifications in the naturally developed proteome. This can be done at the transcription or translation stake. Recently, nucleotide-conjugated nanoparticles are widely studied and analyzed for understanding their efficiency in delivering genetic materials for developing recombinant biocatalysts. However, there are a few challenges associated with these ideas including the reduced loading of biomass and the problems related to mass transmission. Advancements in biotechnology and nanotechnology can be efficiently harnessed for improving the properties of biocatalysts and for enhancing the biofuel producers [106].

4.10.3 Biofuel harvesting

The intrinsic toxicological impacts associated with aliphatic alcoholic products having short chains is one of the leading limiting factors of commercial-scale biofuel production, the primary reason being that the activity of biocatalysts tends to be inhibited at lower volumes of alcohol. Nanotechnological approaches are being extensively studied for better purification of the final product obtained for harvesting better yields. The final biodiesel yields obtained often contain residual water that comes as a by-product of the esterification process. This water needs to be removed because its presence even in trace amounts can lead to the synthesis of glycerin. The potential of nanotechnology can be harnessed through future research and advancements, to overcome these challenges as recently nanomaterial with high porosity have been evaluated for their potential of absorbing hydrophobic molecules [95].

For achieving economical generation of biofuels, the significant financial burdens associated with the production of various essential nanoparticles need to be considered. This capital intensiveness can reduce the practicality of this approach. To make this approach financially viable, alternative strategies for the synthesis of a nanomaterial at lower costs need to be exposed through research advancements. Moreover, the processes need to be improved from a technical perspective as well, to foster the applications of nanotechnology in producing sustainable biofuels. Primary research areas for overcoming the prevalent challenges can be concluded as following [71]:

- Nanoparticles, whose catalytic potential can be regulated, are required to be produced for improving the production processes.
- Nanomaterials that be obtained in financially viable terms need to be studied.
- Improve the synchronization pace of biomass and micro-organisms with the essential enzymes involved in the process and the nanoparticles.
- Develop a comprehensive understanding of the protein conformation and its interaction with the nanomaterials at the molecular scale.
- Evaluate the optimal process conditions for procuring maximum biofuel yield. The opportunity of increasing the bioenergy production by using residual iron containing biomass requires thorough exploration.

4.11 Conclusions

The problem of fossil fuel scarcity is worldwide due to their finite nature and associated problems such as global warming and climate change. This has forced countries all around the globe to look for more sustainable alternatives such as biofuels; biodiesel, bioethanol, and biogas are top priorities. Therefore, biofuel is considered the fuel of the future, and nanotechnology can make interesting contributions by making the production of biofuels easier. Various types of nanocatalysts are emerging as promising ways to become part of the sustainable production of bioenergy. In the biofuel production field, use of immobilized catalysts using nanomaterials as support presents path-breaking research, favoring reuse and recycling of enzymes and even whole cells. Use of nano-supports for catalysts is one technique that can make the production of biofuel economically viable. Despite numerous advantages of nanotechnology, there are certain concerns regarding the use of nanomaterials and nanoparticles, which can produce adverse effects on both the environment and human health. This has led to the stimulation of further studies so that safety measures can be determined so that exposure to nanoparticles could be limited, or problems to human health and the environment could be overcome.

References

1. G. Joshi, J. K. Pandey, S. Rana, and D. S. Rawat. Challenges and opportunities for the application of biofuel. *Renewable and Sustainable Energy Reviews* **79**, 850–866, 2017.

2. F. A. F. Antunes, S. Gaikwad, A. P. Ingle, R. Pandit, J. C. dos Santos, M. Rai, and S. S. da Silva. Bioenergy and biofuels: nanotechnological solutions for sustainable production. In *Nanotechnology for Bioenergy and Biofuel Production* (pp. 3–18). Springer, Cham, 2017.

3. R. W. Whatmore. Nanotechnology—what is it? Should we be worried? *Occupational Medicine* **56**(5), 295–299, 2006.

4. B. S. Sekhon. Nanotechnology in agri-food production: an overview. *Nanotechnology, Science and Applications* **7**, 31, 2014.

5. W. Hannah and P. B. Thompson. Nanotechnology, risk and the environment: a review. *Journal of Environmental Monitoring* **10**(3), 291–300, 2008.

6. A. K. Hussein. Applications of nanotechnology in renewable energies—A comprehensive overview and understanding. *Renewable and Sustainable Energy Reviews* **42**, 460–476, 2015.

7. K. Palaniappan. An Overview of Applications of Nanotechnology in Biofuel Production World. *Applied Sciences Journal* **35**(8), 1305–1311, 2017.

8. S. N. Naik, V. V. Goud, P. K. Rout, and A. K. Dalai. Production of first and second generation biofuels: a comprehensive review. *Renewable and Sustainable Energy Reviews* **14**(2), 578–597, 2010.

9. L. D. Gomez, C. G. Steele-King, and S. J. McQueen-Mason. Sustainable liquid biofuels from biomass: the writing's on the walls. *New Phytologist* **178**(3), 473–485, 2008.

10. C. S. Jones and S. P. Mayfield. Algae biofuels: versatility for the future of bioenergy. *Current Opinion in Biotechnology* **23**(3), 346–351, 2012.

11. D. Y. Leung, X. Wu, and M. K. H. Leung. A review on biodiesel production using catalyzed transesterification. *Applied Energy* **87**(4), 1083–1095, 2010.

12. M. Zabeti, W. M. A. W. Daud, and M. K. Aroua. Activity of solid catalysts for biodiesel production: a review. *Fuel Processing Technology* **90**(6), 770–777, 2009.

13. C. A. Cardona and Ó. J.Sánchez. Fuel ethanol production: process design trends and integration opportunities. *Bioresource Technology* **98**(12), 2415–2457, 2007.

14. N. Rasit, A. Idris, R. Harun, and W. A. W. A. K. Ghani. Effects of lipid inhibition on biogas production of anaerobic digestion from oily effluents and sludges: An overview. *Renewable and Sustainable Energy Reviews* **45**, 351–358, 2015.

15. P. Watkins and P. McKendry. Assessment of waste derived gases as a renewable energy source–Part 1. *Sustainable Energy Technologies and Assessments* **10**, 102–113, 2015.

16. S. Li, H. K. Yoo, M. Macauley, K. Palmer, and J. S. Shih. Assessing the role of renewable energy policies in landfill gas to energy projects. *Energy Economics* **49**, 687–697, 2015.

17. P. S. Nigam and A. Singh. Production of liquid biofuels from renewable resources. *Progress in Energy and Combustion Science* **37**(1), 52–68, 2011.
18. D. Mohan, C. U. Pittman, and P. H. Steele. Pyrolysis of wood/biomass for bio-oil: a critical review. *Energy &Fuels* **20**(3), 848–889, 2006.
19. M. Asadullah. Barriers of commercial power generation using biomass gasification gas: A review. *Renewable and Sustainable Energy Reviews* **29**, 201–215, 2014.
20. A. V. Bridgwater. Review of fast pyrolysis of biomass and product upgrading. *Biomass and Bioenergy* **38**, 68–94, 2012.
21. W. N. R. W. Isahak, M. W. Hisham, M. A. Yarmo, and T. Y. Y. Hin. A review on bio-oil production from biomass by using pyrolysis method. *Renewable and Sustainable Energy Reviews* **16**(8), 5910–5923, 2012.
22. J. Lehto, A. Oasmaa, Y. Solantausta, M. Kytö, and D. Chiaramonti. Review of fuel oil quality and combustion of fast pyrolysis bio-oils from lignocellulosic biomass. *Applied Energy* **116**, 178–190, 2014.
23. S. Murugan and S. Gu. Research and development activities in pyrolysis–contributions from Indian scientific community–a review. *Renewable and Sustainable Energy Reviews* **46**, 282–295, 2015.
24. D. S. Rawat, G. Joshi, B. Y. Lamba, A. K. Tiwari, and P. Kumar. The effect of binary antioxidant proportions on antioxidant synergy and oxidation stability of Jatropha and Karanja biodiesels. *Energy* **84**, 643–655, 2015.
25. H. H. Khoo. Review of bio-conversion pathways of lignocellulose-to-ethanol: sustainability assessment based on land footprint projections. *Renewable and Sustainable Energy Reviews* **46**, 100–119, 2015.
26. J. P. Maity, J. Bundschuh, C. Y. Chen, and P. Bhattacharya. Microalgae for third generation biofuel production, mitigation of greenhouse gas emissions and wastewater treatment: Present and future perspectives–A mini review. *Energy* **78**, 104–113, 2014.
27. R. C. Saxena, D. K. Adhikari, and H. B. Goyal. Biomass-based energy fuel through biochemical routes: a review. *Renewable and Sustainable Energy Reviews* **13**(1), 167–178, 2009.
28. C. Gonzalez-Fernandez, B. Sialve, and B. Molinuevo-Salces. Anaerobic digestion of microalgal biomass: challenges, opportunities and research needs. *Bioresource Technology* **198**, 896–906, 2015.
29. K. Ullah, M. Ahmad, V. K. Sharma, P. Lu, A. Harvey, M. Zafar, and S. Sultana. Assessing the potential of algal biomass opportunities for bioenergy industry: a review. *Fuel* **143**, 414–423, 2015.
30. R. Wirth, G. Lakatos, G. Maróti, Z. Bagi, J. Minárovics, K. Nagy, ... and K. L. Kovács. Exploitation of algal-bacterial associations in a two-stage biohydrogen and biogas generation process. *Biotechnology for Biofuels* **8**(1), 59, 2015.
31. J. C. Escobar, E. S. Lora, O. J. Venturini, E. E. Yáñez, E. F. Castillo, and O. Almazan. Biofuels: environment, technology and food security. *Renewable and Sustainable Energy Reviews* **13**(6–7), 1275–1287, 2009.

32. J. P. Painuly. Barriers to renewable energy penetration; a framework for analysis. *Renewable Energy* **24**(1), 73–89, 2001.
33. L. Lin, Z. Cunshan, S. Vittayapadung, S. Xiangqian, and D. Mingdong. Opportunities and challenges for biodiesel fuel. *Applied Energy* **88**(4), 1020–1031, 2011.
34. F. Passos, M. Solé, J. García, and I. Ferrer. Biogas production from microalgae grown in wastewater: effect of microwave pretreatment. *Applied Energy* **108**, 168–175, 2013.
35. S. Moka, M. Pande, M. Rani, R. Gakhar, M. Sharma, J. Rani, and A. N. Bhaskarwar. Alternative fuels: an overview of current trends and scope for future. *Renewable and Sustainable Energy Reviews* **32**, 697–712, 2014.
36. A. Verbruggen, M. Fischedick, W. Moomaw, T. Weir, A. Nadaï, L. J. Nilsson, … and J.Sathaye. Renewable energy costs, potentials, barriers: Conceptual issues. *Energy Policy* **38**(2), 850–861, 2010.
37. S. Patumsawad. 2nd Generation biofuels: technical challenge and R and D opportunity in Thailand. *J Sustain Energy Environ* (Special Issue), 47–50, 2011.
38. H. Eggert and M. Greaker. Promoting second generation biofuels: does the first generation pave the road? *Energies* **7**, 4430–4445, 2014.
39. M. Ahmed and M. Douek. The role of magnetic nanoparticles in the localization and treatment of breast cancer. *Bio Med Res* **2013**, 1–11, 2013.
40. S. Y.Chen, T. Mochizuki, Y. Abe, M. Toba, and Y. Yoshimura. Production of high-quality biodiesel fuels from various vegetable oils over Ti- incorporated SBA-15 mesoporous silica. *Catalysis Communications* **41**, 136–139, 2013.
41. V. Sajith, C. B. Sobhan, and G. P. Peterson. Experimental Investigations on the Effects of Cerium Oxide Nanoparticle Fuel Additives on Biodiesel. *Advances in Mechanical Engineering* **2**, 581407, 2010.
42. M. Verziu, B. Cojocaru, J. Hu, R. Richards, C. Ciuculescu, P. Filip, and V. I. Parvulescu. Sunflower and rapeseed oil transesterification to biodiesel over different nanocrystalline MgO catalysts. *Green Chemistry* **10**(4), 373–381, 2008.
43. H. Han and Y. Guan. Synthesis of biodiesel from rapeseed oil using $K_2O/\gamma\text{-}Al_2O_3$ as nano-solid-base catalyst. *Wuhan University Journal of Natural Sciences* **14**(1), 75–79, 2009.
44. H. Wang, J. Covarrubias, H. Prock, X. Wu, D. Wang, and S. H. Bossmann. Acid-functionalized magnetic nanoparticle as heterogeneous catalysts for biodiesel synthesis. *The Journal of Physical Chemistry C* **119**(46), 26020–26028, 2015.
45. M. Kaur and A. Ali. Lithium ion impregnated calcium oxide as nano catalyst for the biodiesel production from karanja and jatropha oils. *Renewable Energy* **36**(11), 2866–2871, 2011.
46. F. Qiu, Y. Li, D. Yang, X. Li, and P. Sun. Heterogeneous solid base nanocatalyst: preparation, characterization and application in

biodiesel production. *Bioresource Technology* **102**(5), 4150–4156, 2011.

47. X. Deng, Z. Fang, Y. H. Liu, and C. L. Yu. Production of biodiesel from Jatropha oil catalyzed by nanosized solid basic catalyst. *Energy* **36**(2), 777–784, 2011.
48. C. M. M. Molina. ZnO nanorods as catalyst for biodiesel production from olive oil. M.Sc. Thesis, University of Louisville, 2013.
49. L. L. Mguni, R. Meijboom, and K. Jalama. Biodiesel production over nano-MgO supported ontitania. *International Journal of Chemical, Molecular, Nuclear, Materials and Metallurgical Engineering* 6(4):380–384, 2012
50. M. Feyzi, A. Hassankhani, and H. R. Rafiee. Preparation and characterization of $Cs/Al/Fe_3O_4$ nanocatalysts for biodiesel production. *Energy Conversion and Management* 71, 62–68, 2013.
51. R. Madhuvilakku and S. Piraman. Biodiesel synthesis by TiO_2–ZnO mixed oxide nanocatalyst catalyzed palm oil transesterification process. *Bioresource Technology* **150**, 55–59, 2013.
52. M. B. Alves, F. Medeiros, M. H. Sousa, J. C. Rubim, and P. A. Suarez. Cadmium and tin magnetic nanocatalysts useful for biodiesel production. *Journal of the Brazilian Chemical Society* **25**(12), 2304–2313, 2014.
53. L. Gao, S. Wang, W. Xu, and G. Xiao. Biodiesel production from palm oil over monolithic $KF/\gamma\text{-}Al_2O_3/$honeycomb ceramic catalyst. *Applied Energy* **146**, 196–201, 2015.
54. L. Wen, Y. Wang, D. Lu, S. Hu, and H. Han. Preparation of KF/CaO nanocatalyst and its application in biodiesel production from Chinese tallow seed oil. *Fuel* **89**(9), 2267–2271, 2010.
55. M. Feyzi and L. Norouzi. Preparation and kinetic study of magnetic $Ca/Fe_3O_4@SiO_2$ nanocatalysts for biodiesel production. *Renewable Energy* **94**, 579–586, 2016.
56. A. N. R. Reddy, A. A. Saleh, M. S. Islam, S. Hamdan, and M. A. Maleque. Biodiesel production fromcrude Jatropha oil using a highly active heterogeneous nanocatalyst by optimizing transesterification reaction parameters. *Energy Fuels* **30**, 334–343, 2016.
57. W. J. Goh, V.S. Makam, J. Hu, L. Kang, M. Zheng, S.L. Yoong, C.N.B. Udalagama, and G. Pastorin. Iron oxide filled magnetic carbon nanotube-enzyme conjugates for recycling of amyloglucosidase: toward useful applications in biofuel production process. *Langmuir: The ACS Journal of Surfaces and Colloids* **28**(49), 16864–16873, 2012.
58. M. L. Verma, C.J. Barrow, and M. Puri. Nano Nanobiotechnology as a novel paradigm for enzyme immobilisation and stabilisation with potential applications in biodiesel production. *Applied Microbiology and Biotechnology* **97**(1), 23–39, 2013.
59. F. A. F. Antunes, A. K. Chandel, T. S. S. Mmilessi, J. C. Santos, C. A. Rosa, and S. S. Da Silva. Bioethanol production from sugarcane bagasse by a novel Brazilian pentose fermenting yeast

Scheffersomyces shehatae UFMG-HM 52:2: evaluation of fermentation medium. *Int J Chem Eng* **2014**, 1–8, 2014.

60. M. Rai, J. C. dos Santos, M. F. Soler, P. R. F. Marcelino, L. P. Brumano, A. P. Ingle, S. Gaikwad, A. Gade, and S. S. da Silva. Strategic role of nanotechnology for production of bioethanol and biodiesel. *Nanotechnology Review* **5**(2), 231–250, 2016.

61. R. E. Abraham, M. L. Verma, C. J. Barrow, and M. Puri. Suitability of magnetic nanoparticle immobilised cellulases in enhancing enzymatic saccharification of pre-treated hemp biomass. *Biotechnology Biofuels* **7**, 90, 2014.

62. S. M. Lee, L. H. Jin, J. H. Kim, S. O. Han, H. B. Na, T. Hyeon, ... and J. H. Lee. β-Glucosidase coating on polymer nanofibers for improved cellulosic ethanol production. *Bioprocess and Biosystems Engineering* **33**(1), 141, 2010.

63. M. L. Verma, R. Chaudhary, T. Tsuzuki, C. J. Barrow, and M. Puri. Immobilization of β-glucosidase on a magnetic nanoparticle improves thermostability: application in cellobiose hydrolysis. *Bioresources Technology* **135**, 2–6, 2013b.

64. X. Pan, Z. Fan, W. Chen, Y. Ding, H. Luo, and X. Bao. Enhanced ethanol production inside carbon-nanotube reactors containing catalytic particles. *Nature materials* **6**(7), 507, 2007.

65. J. S. Lupoi and E. A. Smith. Evaluation of nanoparticle-immobilized cellulase for improved yield in simultaneous saccharification and fermentation reactions. *Biotechnology and Bioengineering* **108**, 2835–2843, 2011.

66. W. J. Goh, V. S. Makam, J. Hu, L. Kang, M. Zheng, S. L. Yoong, C. N. Udalagama, and G. Pastorin. Iron oxide filled magnetic carbon nanotube-enzyme conjugates for recycling of amyloglucosidase: toward useful applications in biofuel production process. *Langmuir* **28**(49), 16864–16873, 2012.

67. J. Jordan, C. S. S. Kumar, and C. Theegala. Preparation and characterization of cellulase-bound magnetite nanoparticles. *Journal of Molecular Catalysis B: Enzymatic* **68**, 139–146, 2011.

68. E. Cherian, M. Dharmendirakumar, and G. Baskar. Immobilization of cellulase onto MnO_2 nanoparticles for bioethanol production by enhanced hydrolysis of agricultural waste. *Chinese Journal of Catalysis* **36**(8), 1223–1229, 2015.

69. W. Xie and J. Wang. Immobilized lipase on magnetic chitosan microspheres for transesterification of soybean oil. *Biomass and Bioenergy* **36**, 373–380, 2012.

70. R. Ahmad and M. Sardar. Immobilization of cellulase on TiO_2 nanoparticles by physical and covalent methods: a comparative study, 2014.

71. V. Ivanova, P. Petrova, and J. Hristov. Application in the ethanol fermentation of immobilized yeast cells in matrix of alginate/magnetic nanoparticles, on chitosan-magnetite microparticles

and cellulose-coated magnetic nanoparticles. *International Review of Chemical Engineering* 3(2), 289–299, 2011.

72. A. García, L. Delgado, J.A. Torà, E. Casals, E. González, V. Puntes, X. Font, J. Carrera, and A. Sánchez. Effect of cerium dioxide, titanium dioxide, silver and gold nanoparticles on the of microbial communities intended in wastewater. *Journal of Hazardous Materials* **199–200**, 64–72, 2012.

73. E. Casals, R. Barrena, A. García, E. González, L. Delgado, M. Busquets-Fité, X. Font, J. Arbiol, P. Glatzel, K. Kvashnina, A. Sánchez, and V. Puntes. Programmed Iron Oxide Nanoparticles. Disintegration in Anaerobic Digesters Boosts Biogas Production. *Small* **10**(14), 2801–2808, 2014.

74. L. Su, X. Shi, G. Guo, A. Zhao, and Y. Zhao. Stabilization of sewage sludge in the presence of nanoscale zero-valent iron (nZVI): abatement of odor and improvement of biogas production. *Journal of Material Cycles and Waste Management* **15**(4), 461–468, 2013.

75. Y. Yang, J. Guo, and Z. Hu. Impact of nano zero valent iron (NZVI) on methanogenic activity and population dynamics in anaerobic digestion. *Water Research* **47**(17), 6790–6800, 2013.

76. A. E. Al-Ahmad, S. Hiligsmann, S. Lambert, B. Heinrichs, W. Wannoussa, F. Weekers, F. Delvigne, and P. Thonart. Enhancement of thermophillic anaerobic digestion of methane by metal nanoparticles encapsulated in porous silica, 2013.

77. H. M. Lo, H.Y. Chiu, S.W. Lo, and F.C. Lo. Effects of micro-nano and non-micro-nano MSWI ashes addition on MSW anaerobic digestion. *Bioresource Technology* **114**, 90–94, 2012.

78. A. Nizami and M. Rehan. Towards nanotechnology-based biofuel industry. *Biofuel Research Journal* 5(2), 798–799, 2018.

79. J. García-Martínez. *Nanotechnology for the energy challenge.* John Wiley & Sons, 2010.

80. J. S. Basha and R. B. Anand. Role of nanoadditive blended biodiesel emulsion fuel on the working characteristics of a diesel engine. *Journal of Renewable and Sustainable Energy* 3(2), 023106, 2011.

81. J. Gardy, A. Osatiashtiani, O. Céspedes, A. Hassanpour, X. Lai, A. F. Lee, K. Wilson, and M. Rehan. A magnetically separable $SO_4/$Fe-Al-TiO_2 solid acid catalyst for biodiesel production from waste cooking oil. *Applied Catalysis B: Environmental* **234**, 268–278, 2018.

82. K. H. Kim, O. K. Lee, and E. Y. Lee. Nano-immobilized biocatalysts for biodiesel production from renewable and sustainable resources. *Catalysts* 8(2), 68, 2018.

83. M. Rai and S. S. Da Silva. *Nanotechnology for bioenergy and biofuel production.* Springer, 2017.

84. E. Serrano, G. Rus, and J. Garcia-Martinez. Nanotechnology for sustainable energy. *Renewable and Sustainable Energy Reviews* **13**(9), 2373–2384, 2009.

85. R. W. Lai, K. W. Yeung, M. M. Yung, A. B. Djurišić, J. P. Giesy, and K. M. Leung. Regulation of engineered nanomaterials: current challenges, insights and future directions. *Environmental Science and Pollution Research* **25**(4), 3060–3077, 2018.

86. S. Faisal, F. Yusuf Hafeez, Y. Zafar, S. Majeed, X. Leng, S. Zhao, ... and X. Li. A Review on Nanoparticles as Boon for Biogas Producers—Nano Fuels and Biosensing Monitoring. *Applied Sciences* **9**(1), 59, 2019.

87. A. Zuliani, F. Ivars, and R. Luque. Advances in Nanocatalyst Design for Biofuel Production. *ChemCatChem* **10**(9), 1968–1981, 2018.

88. A. Biswas. Nanotechnology in Biofuels Production: A Novel Approach for Processing and Production of Bioenergy. In *Sustainable Approaches for Biofuels Production Technologies* (pp. 183–193). Springer, Cham, 2019.

89. D. B. Warheit. Hazard and risk assessment strategies for nanoparticle exposures: how far have we come in the past 10 years? *F1000Research*, **7**, 2018.

90. K. M. Rahman, L. Melville, S. I. Huq, and S. K. Khoda. Understanding bioenergy production and optimisation at the nanoscale–a review. *Journal of Experimental Nanoscience* **11**(10), 762–775, 2016.

91. A. Mudhoo and G. Kumar. Catalytic potency of ionic liquid-stabilized metal nanoparticles towards greening biomass processing: Insights, limitations and prospects. *Biochemical Engineering Journal*, 2018.

92. I. Gupta, N. Duran, and M. Rai. Nano-silver toxicity: emerging concerns and consequences in human health. In N. Cioffi and M. Rai (eds.), *Nano-antimicrobials: Progress and Prospects*. Berlin, Germany: Springer, 2012. pp. 525–548.

93. S. Upadhyay, K. Ganguly, and L. Palmberg. Wonders of nanotechnology in the treatment for chronic lung diseases. *J Nanomed Nanotechnol* **6**, 337, 2015.

94. P. V. Asharani, Y. Lianwu, Z. Gong, and S. Valiyaveettil. Comparison of the toxicity of silver, gold and platinum nanoparticles in developing zebrafish embryos. *Nanotoxicology* **5**(1), 43–54, 2011.

95. L. Chen, R. A. Yokel, B. Hennig, and M. Toborek. Manufactured aluminum oxide nanoparticles decrease expression of tight junction proteins in brain vasculature. *J Neuroimmune Pharmacol* **3**(4), 286–295, 2008.

96. Y. J. Kim, H. S. Choi, M. K. Song, D. Y. Youk, J. H. Kim, and J. C. Ryu. Genotoxicity of aluminum oxide (Al2O3) nanoparticle in mammalian cell lines. *Mol Cell Toxicol* **5**, 172–178, 2009.

97. R. Guan, T. Kang, F. Lu, Z. Zhang, H. Shen, and M. Liu. Cytotoxicity, oxidative stress, and genotoxicity in human hepatocyte and embryonic kidney cells exposed to ZnO nanoparticles. *Nanoscale Res Lett* **7**(1), 602, 2012.

98. International Energy Agency. *World Energy Outlook 2011*. Paris: International Energy Agency, 2011.

99. P. Shapira and J. Youtie. The Economic Contributions of Nanotechnology to Green and Sustainable Growth. *International Symposium on Assessing the Economic Impact of Nanotechnology,* 2012.
100. B. Walsh, P. Willis, and A. MacGregor. *A comparative methodology for estimating the economic value of innovation in nanotechnologies. A report for DEFRA.* Oakdene Hollins, Aylesbury, UK, 2010.
101. H. Suttcliffe. A Report on Responsible Research & Innovation. Matter, London, 2012. http://www.matterforall.org/pdf/RRI-Report.pdf (accessed February 20, 2010).
102. E. O'Rourke and M. Morrison. Challenges for Governments in Evaluating Return on Investment from Nanotechnology and its Broader Economic Impact. *Institute of Nanotechnology, United Kingdom,* 2012.
103. M. L. Verma, M. Puri, and C. J.Barrow. Recent trends in nanomaterials immobilised enzymes for biofuel production. *Critical Reviews in Biotechnology* **36**(1), 108–119, 2016.
104. S. Pugh, R. McKenna, R. Moolick, and D. R. Nielsen. Advances and opportunities at the interface between microbial bioenergy and nanotechnology. *The Canadian Journal of Chemical Engineering* **89**(1), 2–12, 2011.
105. C. E. Zhao, P. Gai, R. Song, Y. Chen, J. Zhang, and J. J. Zhu. Nanostructured material-based biofuel cells: recent advances and future prospects. *Chemical Society Reviews* **46**(5), 1545–1564, 2017.
106. Y. C. Lee, K. Lee, and Y. K. Oh. Recent nanoparticle engineering advances in microalgal cultivation and harvesting processes of biodiesel production: a review. *Bioresource Technology* **184**, 63–72, 2015.

Questions

1. What is nanotechnology? Differentiate between various nanomaterials.

2. Define biofuels and explain the three categories they have been classified into, along with examples.

3. What are the conventional methods of biofuel production? Also discuss the related limitations.

4. What kind of conversion processes can be used to produce second- and third-generation biofuels?

5. Nanotechnology is emerging as a promising tool in the biofuel industry. Highlight its importance.

6. What are the limitations and challenges of applying nanotechnology in large-scale processes such as biofuel production?

7. Analyze the role of nanotechnology in biofuel production from a techno-economic perspective.

CHAPTER 5

Fuel and Physical Properties of Biodiesel Components

Gerhard Knothe

5.1   Introduction

Biodiesel is an alternative diesel fuel (DF) derived from vegetable oils or animal fats [1, 2]. Transesterification of an oil or fat with a monohydric alcohol, in most cases methanol, yields the corresponding mono-alkyl esters, which are defined as biodiesel. The successful introduction and commercialization of biodiesel in many countries around the world has been accompanied by the development of standards to ensure high product quality and user confidence. Some biodiesel standards are ASTM D6751 (ASTM stands for American Society for Testing and Materials) and the European standard EN 14214, which was developed from previously existing standards in individual European countries.

The suitability of any material as fuel, including biodiesel, is influenced by the nature of its major as well as minor components arising from production or other sources. The nature of these components ultimately determines the fuel and physical properties. Some of the properties included in standards can be traced to the structure of the fatty esters in the biodiesel. Since biodiesel consists of fatty acid esters, not only the structure of the fatty acids but also that of the ester moiety can influence the fuel properties of biodiesel. The transesterification reaction of an oil or fat leads to a biodiesel fuel corresponding in its fatty acid profile with that of the parent oil or fat. Therefore, biodiesel is largely a mixture of fatty esters with each ester component contributing to the properties of the fuel.

Properties of biodiesel that are determined by the structure of its component fatty esters and the nature of its minor components include ignition quality, cold flow, oxidative stability, viscosity, and lubricity. This chapter discusses the influence of the structure of fatty esters on these properties. Not all of these properties have been included in biodiesel standards, although all of them are essential to proper functioning of the fuel in a diesel engine.

Generally, as the least expensive alcohol, methanol has been used to produce biodiesel. Table 5.1 compares the properties of petro diesel and biodiesel. Biodiesel, in most cases, can therefore be termed the fatty acid methyl esters (FAME) of a vegetable oil or animal fat. However, as mentioned above, both the fatty acid chain and the alcohol functionality contribute to the overall properties of a fatty ester. It is worthwhile to consider the properties imparted by other alcohols yielding fatty acid alkyl esters (FAAE) that could be used for producing biodiesel. Therefore, both structural moieties will be discussed in

Petro diesel	Biodiesel
1. Petro diesel is produced from crude oil by fractional distillation.	1. Biodiesel is converted by a transesterification process from vegetable oil or animal fats.
2. It mostly is a hydrocarbon with carbon atoms varying between 8 and 21.	2. Biodiesels are methyl ester.
3. Petro diesel contains Sulphur, and there is pollution of sulphur or oxides of sulphur.	3. Biodiesel does not contain sulphur and hence reduces pollution.
4. Petroleum diesel contains 95 percent saturated hydrocarbon and 5 percent aromatic compounds, so there is more pollution than biodiesel.	4. Biodiesel's transesterification process contains more oxygen (usually 10 to 12 percent), which results in less pollution. In the presence of esters it has probability of full combustion and reduced pollution.
5. It has more engine peak power than biodiesel.	5. It has less engine peak power than petro diesel.
6. Petro diesel has lower lubricity than biodiesel.	6. It has more lubricity than petro diesel.
7. Petro diesel generally has fewer problems with thickening at low temperature.	7. Biodiesel tends to thicken and "gel up" at low temperatures more readily than petroleum diesel. This is a concern, especially for cold winters.
8. Petro diesel is less chemically active as a solvent.	8. Biodiesel is more chemically active as a solvent than petroleum diesel. As a result, it can be more aggressive to some materials that are normally considered safe for diesel fuel.
9. Petro diesel is more toxic than biodiesel.	9. Biodiesel is much less toxic than petroleum diesel. This can be a real benefit for spill cleanups.

TABLE 5.1   Differences between Petro diesel and Biodiesel

this chapter. Table 5.2 lists fuel properties of neat alkyl esters of fatty acids. Besides the fuel properties discussed here, the heat of combustion (HG) of some fatty compounds [3] is included in Table 5.2, for the sake of underscoring the suitability of fatty esters as fuel with regard to this property.

5.2 Cetane Number and Exhaust Emissions

The cetane number (CN), which is related to the ignition properties, is a prime indicator of fuel quality in the realm of diesel engines. It is conceptually similar to the octane number used for gasoline. Generally, a compound that has a high octane number tends to have a low CN and vice versa. The CN of a DF is related to the ignition delay (ID) time, i.e., the time between injection of the fuel into the cylinder and onset of ignition. The shorter the ID time, the higher the CN, and vice versa.

Standards have been established worldwide for CN determination, e.g., ASTM D613 in the United States, and internationally the International Organization for Standardization (ISO) standard ISO 5165. A long straight-chain hydrocarbon, hexadecane ($C_{16}H_{34}$; trivial name cetane, giving the cetane scale its name) is the high-quality standard on the cetane scale with an assigned CN of 100. A highly branched isomer of hexadecane, 2,2,4,4,6,8,8-heptamethylnonane (HMN), a compound with poor ignition quality, is the low-quality standard with an assigned CN of 15. The two reference compounds on the cetane scale show that CN decreases with decreasing chain length and increasing branching. Aromatic compounds that are present in significant amounts in petrodiesel have low CNs but their CNs increase with increasing size of n-alkyl side chains [12, 13]. The cetane scale is arbitrary, and compounds with CN > 100 or CN < 15 have been identified. The American standard for petrodiesel (ASTM D975) prescribes a minimum CN of 40, while the standards for biodiesel prescribe a minimum of 47 (ASTM D6751) or 51 (European standard EN 14214). Due to the high CNs of many fatty compounds, which can exceed the cetane scale, the lipid combustion quality number for these compounds has been suggested [14].

The use of biodiesel reduces most regulated exhaust emissions from a diesel engine. The species reduced include carbon monoxide, hydrocarbons, and particulate matter (PM). Nitrogen oxide (NO_x) emissions are slightly increased, however. When blending biodiesel with petrodiesel, the effect of biodiesel is approximately linear to the blend level. A report summarizing exhaust emissions tests with biodiesel is available [15], and other summaries are given in Refs. [16, 17].

The structure of the fatty esters in biodiesel affects the levels of exhaust emissions. When using a 1991-model, 6-cylinder, 345-bhp (257-kW), direct-injection, turbocharged, and intercooled diesel

Trivial (systematic) name; acronym[b]	MP[c] (°C)	BP[c,d] (°C)	Cetane no.	Viscosity[e]	HG[f], (kcal/mol)
Caprylic (Octanoic); 8:0		237	18		
Methyl ester		193	33.6 (98.6)[g]	0.99[j]; 1.19[k]	1313
Ethyl ester	−43.1	208.5		1.37 (25°)[j]	1465
Capric (Decanoic); 10:0	31.5	270	47.6 (98.0)[g]		1453.07 (25°)
Methyl ester		224	47.2 (98.1)[g]	1.40[j]; 1.72[k]	1625
Ethyl ester	−20	243–5	51.2 (99.4)[g]	1.99 (25°)[j]	1780
Lauric (Dodecanoic); 12:0	44	131[1]			1763.25 (25°)
Methyl ester	5	266[766]	61.4 (99.1)[g]	1.95[j]; 2.43[k]	1940
Ethyl ester	−1.8fr	163[25]		2.88[j]	2098
Myristic (Tetradecanoic); 14:0	58	250.5[100]			2073.91 (25°)
Methyl ester	18.5	295[751]	66.2 (96.5)[g]	2.69[j]	2254
Ethyl ester	12.3	295	66.9 (99.3)[g]		2406
Palmitic (Hexadecanoic); 16:0	63	350			2384.76 (25°)
Methyl ester	30.5	415–8[747]	74.5 (93.6)[g]; 85.9[j]	3.60[j]; 4.38[k]	2550
Ethyl ester	19.3/24	191[10]	93.1[i]		2717
Propyl ester	20.4	190[12]	85.0[i]		
Isopropyl ester	13–4	160[2]	82.6[i]		
Butyl ester	16.9		91.9[i]		
2-Butyl ester			84.8[i]		
Isobutyl ester	22.5, 28.9	199[5]	83.6[i]		
Stearic (Octadecanoic); 18:0	71	360d	61.7[h]		2696.12 (25°)

Trivial (systematic) name; acronym[b]	MP[c] (°C)	BP[c,d] (°C)	Cetane no.	Viscosity[e]	HG[f], (kcal/mol)
Methyl ester	39	442–3[747]	86.9 (92.1)[g]; 101[i]	4.74[j]	2859
Ethyl ester	31–33.4	199[10]	76.8[h]; 97.7[i]		3012
Propyl ester			69.9[h]; 90.9[i]		
Isopropyl ester			96.5[i]		
Butyl ester	27.5	343	80.1[h]; 92.5[i]		
2-Butyl ester			97.5[i]		
Isobutyl ester			99.3[i]		
Palmitoleic (9(Z)-Hexadecanoic); 16:1					
Methyl ester			51.0[i]		2521
Oleic (9(Z)-Octadecanoic); 18:1	16	286[100]	46.1[h]		2657.4 (25°)
Methyl ester	−20	218.5[20]	55[h]; 59.3[i]	3.73[j]; 4.51[k]	2828
Ethyl ester		216–7[151]	53.9[h]; 67.8[i]	5.50 (25°)[j]	
Propyl ester			55.7[h]; 58.8[i]		
Isopropyl ester			86.6[i]		
Butyl ester			59.8[h]; 61.6[i]		
2-Butyl ester			71.9[i]		
Isobutyl ester			59.6[i]		
Linoleic (9Z,12Z-Octadecadienoic); 18:2	−5	229–30[16]	31.4[h]		
Methyl ester	−35	215[20]	42.2[h]; 38.2[i]	3.05[i]; 3.65[k]	2794
Ethyl ester		270–5[180]	37.1[h]; 39.6[i]		

TABLE 5.2 Properties of Fatty Acids and Esters[a] (Continued)

Propyl ester			40.6[h]; 44.0[i]		
Butyl ester			41.6[h]; 53.5[i]		
Linolenic (9Z,12Z, 15Z-Octadecatrienoic); 18:3	−11	230–2[17]		20.4[h]	
Methyl ester					
Ethyl ester	−57/−52	109[0.018]	20.6[g]; 22.7[i]	2.65[j]; 3.14[k]	2750
Propyl ester		174[2.5]	26.7[h]		
Butyl ester			26.8[h]		
Ricinoleic (12-Hydroxy-9Z-octadecenoic); 18:1, 12-OH	5.5	245[10]			
Methyl ester		225–7[15]		15.44[k]	
Erucic (13Z-Docosenoic); 22:1	33–4	265[15]			
Methyl ester		221–2[5]		5.91[j]	3454
Ethyl ester		229–30[5]			

[a] Adapted from Ref. [4].
[b] The numbers denote the number of carbons and double bonds. For example, in oleic acid, 18:1 stands for 18 carbons and 1 double bond.
[c] Melting point and boiling point data are from Refs. [5] and [6].
[d] Superscripts denote pressure (mm Hg) at which the boiling point was determined.
[e] Viscosity values determined at 40°C, unless indicated otherwise.
[f] HG values are from Refs. [3] and [5].
[g] Number in parentheses indicates purity (%) of the material used for CN determination as given in Ref. [7].
[h] Ref. [8].
[i] Ref. [9].
[j] Dynamic viscosity (mPa · s = cP), Ref. [10].
[k] Kinematic viscosity (mm2/s = cSt), Ref. [11].

TABLE 5.2 Properties of Fatty Acids and Esters[a] (Continued)

engine, NO_x exhaust emission increased with increasing number of double bonds and decreasing chain length for saturated chains [18]. Although often a trade-off is observed between NO_x and PM exhaust emissions, no trade-off has been observed in this work when varying the chain length [18]. The CN and density were correlated with emission levels [18]. However, emissions are likely affected by the technology level of the engine. When conducting tests on a 2003-model, 6-cylinder, 14–L, direct-injection, turbocharged, intercooled diesel engine with exhaust gas recirculation (EGR), no chain length effect has been observed for NO_x exhaust emissions, although the level of saturation still played a significant role [19]. PM exhaust emissions were reduced to levels close to the US 2007 regulations required for ultra-low-sulfur petrodiesel fuel. Also, PM levels were lower than those for neat hydrocarbons which would be enriched in "clean" petrodiesel fuel [19]. In both studies [18, 19], NO_x emissions of the saturated esters were slightly below those of the reference petrodiesel fuel.

For petrodiesel fuel, higher CNs have been correlated with reduced NO_x exhaust emissions [20]. This correlation has led to efforts to improve the CN of biodiesel fuels by using additives known as cetane improvers [8]. Despite the inherent relatively high CNs of fatty compounds, NO_x exhaust emissions usually increase slightly when operating a diesel engine on biodiesel, as mentioned above. The relationship between the CN and engine emissions is complicated by many factors, including the technology level of the engine. Older, lower-injection pressure engines are generally very sensitive to CN, with increased CN causing significant reductions in NOx emissions, due to shorter ID times and the resulting lower average combustion temperatures. More modern engines that are equipped with injection systems that control the rate of injection are not very sensitive to CN [21–23].

Historically, the first CN tests were carried out on palm oil ethyl esters [24, 25], which have a high CN, a result confirmed by later studies on many other vegetable oil-based DFs and individual fatty compounds. The influence of the compound structure on CNs of fatty compounds has been discussed in more recent literature [26], with the predictions made in that paper bein2 confirmed by practical cetane tests [7–9, 13]. CNs of neat fatty compounds are given in Table 5.2. In summary, the results show that CNs decrease with increasing unsaturation and increase with increasing chain length, i.e., uninterrupted CH_2 moieties. However, branched esters derived from alcohols such as isopropanol have CNs competitive with methyl or other straight-chain alkyl esters [9, 27]. Thus, one long, straight chain suffices to impart a high CN, even if the other moiety is branched. Branched esters are of interest because they exhibit improved low-temperature properties.

Recently, cetane studies on fatty compounds have been conducted using the Ignition Quality Tester™ (IQT) [9]. The IQT is a further, automated development of a constant volume combustion apparatus (CVCA) [28, 29]. The CVCA was originally developed for determining CNs more rapidly with greater experimental ease, better reproducibility, reduced use of fuel, and therefore less cost than the ASTM D613 method utilizing a cetane engine. The IQT method, which is the basis of ASTM D6890, was shown to be reproducible and the results competitive with those derived from ASTM D613. Some results from the IQT are included in Table 5.2. For the IQT, ID and CN are related by the following equation [9]:

$$CN_{IQT} = 83.99 \, (ID - 1.512)^{-0.658} + 3.547 \qquad (5.1)$$

In the recently approved method ASTM D6890, which is based on this technology, only ID times of 3.6–5.5 ms [corresponding to 55.3–40.5 DCN (derived CN)] are covered as the precision may be affected outside that range. However, for fatty compounds, the results obtained by using the IQT are comparable to those obtained by other methods [9]. Generally, the results of cetane testing for compounds with lower CNs, such as more unsaturated fatty compounds, show better agreement over various related literature references than the results for compounds with higher CNs, because of the nonlinear relationship [see Eq. (5.1)] between the ID time and the CN, which was observed previously [30]. Thus, small changes at shorter ID times result in greater changes in CN than at longer ID times. This would indicate a leveling-off effect on emissions such as NOx, as discussed above, once a certain ID time with corresponding CN has been reached as the formation of certain species depend on the ID time. However, for newer engines, this aspect must be modified as discussed above.

5.3   Cold-Flow Properties

One of the major problems associated with the use of biodiesel is poor low-temperature flow properties, documented by relatively high cloud points (CPs) and pour points (PPs) [1, 2]. The CP, which usually occurs at a higher temperature than the PP, is the temperature at which a fatty material becomes cloudy due to the formation of crystals and solidification of saturates. Solids and crystals rapidly grow and agglomerate, clogging fuel lines and filters and causing major operability problems. With decreasing temperature, more solids form and the material approaches the PP, the lowest temperature at which the material will still flow. Saturated fatty compounds have significantly higher melting points than unsaturated fatty compounds (Table 5.2), and in a mixture, they crystallize at higher temperatures than the unsaturates. Thus, biodiesel fuels derived from fats or oils with significant amounts of saturated fatty compounds will display higher CPs and PPs.

Besides the CP (ASTM D2500) and PP (ASTM D97) tests, two test methods for the low-temperature flow properties of petrodiesel exist, namely, the low-temperature flow test (LTFT) (used in North America; e.g., ASTM D4539) and cold filter plugging point (CFPP) (used outside North America; e.g., the European standard EN 116). These methods have also been used to evaluate biodiesel and its blends with No. 1 and 2 petrodiesel. Low-temperature filterability tests were stated to be necessary because of better correlation with operability tests than the CP or PP test [31]. However, for fuel formulations containing at least 10 vol% methyl esters, both LTFT and CFPP are linear functions of the CP [32]. Additional statistical analysis has shown a strong 1:1 correlation between LTFT and CP [32].

Several approaches to low-temperature problems of esters have been investigated, including blending with petrodiesel, winterization, additives, branched-chain esters, and bulky substituents in the chain. The latter approach may be considered a variation of the additive approach, as the corresponding compounds have been investigated in biodiesel at additive levels. Blending of esters with petrodiesel will not be discussed here.

Numerous, usually polymeric, additives were synthesized and reported mainly in the patent literature to have the effect of lowering the PP or sometimes even the CP. A brief overview of such additives has been presented [33]. Similarly, the use of fatty compound-derived materials with bulky moieties in the chain [34] at additive levels has been investigated. The idea associated with these materials is that the bulky moieties in these additives would destroy the harmony of the crystallizing solids. The effect of some additives appears to be limited because they more strongly affect the PP than the CP or they have only a slight influence on the CP. The CP, however, is more important than the PP for improving low-temperature flow properties [35].

The use of branched esters such as isopropyl, isobutyl, and 2-butyl esters instead of methyl esters [36, 37] is another approach for improving the low-temperature properties of biodiesel. Branched esters have lower melting points in the neat form (Table 5.2). These esters showed a lower TCO (crystallization onset temperature), as determined by differential scanning calorimetry (DSC) for the isopropyl esters of soybean oil (SBO) by 7–11°C and for the 2-butyl esters of SBO by 12–14°C [36]. The CPs and PPs were also lowered by the branched-chain esters. For example, the CP of isopropyl soyate was given as −9°C [7] and that of 2-butyl soyate as −12°C [36]. In comparison, the CP of methyl soyate is around 0°C [32]. However, in terms of economics, only isopropyl esters appear attractive as branched-chain esters, although even they are more expensive than methyl esters.

Branching in the ester chain does not have any negative effect on the CN of these compounds, as discussed above.

Winterization [35, 38, 39] is based on the lower melting points of unsaturated fatty compounds than saturated compounds (Table 5.2). This method removes by filtration the solids formed during the cooling of the vegetable oil esters, leaving a mixture with a higher content of unsaturated fatty esters and thus with lower CP and PP. This procedure can be repeated to further reduce the CPs and PPs. Saturated fatty compounds, which have higher CNs (Table 5.2) than unsaturated fatty compounds, are among the major compounds removed by winterization. Thus the CN of biodiesel decreases during winterization. Loss of material was reduced when winterization was carried out in the presence of cold-flow improvers or solvents such as hexane and isopropanol [39].

In other work [40], tertiary fatty amines and amides have been reported to be effective in enhancing the ignition quality of biodiesel without negatively affecting the low-temperature properties. Also, saturated fatty alcohols of chain lengths $>C_{12}$ increased the PP substantially. Ethyl laurate weakly decreased the PP.

5.4   Oxidative Stability

Oxidative stability of biodiesel has been the subject of considerable research [41–62]. This issue affects biodiesel primarily during extended storage. The influence of parameters such as presence of air, heat, traces of metal, antioxidants, and peroxides as well as nature of the storage container was investigated in the aforementioned studies. Generally, factors such as the presence of air, elevated temperatures, or the presence of metals facilitate oxidation. Studies performed with the automated oil stability index (OSI) method have confirmed the catalyzing effect of metals on oxidation; however, the influence of the compound structure of the fatty esters, especially unsaturation as discussed below, was even greater [52]. Numerous other methods, including not only wet-chemical ones such as the acid value and peroxide value, but also pressurized differential scanning calorimetry, nuclear magnetic resonance (NMR), and so forth, have been applied in oxidation studies of biodiesel.

Two simple methods for assessing the quality of stored biodiesel are the acid value and viscosity since both increase continuously with increasing fuel degradation, i.e., deteriorating fuel quality. The peroxide value is less suitable because it reaches a maximum and then can decrease again due to the formation of secondary oxidation products [48].

Autoxidation occurs due to the presence of double bonds in the chains of many fatty compounds. Autoxidation of unsaturated fatty compounds proceeds with different rates, depending on the number and position of double bonds [63]. Especially the positions allylic to double bonds are susceptible to oxidation. The bis-allylic positions in common polyunsaturated fatty acids, such as linoleic acid (double bonds at .C-9 and .C-12, giving one bis-allylic position at C-11) and linolenic acid (double bonds at .C-9, .C-12, and C-15, giving two bis-allylic positions at C-11 and C-14), are even more prone to autoxidation than the allylic positions. The relative rates of oxidation given in the literature [63] are 1 for oleates (methyl, ethyl esters), 41 for linoleates, and 98 for linolenates. This is essential because most biodiesel fuels contain significant amounts of esters of oleic, linoleic, or linolenic acids, which influence the oxidative stability of the fuels. The species formed during the oxidation process cause the fuel to eventually deteriorate.

A European standard (EN 14112; Rancimat method) for oxidative stability has been included in the American and European biodiesel standards (ASTM D6751 and EN 14214). Both biodiesel standards call for determining oxidative stability at 110ᵧC; however, EN 14214 prescribes a minimum induction time of 6 h by the Rancimat method while ASTM D6751 prescribes 3 h. The Rancimat method is nearly identical to the OSI method, which is an AOCS (American Oil Chemists' Society) method.

Besides preventing exposure of the fatty material to air, adding antioxidants is a common method to address the issue of oxidative stability. Common antioxidants are synthetic materials such as tert-butylhydroquinone (TBHQ), butylated hydroxytoluene (BHT), butylated hydroxyanisole (BHA), and propyl gallate (PG) as well as natural materials such as tocopherols. Antioxidants delay oxidation but do not prevent it, as oxidation will commence once the antioxidant in a material has been consumed.

5.4.1 Iodine value

The iodine value (IV) has been included in the European biodiesel standards to purportedly address the issue of oxidative stability and the propensity of the oil or fat to polymerize and form engine deposits. The IV is a measure of the total unsaturation of a fatty material measured in grams of iodine per 100 g of sample when formally adding iodine to the double bonds. An IV of 120 has been specified in EN 14214 and 130 in EN 14213, which would largely exclude vegetable oils such as soybean and sunflower oils as biodiesel feedstock. Thus the IV has not been included in biodiesel standards in the United States and Australia, and is limited to 140 in the South African standard (which would permit sunflower and soybean oils); the provisional Brazilian standard requires that it only be noted.

The IV of a vegetable oil or animal fat is almost identical to that of the corresponding methyl esters; however, the IV of alkyl esters decreases with higher alcohols used in their production since the IV is molecular weight dependent. For example, the IV of methyl, ethyl, propyl, and butyl linoleate is 172.4, 164.5, 157.4, and 150.8, respectively [64].

The use of the IV of a mixture for such purposes does not take into consideration that an infinite number of fatty acid profiles can yield the same IV and that different fatty acid structures can give the same IV, although the propensity for oxidation can differ significantly [64]. Other, new structure indices termed allylic and bis-allylic position equivalents (APE and BAPE), which are based on the number of such positions in a fatty acid chain and are independent of molecular weight, are likely more suitable than the IV [64]. The BAPE index distinguishes mixtures having nearly identical IV correctly by their OSI times. Note that the BAPE index is the decisive index compared to the APE because it relates to the more reactive bis-allylic positions. Engine performance tests with a mixture of vegetable oils of different IVs did not yield results that would have justified a low IV [65, 66]. No relationship between the IV and oxidative stability has been observed in another investigation on biodiesel with a wide range of IV [52].

5.5   Viscosity

Viscosity affects the atomization of a fuel upon injection into the combustion chamber and, thereby, ultimately the formation of engine deposits. The higher the viscosity, the greater the tendency of the fuel to cause such problems. The viscosity of a transesterified oil, i.e., biodiesel, is about an order of magnitude lower than that of the parent oil [1, 2]. High viscosity is the major fuel property why neat vegetable oils have been largely abandoned as alternative DF. Kinematic viscosity has been included in most biodiesel standards. It can be determined by standards such as ASTM D445 or ISO 3104. The difference in viscosity between the parent oil and the alkyl ester derivatives can be used in monitoring biodiesel production [67]. The effect on viscosity of blending biodiesel and petrodiesel has also been investigated [68], and an equation has been derived, which allows calculating the viscosity of such blends.

The prediction of viscosity of fatty materials has received considerable attention in the literature. Viscosity values of biodiesel/mixtures of fatty esters have been predicted from the viscosities of the individual components by a logarithmic equation for dynamic viscosity [10]. Viscosity increases with chain length (number of carbon atoms) and with increasing degree of saturation. This holds also for the alcohol moiety as the viscosity of ethyl esters is slightly higher than that of methyl esters [11]. Factors such as double bond configuration influence viscosity (cis double bond configuration giving a lower viscosity than the trans configuration), while the double bond position affects viscosity less [11]. Thus, a feedstock such as used frying oils, which is more saturated and contains some amounts of trans fatty acid chains, has a higher viscosity than its parent oil. Branching in the ester moiety, however, has little or no influence on viscosity, again showing that this is a technically promising approach for improving low-temperature properties without significantly affecting other fuel properties. Values for dynamic viscosity and kinematic viscosity of neat fatty acid alkyl esters are included in Table 5.2.

5.6   Lubricity

With the advent of low-sulfur petroleum-based DFs, the issue of DF lubricity is becoming increasingly important. Desulfurization of petrodiesel reduces or eliminates the inherent lubricity of this fuel, which is essential for proper functioning of vital engine components such as fuel pumps and injectors. Several studies [10, 11, 67–82] on the lubricity of biodiesel or fatty compounds have shown a beneficial effect of these materials on the lubricity of petrodiesel, particularly low-sulfur petrodiesel fuel. Adding biodiesel at low levels (1–2%) restores the lubricity to low-sulfur petroleum-derived DFs. However, the lubricity-enhancing effect of biodiesel at low blend levels is mainly caused by minor components of biodiesel such as free fatty acids and monoacylglycerols [83], which have free COOH and OH groups. Other studies [84, 85] also point out the beneficial effect of minor components on biodiesel lubricity, but these studies do not fully agree on the responsible species [83–85]. Thus, biodiesel is required at 1–2% levels in low-lubricity petrodiesel, in order for the minor components to be effective lubricity enhancers [83]. At higher blend levels, such as 5%, the esters are sufficiently effective without the presence of minor components.

While the length of a fatty acid chain does not significantly affect lubricity, unsaturation enhances lubricity slightly; thus an ester such as methyl linoleate or methyl linolenate improves lubricity more than methyl stearate [80, 83]. In accordance with the above observation on the effect of free OH groups on lubricity, castor oil displayed better lubricity than other vegetable oil esters [75, 80, 81]. Ethyl esters have improved lubricity compared to methyl esters [75].

Standards for testing DF lubricity use the scuffing load ball-on-cylinder lubricity evaluator (SLBOCLE) (ASTM D6078) or the high-frequency reciprocating rig (HFRR) (ASTM D6079; ISO 12156). Lubricity has not been included in biodiesel standards despite the definite advantage of biodiesel over petrodiesel with respect to this fuel property. However, the HFRR method has been included in the petrodiesel standards ASTM D975 and EN 590.

5.7   Outlook

The fuel properties of biodiesel are strongly influenced by the properties of the individual fatty esters as well as those of some minor components. Both moieties, the fatty acid and alcohol, have considerable influence on fuel properties such as CN, with relation to combustion and exhaust emissions, cold flow, oxidative stability, viscosity, and lubricity. It therefore appears reasonable to enrich (a) certain fatty ester(s) with desirable properties in the fuel, in order to improve the properties of the whole fuel. For example, from the presently available data, it appears that isopropyl esters have better fuel properties than methyl esters. The major disadvantage is the higher price of isopropanol in comparison to methanol, besides modifications needed for the transesterification reaction. Similar observations likely hold for the fatty acid moiety. It may be possible in the future to improve the properties of biodiesel by means of genetic engineering of the parent oils, which could eventually lead to a fuel enriched with (a) certain fatty acid(s), possibly oleic acid, that exhibits a combination of improved fuel properties.

References

1. G. Knothe, J. Krahl, and J. Van Gerpen (Eds.). *The Biodiesel Handbook*, Champaign, IL: AOCS Press, 2005.
2. M. Mittelbach and C. Remschmidt. Biodiesel—*The Comprehensive Handbook*, Graz, Austria: Martin Mittelbach, 2004.
3. B. Freedman and M. O. Bagby. Heats of combustion of fatty esters and triglycerides, *Journal of the American Oil Chemists' Society* **66**, 1601–1605 (1989).
4. G. Knothe. Dependence of biodiesel fuel properties on the structure of fatty acid alkyl esters, *Fuel Processing Technology* **86**, 1059–1070, 2005.
5. R. C. Weast (Ed.). *Handbook of Chemistry and Physics*, 66th ed., Boca Raton, FL: CRC Press, p. D272, 1986.
6. F. D. Gunstone, J. L. Harwood, and F. B. Padley. *The Lipid Handbook*, London: Chapman & Hall, 1994.
7. W. E. Klopfenstein. Effect of molecular weights of fatty acid esters on cetane numbers as diesel fuels, *Journal of the American Oil Chemists' Society* **62**, 1029–1031, 1985.
8. G. Knothe, M. O. Bagby, and T. W. Ryan III. Cetane Numbers of Fatty Compounds: Influence of Compound Structure and of Various Potential Cetane Improvers, SAE Technical Paper Series 971681, In: *State of Alternative Fuel Technologies*, SAE Publication SP-1274, Warrendale, PA, pp. 127–132, 1997.
9. G. Knothe, A. C. Matheaus, and T. W. Ryan III. Cetane numbers of branched and straight-chain fatty esters determined in an ignition quality tester, *Fuel* **82**, 971–975, 2003.
10. C. A. W. Allen, K. C. Watts, R. G. Ackman, and M. J. Pegg. Predicting the viscosity of biodiesel fuels from their fatty acid ester composition, *Fuel* **78**, 1319–1326, 1999.

11. G. Knothe and K. R. Steidley. Kinematic viscosity of biodiesel fuel components. Influence of compound structure and comparison to petrodiesel fuel components, *Fuel* **84**, 1059–1065, 2005.
12. A. D. Puckett and B. H. Caudle. Ignition qualities of hydrocarbons in the diesel-fuel boiling range, *U.S. Bureau of Mines Information Circular*, No. 7474, 14 pp., 1948.
13. P. Q. E. Clothier, B. D. Aguda, A. Moise, and H. Pritchard. How do diesel-fuel ignition improvers work? *Chemical Society Reviews* **22**, 101–108, 1993.
14. B. Freedman, M. O. Bagby, T. J. Callahan, and T. W. Ryan III. Cetane Numbers of Fatty Esters, Fatty Alcohols and Triglycerides Determined in a Constant Volume Combustion Bomb, SAE Technical Paper Series 900343, 1990.
15. U.S. Environmental Protection Agency. A Comprehensive Analysis of Biodiesel Impacts on Exhaust Emissions, Draft Technical Report EPA420-P-02-00, October 2002; see also www.epa.gov/oms/models/analysis.p02001.pdf (accessed January 2007).
16. R. L. McCormick and T. L. Alleman. Effect of biodiesel fuel on pollutant emissions from diesel engines, In: *The Biodiesel Handbook*, Knothe, G., Van Gerpen, J., and Krahl, J., (Eds.), Champaign, IL: AOCS Press, pp. 165–174, 2005.
17. J. Krahl, A. Munack, O. Schröder, H. Stein, and J. Bünger. Influence of biodiesel and different petrodiesel fuels on exhaust emissions and health effects, In: *The Biodiesel Handbook*, Knothe, G., Van Gerpen, J., and Krahl, J. (Eds.), Champaign, IL: AOCS Press, pp. 175–182, 2005.
18. R. L. McCormick, M. S. Graboski, T. L. Alleman, and A. M. Herring. Impact of biodiesel source material and chemical structure on emissions of criteria pollutants from a heavy-duty engine, *Environmental Science and Technology* **35**, 1742–1747, 2001.
19. G. Knothe, C. A. Sharp, and T. W. Ryan III. Exhaust emissions of biodiesel, petrodiesel, neat methyl esters, and alkanes in a new technology engine, *Energy & Fuels* **20**, 403–408, 2006.
20. N. Ladommatos, M. Parsi, and A. Knowles. The effect of fuel cetane improver on diesel pollutant emissions, *Fuel* **75**, 8–14, 1996.
21. R. L. Mason, A. C. Matheaus, T. W. Ryan III, R. A. Sobotowski, J. C. Wall, C. H. Hobbs et al. EPA HDEWG Program—Statistical Analysis, SAE Technical Paper Series 2001-01-1859, In: *Diesel and Gasoline Performance and Additives*, SAE Publication SP-1551, Warrendale, PA, 2001.
22. A. C. Matheaus, G. D. Neely, T. W. Ryan III, R. A. Sobotowski, J. C. Wall, C. H. Hobbs, G. W. Passavant, and T. J. Bond. EPA HDEWG Program—Engine Test Results, SAE Technical Paper Series 2001-01-1858, In: *Diesel and Gasoline Performance and Additives*, SAE Publication SP-1551, Warrendale, PA, 2001.
23. R. A. Sobotowski, J. C. Wall, C. H. Hobbs, A. C. Matheaus, R. L. Mason, T. W. Ryan III, et al. EPA HDEWG Program—Test Fuel

Development, SAE Technical Paper Series 2001-01-1857, In: *Diesel and Gasoline Performance and Additives*, SAE Publication SP-1551, Warrendale, PA, 2001.

24. M. van den Abeele. Palm oil as raw material for the production of a heavy motor fuel (L'huile de palme. Matière première pour la préparation d'un carburant lourd utilisable dans les moteurs à combustion interne.), *Bulletin Agricole du Congo Belge* **33**, 3–90, 1942.

25. G. Knothe. Historical perspectives on vegetable oil-based diesel fuels, *INFORM* **12**, 1103–1107, 2001.

26. K. J. Harrington. Chemical and physical properties of vegetable oil esters and their effect on diesel fuel performance, *Biomass* **9**, 1–17, 1986.

27. Y. Zhang and J. H. Van Gerpen. Combustion Analysis of Esters of Soybean Oil in a Diesel Engine, SAE Technical Paper Series 960765, In: *Performance of Alternative Fuels for SI and CI Engines*, SAE Publication SP-1160, Warrendale, PA, pp. 1–15, 1996.

28. T. W. Ryan III and B. Stapper. Diesel Fuel Ignition Quality as Determined in a Constant Volume Combustion Bomb, SAE Technical Paper Series 870586, 1987.

29. A. A. Aradi and T. W. Ryan III. Cetane Effect on Diesel Ignition Delay Times Measured in a Constant Volume Combustion Apparatus, SAE Technical Paper Series 952352, In: *Emission Processes and Control Technologies in Diesel Engines*, SAE Publication SP-1119, Warrendale, PA, pp. 43–56, 1995.

30. L. N. Allard, G. D. Webster, N. J. Hole, T. W. Ryan III, D. Ott, and C. W. Fairbridge. Diesel Fuel Ignition Quality as Determined in the Ignition Quality Tester (IQT), SAE Technical Paper Series 961182, 1996.

31. K. Owen and T. Coley. *Automotive Fuels Reference Book*, Second ed., Warrendale, PA: SAE, 1995.

32. R. O. Dunn and M. O. Bagby. Low-temperature properties of triglyceride-based diesel fuels: Transesterified methyl esters and petroleum middle distillate/ester blends, *Journal of the American Oil Chemists' Society* **72**, 895–904, 1995.

33. G. Knothe, R. O. Dunn, and M. O. Bagby. Biodiesel: The use of vegetable oils and their derivatives as alternative diesel fuels, ACS Symposium Series 666, In: *Fuels and Chemicals from Biomass*, American Chemical Society, Washington, DC, pp. 172–208, 1997.

34. G. Knothe, R. O. Dunn, M. W. Shockley, and M. O. Bagby. Synthesis and characterization of some long-chain diesters with branched or bulky moieties, *Journal of the American Oil Chemists' Society* **77**, 865–871, 2000.

35. R. O. Dunn, M. W. Shockley, and M. O. Bagby. Improving the low-temperature properties of alternative diesel fuels: Vegetable-oil derived methyl esters, *Journal of the American Oil Chemists' Society* **73**, 1719–1728, 1996.

36. I. Lee, L. A. Johnson, and E. G. Hammond. Use of branched-chain esters to reduce the crystallization temperature of biodiesel, *Journal of the American Oil Chemists' Society* **72**, 1155–1160, 1995.

37. T. A. Foglia, L. A. Nelson, R. O. Dunn, and W. Marmer. Low-temperature properties of alkyl esters of tallow and grease, *Journal of the American Oil Chemists' Society* **74**, 951–955, 1997.

38. I. Lee, L. A. Johnson, and E. G. Hammond. Reducing the crystallization temperature of biodiesel by winterizing methyl soyate, *Journal of the American Oil Chemists' Society* **73**, 631, 1996.

39. R. O. Dunn, M. W. Shockley, and M. O. Bagby. Winterized Methyl Esters from Soybean Oil: An Alternative Diesel Fuel with Improved Low-Temperature Flow Properties, SAE Technical Paper Series 971682, In: *State of Alternative Fuel Technologies*, SAE Publication SP-1274, Warrendale, PA, p. 133, 1997.

40. S. Stournas, E. Lois, and A. Serdari. Effects of fatty acid derivatives on the ignition quality and cold flow of diesel fuel, *Journal of the American Oil Chemists' Society* **72**, 433–437, 1995.

41. L. M. Du Plessis. Plant oils as diesel fuel extenders: Stability tests and specifications on different grades of sunflower seed and soyabean oils, *CHEMSA* **8**, 150–154, 1982.

42. L. M. Du Plessis, J. B. M. de Villiers, and W. H. van der Walt. Stability studies on methyl and ethyl fatty acid esters of sunflowerseed oil, *Journal of the American Oil Chemists' Society* **62**, 748–752, 1985.

43. P. Bondioli, A. Gasparoli, A. Lanzani, E. Fedeli, S. Veronese, and M. Sala. Storage stability of biodiesel, *Journal of the American Oil Chemists' Society* **72**, 699–702, 1995.

44. P. Bondioli and L. Folegatti. Evaluating the oxidation stability of biodiesel. An experimental contribution, *Rivista Italiana delle Sostanze Grasse* **73**, 349–353, 1996.

45. N. M. Simkovsky and A. Ecker. Influence of light and contents of tocopherol on the oxidative stability of fatty acid methyl esters (Einfluß von Licht und Tocopherolgehalt auf die Oxidationsstabilität von Fettsäuremethylestern.), *Fett/Lipid* **100**, 534–538, 1998.

46. J. C. Thompson, C. L. Peterson, D. L. Reece, and S. M. Beck. Two-year storage study with methyl and ethyl esters of rapeseed, *Transactions of the American Society of Agricultural Engineers* **41**, 931–939, 1998.

47. N. M. Simkovsky and A. Ecker. Effect of antioxidants on the oxidative stability of rapeseed oil methyl esters, *Erdoel, Erdgas, Kohle* **115**, 317–318, 1999.

48. M. Canakci, A. Monyem, and J. Van Gerpen. Accelerated oxidation processes in biodiesel, *Transactions of the American Society of Agricultural Engineers* **42**, 1565–1572, 1999.

49. R. O. Dunn. Analysis of oxidative stability of methyl soyate by pressurized-differential scanning calorimetry, *Transactions of the American Society of Agricultural Engineers* **43**, 1203–1208, 2000.

50. M. Mittelbach and S. Gangl. Long storage stability of biodiesel made from rapeseed and used frying oil, *Journal of the American Oil Chemists' Society* **78**, 573–577, 2001.
51. R. O. Dunn. Effect of oxidation under accelerated conditions on fuel properties of methyl soyate (biodiesel), *Journal of the American Oil Chemists' Society* **79**, 915–920, 2002.
52. G. Knothe and R. O. Dunn. Dependence of oil stability index of fatty compounds on their structure and concentration and presence of metals, *Journal of the American Oil Chemsits' Society* **80**, 1021–1026, 2003.
53. S. Schober and M. Mittelbach. The impact of antioxidants on biodiesel oxidation stability, *European Journal of Lipid Science and Technology* **106**, 382–389, 2004.
54. P. Bondioli, A. Gasparoli, L. della Bella, S. Tagliabue, F. Lacoste, and L. Lagardere. The prediction of biodiesel storage stability. Proposal for a quick test, *European Journal of Lipid Science and Technology* **106**, 822–830, 2004.
55. T. Dittmar, B. Ondruschka, J. Haupt, and M. Lauterbach. Improvement of the oxidative stability of fatty acid methyl esters with antioxidants—Limitations of the Rancimat test (Verbesserung der Oxidationsstabilität von Fettsäuremethylestern mit Antioxidantien—Grenzen des Rancimat-Tests.), *Chemie Ingenieur Technik* **76**, 1167–1170, 2004.
56. O. Falk and R. Meyer-Pitroff. The effect of fatty acid composition on biodiesel oxidative stability, *European Journal of Lipid Science and Technology* **106**, 837–843, 2004.
57. R. O. Dunn. Oxidative stability of soybean oil fatty acid methyl esters by oil stability index (OSI), *Journal of the American Oil Chemists' Society* **82**, 381–387, 2005.
58. R. O. Dunn. Effect of antioxidants on the oxidative stability of methyl soyate (biodiesel), *Fuel Processing Technology* **86**, 1071–1085, 2005.
59. J. Polavka, J. Paligová, J. Cvengroš, and P. Simon. Oxidation stability of methyl esters studied by differential thermal analysis and Rancimat, *Journal of the American Oil Chemists' Society* **82**, 519–524 (2005).
60. R. O. Dunn. Oxidative stability of biodiesel by dynamic mode pressurized-differential scanning calorimetry (P-DSC), *Transactions of the American Society of Agricultural and Biological Engineers* **49**, 1633–1641, 2006.
61. H. L. Fang and R. L. McCormick. Spectroscopic Study of Biodiesel Degradation Pathways, SAE Technical Paper Series 2006-01-3300, 2006.
62. G. Knothe. Analysis of oxidized biodiesel by 1H-NMR and effect of contact area with air, *European Journal of Lipid Science and Technology* **108**, 493–500, 2006.
63. E. N. Frankel. *Lipid Oxidation*, Brigdwater, England: The Oily Press, 2005.

64. G. Knothe. Structure indices in FA chemistry. How relevant is the iodine value? *Journal of the American Oil Chemists' Society* **79**, 847–854, 2002.
65. H. Prankl, M. Wörgetter, and J. Rathbauer. Technical performance of vegetable oil methyl esters with a high iodine number, In: *Biomass, Proceedings of the 4th Biomass Conference of the Americas*, 1999, Oakland, California, pp. 805–810.
66. H. Prankl and M. Wörgetter. Influence of the iodine number of biodiesel to the engine performance, In: *Liquid Fuels and Industrial Products from Renewable Resources, Proceedings of the 3rd Liquid Fuel Conference*, Cundiff, J. S., et al. (Eds.), St Joseph, MI: ASAE, pp. 191–196, 1996.
67. P. De Filippis, C. Giavarini, M. Scarsella, and M. Sorrentino. Transesterification processes for vegetable oils: A simple control method of methyl ester content, *Journal of the American Oil Chemists' Society* **71**, 1399–1404, 1995.
68. M. E. Tat and J. H. Van Gerpen. The kinematic viscosity of biodiesel and its blends with diesel fuel, *Journal of the American Oil Chemists' Society* **76**, 1511–1513, 1999.
69. J. A. Waynick. Evaluation of the stability, lubricity, and cold flow properties of biodiesel fuel, In: *Proceedings of the 6th International Conference on Stability and Handling of Liquid Fuels*, 1997, Vancouver, BC, Canada.
70. J. H. Van Gerpen, S. Soylu, and M. E. Tat. Evaluation of the lubricity of soybean oil-based additives in diesel fuel, In: *Proceedings of 1999 ASAE/CSAE-SCGR Annual International Meeting*, Paper No. 996134, Toronto, Canada, 1999.
71. D. Karonis, G. Anostopoulos, E. Lois, S. Stournas, F. Zannikos, and A. Serdari. Assessment of the Lubricity of Greek Road Diesel and the Effect of the Addition of Specific Types of Biodiesel, SAE Technical Paper Series 1999-01-1471, 1999.
72. L.G. Schumacher and J. Van Gerpen. Engine Oil Analysis of Diesel Engines Fueled with 0, 1, 2 and 100 Percent Biodiesel, In: ASAE Meeting Presentation 006010, *ASAE International Meeting*, 2000, Midwest Express Center, Milwaukee, WI.
73. G. Anastopoulos, E. Lois, A. Serdari, F. Zanikos, S. Stornas, and S. Kalligeros. Lubrication properties of low-sulfur diesel fuels in the presence of specific types of fatty acid derivatives, *Energy & Fuels* **15**, 106–112, 2001.
74. G. Anastopoulos, E. Lois, D. Karonis, F. Zanikos, and S. Kalligeros. A preliminary evaluation of esters of monocarboxylic fatty acid on the lubrication properties of diesel fuel, *Industrial & Engineering Chemistry Research* **40**, 452–456, 2001.
75. D. C. Drown, K. Harper, and E. Frame. Screening vegetable oil alcohol esters as fuel lubricity enhancers, *Journal of the American Oil Chemists' Society* **78**, 579–584, 2001.

76. C. Kajdas and M. Majzner. The Influence of Fatty Acids and Fatty Acids Mixtures on the Lubricity of Low-Sulfur Diesel Fuels, SAE Technical Paper Series 2001-01-1929, 2001.

77. L. G. Schumacher and B. T. Adams. Using biodiesel as a lubricity additive for petroleum diesel fuel, *ASAE Paper*, No. 02-6085, 2002.

78. A. K. Agarwal, J. Bijwe, and L. M. Das. Effect of biodiesel utilization of wear of vital parts in compression ignition engine, *Transactions of the American Society of Mechanical Engineers (Journal of Engineering for Gas Turbines and Power)* **125**, 604–611, 2003.

79. A. K. Agarwal, J. Bijwe, and L. M. Das. Wear assessment in a biodiesel fueled compression ignition engine, *Transactions of the American Society of Mechanical Engineers (Journal of Engineering for Gas Turbines and Power)* **125**, 820–826, 2003.

80. D. P. Geller and J. W. Goodrum. Effects of specific fatty acid methyl esters on diesel fuel lubricity, *Fuel* **83**, 2351–2356, 2004.

81. J. W. Goodrum and D. P. Geller. Influence of fatty acid methyl esters from hydroxylated vegetable oils on diesel fuel lubricity, *Bioresource Technology* **96**(7), 851–855, 2005.

82. A. K. Bhatnagar, S. Kaul, V. K. Chhibber, and A. K. Gupta. HFRR studies on methyl esters of nonedible vegetable oils, *Energy & Fuels* **20**(3), 1341–1344, 2006.

83. G. Knothe and K. R. Steidley. Lubricity of components of biodiesel and petrodiesel. The origin of biodiesel lubricity, *Energy & Fuels* **19**, 1192–1200, 2005.

84. G. Hillion, X. Montagne, and P. Marchand. Methyl esters of plant oils used as additives or organic fuel. (Les esters méthyliques d'huiles végétales: additif ou biocarburant?), *Oleagineux, Corps Gras, Lipides* **6**, 435–438, 1999.

85. J. Hu, Z. Du, C. Liu, and E. Min. Study on the lubrication properties of biodiesel as fuel lubricity enhancers, *Fuel* **84**, 1601–1606, 2005.

Questions

1. What is cetane number, and how does the structure of fatty acids in biodiesel affect exhaust emissions of an engine?

2. What are the working problems of biodiesel in an engine at low temperature? How are these problems solved?

3. What is autooxidation? How is oxidation stability in oil maintained?

4. How does fuel viscosity create engine problems? How has biodiesel viscosity been standardized?

5. What is lubricity? How is biodiesel lubricity maintained?

CHAPTER **6**

Processing of Vegetable Oils as Biodiesel and Engine Performance

Ahindra Nag

6.1 Introduction

Processing of vegetable oils as biodiesel [1, 2] and its engine performance is very challenging. From an environmental point of view, diesel engines are a major source of air pollution. Exhaust gases from diesel engines contain oxides of nitrogen, carbon monoxide, organic compounds consisting of unburned or partially burned hydrocarbons, and particulate matter (consisting primarily of soot).

Interest in clean burning fuels is growing worldwide, and reduction in exhaust emissions from diesel engines is of utmost importance. It is widely recognized that alternative diesel fuels produced from vegetable oils and animal fats can reduce exhaust emissions from compression ignition (CI) engines, without significantly affecting engine performance. But reducing pollutant emissions from diesel engines requires a detailed knowledge of the combustion process. However, the complex nature of the combustion process in an engine makes it difficult to understand the events occurring in the combustion chamber that determine the emission of exhaust gases.

Dr. Rudolf Diesel [3], the inventor of the CI engine, used peanut oil in one of his engines for a demonstration at the Paris exhibition in 1900. Then there was considerable interest in the use of vegetable oils as fuel in diesel engines.

Several studies have reported the effects of fuel and engine parameters on diesel exhaust emissions. Chowdhury [4] claims to have successfully used raw vegetable oils in diesel engines. He observed that no major changes were necessary in the engine, but the engine could not be run for more than 4 h. The performance and economic aspects of vegetable oil were also discussed.

Barve and Amurthe [5] cite an example of using groundnut oil as fuel in a diesel engine generator set (103 kW) of a local water pump house. They claimed that the power output and fuel consumption were very much comparable with certified diesel fuel. Weibe and Nowakowska [6] have reported the use of palm oil as a motor fuel. The performance was found satisfactory with higher fuel consumption. Fang [7] has reported that soybean and castor oil blended with diesel fuel burns adequately in a small diesel engine. Engelman et al. [8] have presented data on the performance of soybean diesel oil blends compared with diesel fuel. Results from a short-duration test showed that the use of blends was feasible in the diesel engine; but in fact, in the long-term, test problems associated with lubrication, sticking piston rings, and injector atomization patterns contributed to mechanical difficulties in the engine. Cruze et al. [9] have found that atomization of the fuel by the injector, in some cases, has caused delayed ignition characteristics and reduced efficiency of mechanical power production, compared to diesel fuel. Pryde [10] has stated that raw vegetable oil has had no great promise for engine tests and that modified oil esters were required for further engine tests. Bruwer [11] has reported that even without modification, nine diesel engines started and operated almost normally on sunflower oil and delivered power equal to that of diesel fuel. Brake thermal efficiency and maximum engine power were 3 percent lower, while the specific fuel consumption was 10 percent higher than that of diesel fuel. The bench test, however, showed that atomization of 100 percent sunflower oil was much poorer than diesel but could be improved by reducing the viscosity of oil. Energy wise, sunflower oil was favorable for running diesel engines for a shorter duration.

Baranescu et al. [12] have conducted tests on a turbocharged engine, using mixtures of sunflower oil in 25 percent, 50 percent, and 75 percent with diesel fuel. They have concluded that the use of sunflower oil blended with diesel brought modification in the fuel injection process that mainly included an increase in injection pressure and a longer ignition duration. These effects led to longer combustion duration. Cold-temperature operation was very critical due to high viscosity that caused fuel system problems such as starting failure, unacceptable emission levels, and injection pump failure. Engine shutdown for a long duration accelerated gum formation, where the fuel contacted the bare metal. This might further impair the engine or injection system.

Wagner et al. [13] have conducted tests on a number of diesel engines with different blends of winter rape and safflower oil with diesel fuel. The following specific conclusions were drawn from the results obtained:

- High viscosity and tendency to polymerize within the cylinder were major physical and chemical problems.
- Attempt to reduce the viscosity of the oil by preheating the fuel by increasing the temperature of the fuel at the injector to the required value was not successful.
- Short-term engine performance showed power output and fuel consumption equivalent to diesel fuel.
- Severe engine damage occurred within a very short duration when the test was conducted for maximum power with varying engine rpm (revolutions per minute).
- A blend of 70 percent winter rape with 30 percent diesel was successfully used for 50 h. No adverse effect was noted.
- A diesel injector pump when run for 154 h with safflower oil had no abnormal wear, gumming, or corrosion.

Borgelt et al. [14] have conducted tests on three diesel engines containing 25–75 percent and 50–50 percent soybean oil and diesel. The engines were operated under 50 percent load for 1000 horsepower (HP); the output ranged from 2.55 to 2.8 kW. Thermal efficiency ranged from 19.3 percent to 20 percent. Engine performances were not significantly different. Carbon deposit increased with increased percentage of soybean oil. Thus, Borgelt et al. concluded that use of 25 percent or less soybean oil caused negligible changes in engine performance.

Barsic and Humke [15] performed a study in which blends of unrefined peanut and sunflower with diesel fuel (50–50 percent) were used in a single-cylinder engine. The engine produced equivalent power or a minor increase (6 percent) with vegetable oils and blends, with a 20 percent increase in specific fuel consumption. Performance tests at equal energy showed that the power level remained constant or decreased slightly, thermal efficiency decreased slightly, and the exhaust temperature increased with an increase in the percentage of vegetable oil in the fuel. Exhaust emission at equal energy input showed slightly higher NO_x for vegetable oils and their blends. Unburned hydrocarbon emission was about 50 percent higher than pure diesel fuel because the injection system was not optimized for more viscous fuels. Ziejewski et al. [16] reported the results of an endurance test using a 25–75 percent blend of alkali-refined sunflower oil with diesel and 25–75 percent blend of safflower oil with diesel on a volume basis. The major problems experienced were premature injection, determination of nozzle performance, and heavier carbon deposits in the piston ring grooves. There was no significant problem with engine operation when the blend of safflower oil was used. That investigation revealed that chemical differences between vegetable oil and diesel had a very important influence on long-term engine performance. Bhattacharya et al. [17] have reported that a blend of 50 percent rice bran oil with diesel could be a supplementary fuel for their 10-bhp CI engine. No significant difference in the brake thermal efficiency was reported.

Samson et al. [18] have reported the use of tallow and stillingia oil in 25–75 percent and 50–50 percent blends by mass with diesel. The fuel properties of the blends were found to be within the limits proposed for diesel. The heat of combustion appeared to decrease. Specific gravity and kinematic viscosity increased with the increase in concentration of oil. Dunn et al. [19] conducted the test on rubber seed oil blended with diesel in 25 percent, 50 percent, 75 percent, and 100 percent in an air-cooled engine with 4.9 kW at 3600 rpm. Higher specific fuel consumption and slightly higher thermal efficiency were observed. But, carbon deposits were heavier than those for pure diesel fuel.

Samga [20] conducted a test on a water-cooled single-cylinder diesel engine, using hone oil (ken seed oil). He concluded that hone oil gave acceptable performance, smooth running, and ease in starting without preheating. The exhaust temperature and specific fuel consumption were higher than those for diesel. The partial-load efficiency was lower, but full-load efficiency was better than with diesel fuel.

Auld et al. [21] evaluated the potential yield and fuel qualities of winter rape, safflower, and sunflower as sources of fuel for diesel engines. Vegetable oils contained 94–95 percent heat value of diesel fuel, but were 11.1–17.6 times more viscous and also 7–9 percent heavier than diesel fuel. Viscosities of vegetable oils were closely related to fatty acid chain length and number of unsaturated bonds. During short-term engine tests, all vegetable oils produced power comparable to that of diesel, and the thermal efficiency was 1.8–2.8 percent higher than that of diesel. Based on the results, they concluded that vegetable oil as fuel should be selected on identification of the crop species that produced the most optimum yield of fuel quality vegetable oils.

Ryan et al. [22] have tested four different types of vegetable oils (soybean, sunflower, cottonseed, and peanut) in at least three different stages of processing. All the oils were characterized according to their physical and chemical properties. The spray characteristics of oils were determined at different fuel temperatures, using a high-pressure, high-temperature injection bomb, and high-speed motion picture camera. The injection study pointed out that vegetable oil behaved differently from diesel fuel. Normally, as the viscosity decreased, the penetration rate decreased and the spray cone angle increased. Using vegetable oils, however, increased the penetration rate, and increasing the temperature of the oil from 45°C to 145°C reduced the cone angle and decreased the viscosity. Engine test results, based on the specific energy, showed that degummed soybean performed as well as the base fuel, but performance of the deodorized sunflower was the worst of those tested with an energy consumption 10 percent higher than the base fuel. Vegetable oils had a much smaller premixed combustion stage, with the diffused stage of combustion being flatter for sunflower and soybean oil than for the diesel fuel. Engine inspection showed that heating of the oil reduced the carbon deposit problem. It was concluded that deposits and overall durability were related to viscosity differences and the chemical structures of the other oil as compared to diesel fuel.

Mathur and Das [23] have conducted tests on diesel engines, using blends of mahua and neem oil with diesel. Results showed that neem oil could be substituted for up to 35 percent with marginal reduction in efficiency and power output. Mahua oil with diesel had exhaust characteristics similar to those of diesel. Further, savings in the diesel fuel through the use of both these nonedible oils outweighed the demerits of a marginal drop in efficiency and a slight loss in power output.

Goering et al. [24] conducted tests on a diesel engine using a hybrid fuel formed by micro-emulsion of aqueous ethanol in soybean oil. The test data were compared with the data from a baseline test on diesel fuel. The nonionic emulsion produced the same power as diesel fuel, with 19 percent lower heating value. Brake specific fuel consumption (BSFC) was 16 percent higher, and the brake thermal efficiency was 6 percent higher, with diesel at full power. Diesel knock for the hybrid fuel was not worse than for diesel fuel; thus the low cetane number of the hybrid fuel was not reflected in engine performance. Hybrid fuels were less volatile than ethanol and thus safer. The effect of hybrid fuel on the engine durability was unknown.

6.2 Processing of Vegetable Oils to Biodiesel

Different techniques adopted for converting vegetable oils to biodiesel are (a) degumming of vegetable oils, (b) transesterification by acid or alkali, and (c) enzymatic transesterification.

6.2.1 Degumming of vegetable oils

Degumming is an economical chemical process involving acid treatment to improve the viscosity and cetane number up to a certain limit so that the blends of nonedible oils with diesel can be used satisfactorily in a diesel engine. It is a very simple process by which the gum of the vegetable oil is removed to decrease the viscosity of oil by using an appropriate acid that can be optimized for reduction in viscosity. The quantity of acid and the duration of the process are very important to obtain optimum results. Compared to transesterification, the process of degumming is simple, very easy, and less costly, and the reduction in viscosity of vegetable oil is very small.

Nag et al. [25] degummed karanja, putranjiva, and jatropha oils by phosphoric acid treatment. Before degumming the oils, the fuel properties of three oils have been measured and compared with diesel (Table 6.1). Acid concentrations of 1 percent, 2 percent, 3 percent, 4 percent, and 5 percent were used at 40°C with vigorous stirring. The stirring was continued for 10 min after adding the acid. After stirring, the mixtures were held for 1 week to complete the reactions and to settle the gum materials. Then the mixtures were filtered through a packed bed filled with charred sawdust. Viscosities of the filtrate were then measured.

Performance and emission measurement. After studying the properties of the jatropha, karanja, and putranjiva oils, they were degummed. In this context, the Ricardo variable-compression engine (Ricardo & Co. Engineers Ltd., England, single cylinder, 3-in bore, 35/8 in stroke) was run with 10 percent, 20 percent, 30 percent, and 40 percent blends of degummed karanja, jatropha, and putranjiva oils with diesel at different loads (0–2.7 kW) and different timings (45°, 40°, 35°, and 30° bTDC [before top dead center]). To measure emissions, an automotive exhaust monitor (model PEA205) and smoke meter (model OMS103, Indus Scientific Pvt. Ltd., India) were used.

Properties	Karanja	Jatropha	Putranjiva	Diesel
Viscosity in cSt (at 40°C)	43.67	35.38	37.62	5.032
Cetane number	29.9	33.7	31.3	46.3
Calorific value (kJ/kg)	36,258	38,833	39,582	42,707
Pour point (°C)	5	2	−3	−12
Specific gravity at 25°C	0.932	0.916	0.918	0.834
Flash point (°C)	215	280	48	78
Fire point (°C)	235	291	53	85
Carbon residue (%)	1.4	0.2	0.9	0.1

TABLE 6.1 Fuel Properties of Three Nonedible Oils and Diesel

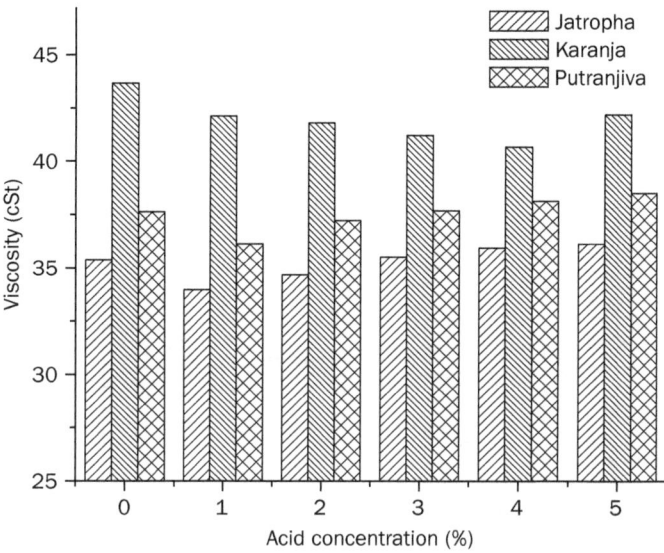

FIGURE 6.1   Viscosity versus acid concentration of jatropha, karanja, and putranjiva oils at 40°C.

Degumming by acid treatment lowers the viscosity. Viscosities of karanja, jatropha, and putranjiva oils degummed at 40°C and at various acid concentrations are shown in Figure 6.1. Karanja oil with 4 percent acid treatment had the lowest viscosity, whereas jatropha and putranjiva oils both had the lowest viscosities with 1% acid treatment.

Effect of timing.   By observing the performance data at various timings (45°, 40°, and 35° bTDC) in Figure 6.2, it was concluded that at 45° bTDC timing, the nonedible karanja, jatropha, and putranjiva oils gave the highest yields, whereas at 40° bTDC timing, diesel gave the highest yield. That may have been due to the different ignition temperatures of the nonedible oils from diesel.

Performance of various blends.   Performances of blends of degummed vegetable oil with diesel are shown in Figures 6.3 and 6.4. The 20 percent blends of jatropha, karanja, and putranjiva oils with diesel gave quite satisfactory performance related to BSFC and brake thermal efficiency (hbt). Beyond the 20 percent blends, the cetane numbers and viscosities of the blends were not so effective.

Comparison of the performance of blends.   As per Figures 6.5 and 6.6, engine performance using jatropha and karanja oils was better than diesel but the use of putranjiva oil gave reverse results at all loads, although the results were more or less the same. Degummed karanja

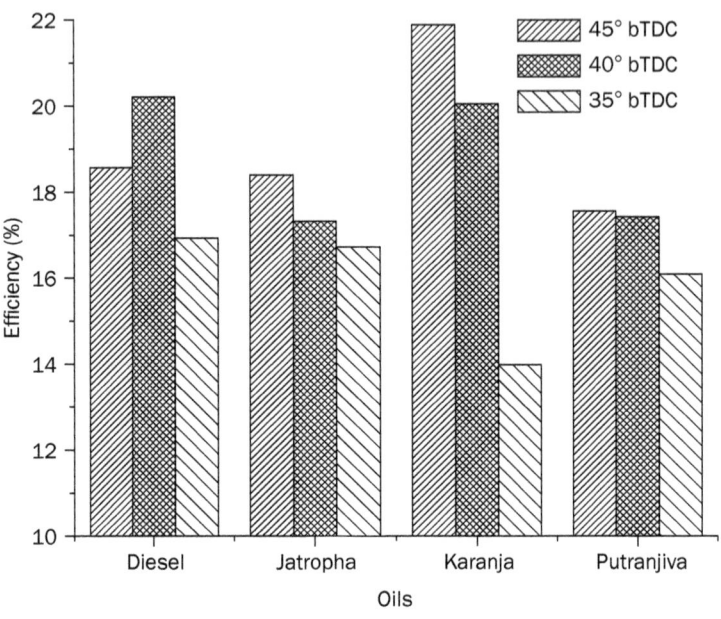

Figure 6.2 Brake thermal efficiency at various timings of diesel and 20 percent vegetable oil blends at 1-kW brake power, 1200 rpm, and 20 compression ratio.

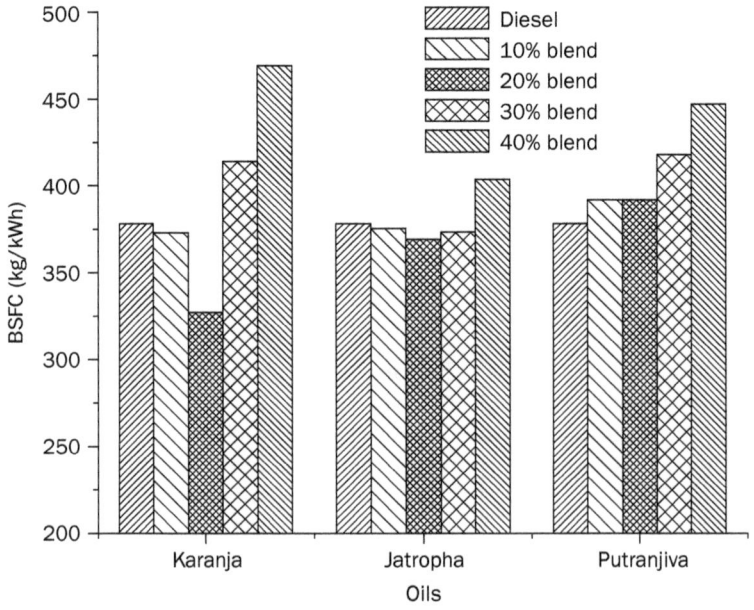

Figure 6.3 Brake specific fuel consumption versus vegetable oils and diesel blends at 1200 rpm, 45° bTDC, 20 compression ratio, and 1.4-kW brake power.

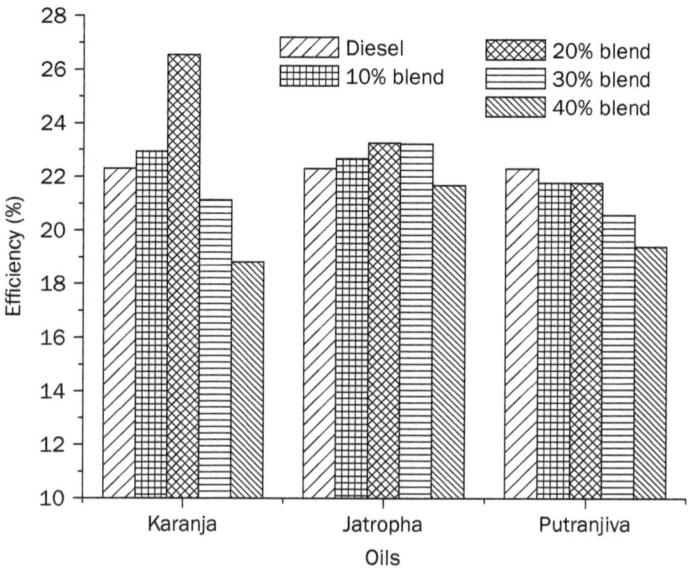

Figure 6.4 Brake thermal efficiency versus brake horsepower of vegetable oil and diesel blends at 1200 rpm, 45° bTDC, 20 compression ratio, and 1.4-kW brake power.

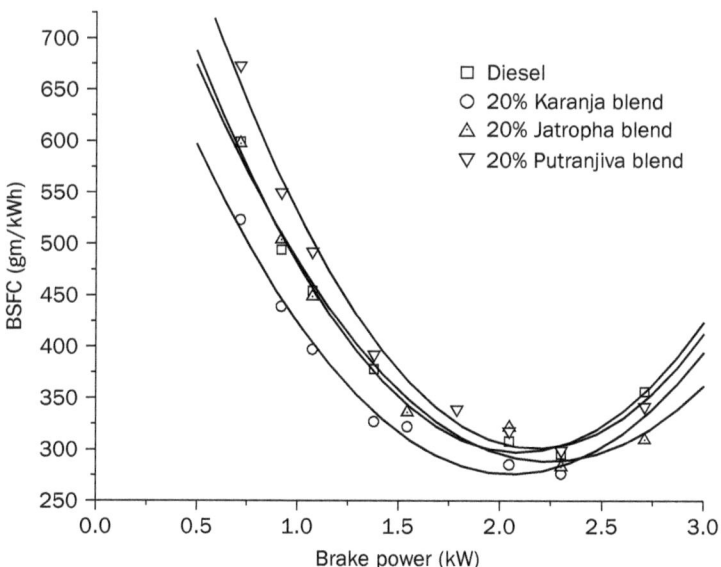

Figure 6.5 Brake specific fuel consumption versus brake power of diesel, 20 percent karanja oil, jatropha oil, and putranjiva oil blends at 1200 rpm, 45° bTDC, and 20 compression ratios.

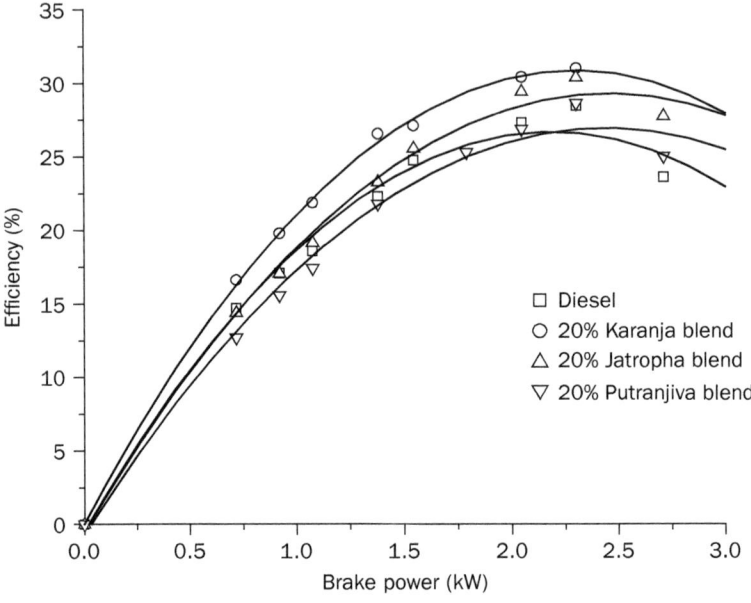

FIGURE 6.6 Brake thermal efficiency versus brake power of diesel, 20 percent karanja oil, 20 percent jatropha oil, and 20 percent putranjiva oil blends at 1200 rpm, 45° bTDC, and 20 compression ratio.

oil blends gave better performance, but at high loads, the performance of jatropha oil blends was better in comparison to the performance of karanja oil blends. The performance data showed that all three vegetable oils could be used as alternative fuels for diesel engines.

Effect of loads on emissions of vegetable oil blends and comparison. As per Figures 6.7 and 6.8, it is interesting to note that for the karanja, jatropha, and putranjiva oils, in every case, smoke and particulates decreased, which was very favorable in terms of their environmental impact on human beings. The rate of increase in smoke and particulate generation with the load of jatropha oil, in comparison to karanja and putranjiva oils, was very low. It is very interesting to observe that although the particulates and smoke for all the oils decreased, jatropha oil blends gave the highest reduction.

In Figures 6.9 and 6.10, the CO, CO_2, NO_x, and HC (hydrocarbon) emissions for the three nonedible oils were less in comparison to diesel at high loads. However, at low loads, emissions from the nonedible oils are almost parallel to diesel. Because of the higher ignition temperature of nonedible oils than diesel, the better combustion of these oils gave less exhaust emissions.

Thus, degumming is an economic chemical process for a 20 percent blend of karanja, jatropha, and putranjiva oils with diesel to have

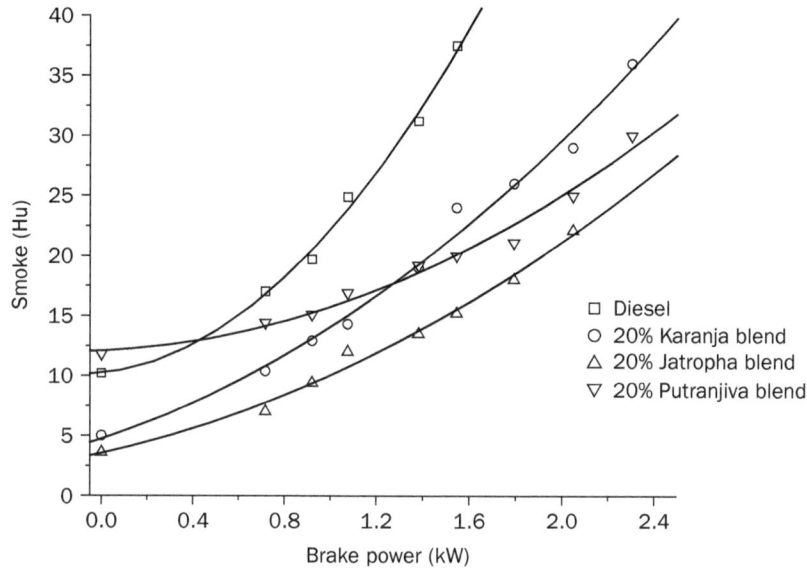

Figure 6.7 Smoke versus brake power of diesel, 20 percent karanja oil, 20 percent jatropha oil, and 20 percent putranjiva oil blends at 1200 rpm, 45° bTDC, and 20 compression ratio.

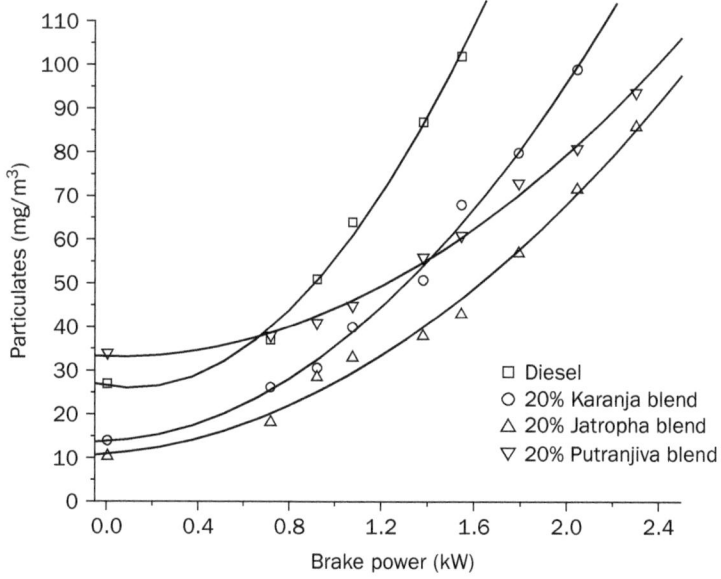

Figure 6.8 Particulates versus brake power of diesel, 20 percent karanja oil, 20 percent jatropha oil, and 20 percent putranjiva oil blends at 1200 rpm, 45° bTDC, and 20 compression ratio.

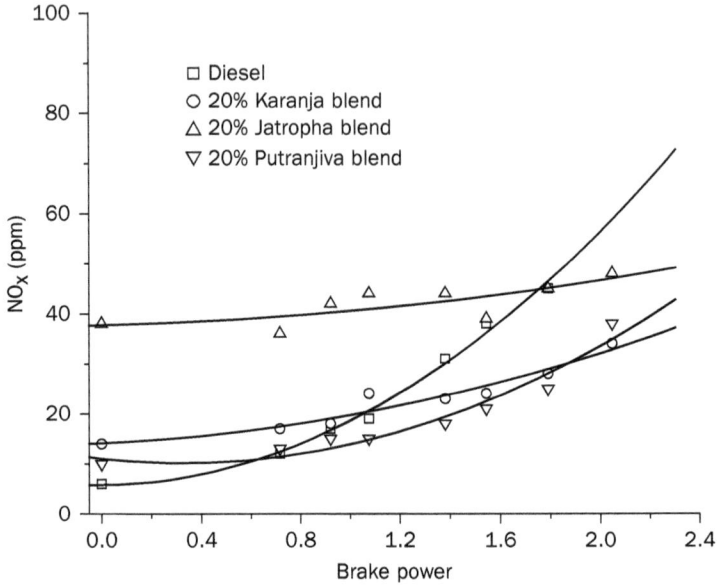

Figure 6.9 Nitrogen oxide versus brake power of diesel, 20 percent karanja oil, 20 percent jatropha oil, and 20 percent putranjiva oil blends at 1200 rpm, 45° bTDC, and 20 compression ratio.

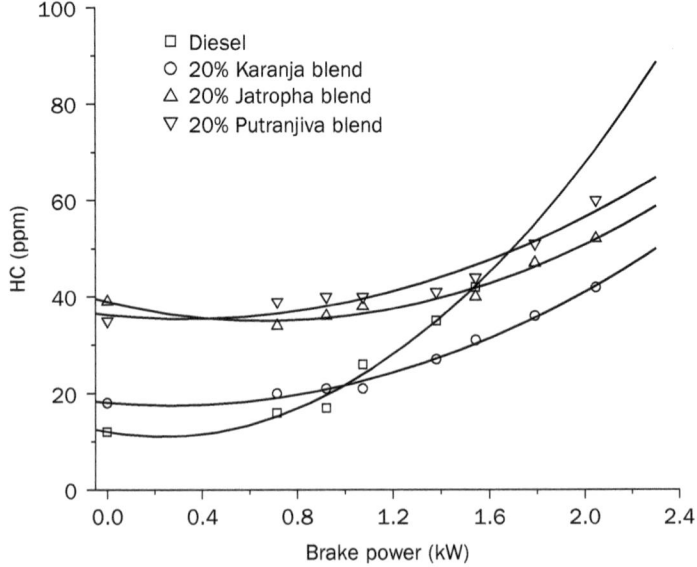

Figure 6.10 Unburnt hydrocarbon versus brake power of diesel, 20 percent karanja oil, 20 percent jatropha oil, and 20 percent putranjiva oil blends at 1200 rpm, 45° bTDC, and 20 compression ratio.

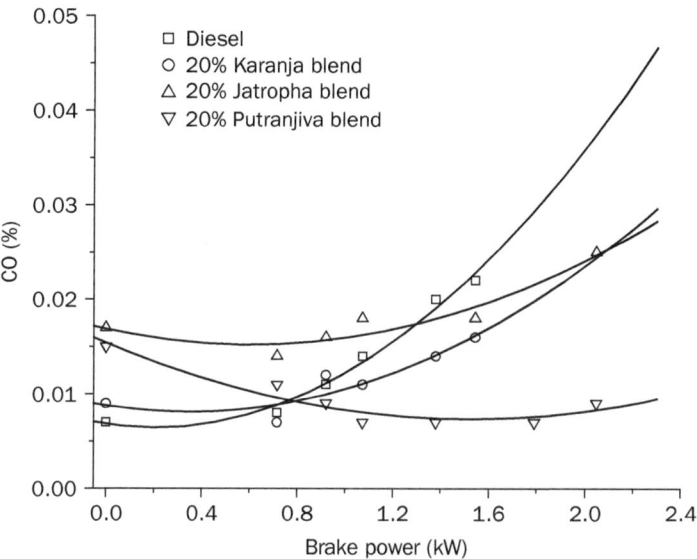

FIGURE 6.11 Carbon monoxide versus brake power of diesel, 20 percent karanja oil, 20 percent jatropha oil, and 20 percent putranjiva oil blends at 1200 rpm, 45° bTDC, and 20 compression ratio.

very satisfactory results. The degumming method, therefore, offers a potential low-cost method with simple technology for producing an alternative fuel for CI engines. Out of the three nonedible oils, jatropha oil was the most promising to yield good performance and emissions at high loads in all respects. Comparing CO, CO_2, NO_x, HC, smoke, and particulate emissions from using the three nonedible oils, jatropha oil was very encouraging (see Fig. 6.11). Considering the above-mentioned points, it can be concluded that the diesel engine can be run very satisfactorily using a 20 percent blend of vegetable oil with diesel at 45° bTDC, 1200 rpm, and 20 compression ratios. Any diesel engine can be operated with a 20 percent blend of degummed vegetable oils as a prime mover for agriculture purposes without any modification of the engine.

6.2.2 Transesterification of vegetable oils by acid or alkali

Goering et al. [24] have suggested that vegetable oils are too viscous for prolonged use in direct-injected diesel engines, which has led to poor fuel atomization and inefficient mixing with air, contributing to incomplete combustion. These chemical and physical properties caused vegetable oils to accumulate and remain as charred deposits when they contacted engine cylinder walls. The problem of charring and deposits of oils on the injector and cylinder wall can be overcome by better esterification of the oil to reduce the viscosity and remove glycerol.

Acid-catalyzed alcoholysis of triglycerides (TG) can be used to produce alkyl esters for a variety of traditional applications and for potentially large markets in the biodiesel fuel industry [26]. It can overcome some of the shortcomings of traditional base catalysis for producing alkyl esters. A significant disadvantage of base catalysts is their inability to esterify free fatty acids (FFA). These FFA are present at about 0.3 wt% in refined soybean oil and at significantly higher concentrations in waste greases, due to hydrolysis of the oil with water to produce FFA. The FFA react with soluble bases to form soaps through the saponification reaction mechanism. The soap forms emulsions and makes recovery of methyl esters (ME) difficult. Saponification consumes the base catalyst and reduces product yields. The use of alkaline catalysts requires that the oil reagent be dry and contain less than about 0.3 wt% FFA [27, 28].

Acid catalysts can handle large amounts of FFA and are commonly used to esterify FFA in fat or oil feedstock prior to base-catalyzed FFA alcoholysis to ME [29]. Though it solves FFA problems, it adds additional reaction and cleanup steps that increase batch times, catalyst cost, and waste generation.

Generally, acid-catalyzed methanolysis of TG is carried out at temperatures at or below that of methanol reflux (65°C). Using sulfuric acid catalysis under reflux conditions, Harrington and D'Arcy-Evans [30] first explored the feasibility of in situ transesterification, using homogenized whole sunflower seeds as a substrate. Using reflux conditions, a 560-fold molar excess of methanol and a 12-fold molar excess of sulfuric acid relative to the number of moles of triacylglycerol (TAG) were used. They observed ester production, with yields up to 20 percent greater than in the transesterification of preextracted oil, and suggested that this was an effect of the water content of the seeds, an increased extractability of some seed lipids under acidic conditions, and also the transesterification of seed-hull lipids.

Stern et al. [31] have developed a process to prepare ethyl esters for use as a diesel fuel substitute from various vegetable oils using hydrated ethyl alcohol and crude vegetable oil, with sulfuric acid as a catalyst. Ethyl ester of 98 percent purity with a very low acidity has been reported.

Schwab et al. [32] have compared acid and base catalysts and confirmed that, although base catalysts performed well at lower temperatures, acid catalysis requires higher temperatures. Liu [33] has compared the influence of acid and base catalysts on yield and purity of the product, and suggested that an acid catalyst is more effective for alcoholysis if the vegetable oil contains more than 1 percent FFA.

Goff et al. [34] have conducted acid-catalyzed alcoholysis of soybean oil using sulfuric, hydrochloric, formic, acetic, and nitric acids, which were evaluated at 0.1 and 1 wt% loadings at temperatures of 100ᵧC and 120ᵧC in sealed ampoules, and observed sulfuric acid was effective. Kinetic studies at 100ᵧC with 0.5 wt% sulfuric acid catalyst

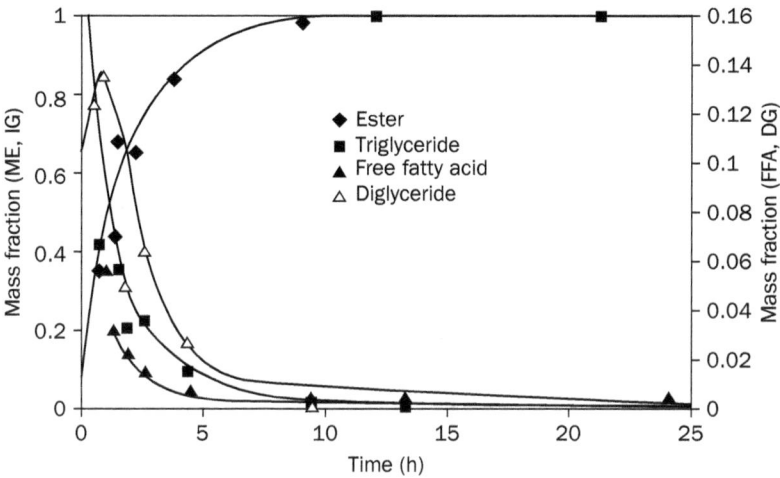

FIGURE 6.12   Kinetics of 0.5 wt% sulfuric acid catalyst at 100°C and 9:1 methanol-TG molar ratio. (*Used with permission from Goff et al. [34].*)

and 9 times methanol stoichiometry provided more than 99 wt% conversion of TG in 8 h, and with less than 0.8 wt% FFA concentration in less than 4 h (see Fig. 6.12).

Base catalysts are generally preferred to acid catalysts because they lead to faster reactions [35]. Base catalysts generally used in transesterification reactions are NaOH, KOH, and their alkoxides. KOH is preferred to other bases because the end reaction mixture can be neutralized with phosphoric acid, which produces potassium phosphate, a well-known fertilizer [36].

Darnoko et al. [37] explained transesterification of palm oil with methanol and KOH as a catalyst by the following three-step reaction sequence:

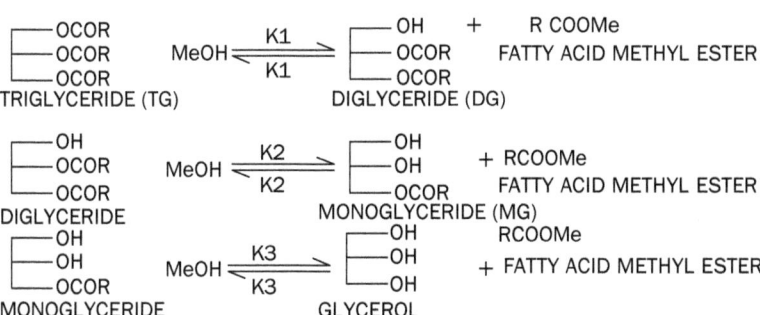

Knothe et al. [38] have reported optimal conditions of a 1 wt% KOH catalyst at 69°C and 7:1 alcohol—vegetable oil molar ratio gave 97.7 percent conversions in 18 min, when KOH was used with high-purity feedstocks.

Freedman et al. [39] have studied transesterification of sunflower oil and soybean oil with the reaction variables (a) molar ratio of alcohol to vegetable oil, (b) type of alcohol (methanol, ethanol, and t-butanol), (c) type of catalyst (acidic and alkali), and (d) reaction temperature (60°C, 45°C, and 32°C). They have suggested that esterification was 90 to 98 percent completed at the respective molar ratio of methanol to sunflower oil 4:1 and 6:1. All three alcohols produced high yields of esters. Alkaline catalysts were generally much more effective than acid catalysts. The reaction was performed successfully at both 45°C and 60°C in 4 h, with the production of 97 percent of ME.

Kruclen et al. [40] have presented a process for conversion of a high-melting point palm oil fraction into ethyl esters, which could be used as a diesel fuel substitute. The amount of catalyst used (KOH) was 0.1–1 percent, and the reaction was completed rapidly at 80°C with yields of 80–94 percent, depending on the concentration of catalysts. The specific gravity of ethyl ester varied from 0.847 to 0.864 with kinematic viscosity of 4.4–4.6 cSt at 40°C.

Gelbard et al. [41] have determined the yield of transesterification of rapeseed oil with methanol and base by ^{1}H-NMR (nuclear magnetic resonance) spectroscopy. The relevant signals chosen for integration are those of methoxy groups in ME at 3.7 ppm (parts per million) (singlet) and of the α-carbonyl methylene groups present in all fatty ester derivatives at 2.3 ppm. The latter appears as a triplet, so accurate measurements require good separation of this multiple at 2.1 ppm, which is related to allylic protons.

Chadha et al. [42] have studied base-catalyzed transesterification of monoglycerides from pongamia oil. They separated monoglyceride fractions (MG) by column chromatography and then characterized the fractions by ^{1}H-NMR spectroscopy in deuterated chloroform (CDCl3) and tetramethylsilane (TMS) (see Fig. 6.13). They explain that 1- or 2-MG are positional isomers. Consequently, in 1-MG, the methylene protons at C-1 and C-3 are magnetically nonequivalent, due to four double doublets, which are observed in the spectra. But 2-MG, on the other hand, are symmetrical, and C-1 and C-3 methylene protons are magnetically equivalent and appear as a multiplate.

6.2.3 Enzymatic transesterification of vegetable oils

Enzymatic transesterification of TG by lipases (3.1.1.3) is a good alternative over a chemical process due to its eco-friendly, selective nature and low temperature requirement. Lipases break down the TAG into FFA and glycerol that exhibits maximum activity at the oil–water

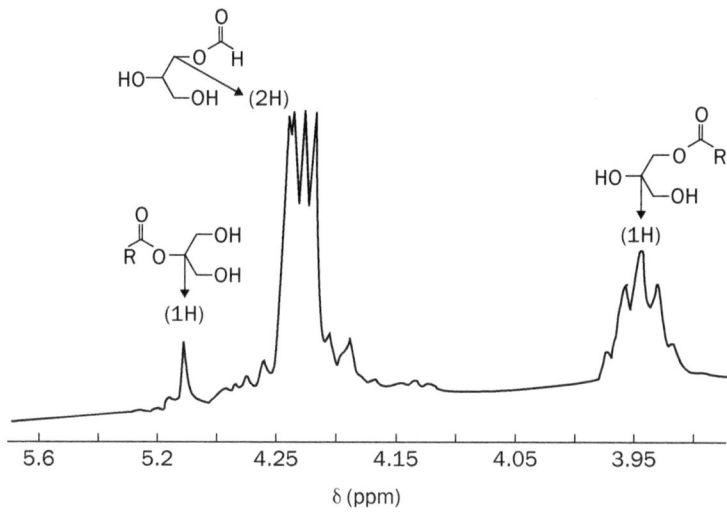

FIGURE 6.13 Characteristic ^{1}H-NMR signals of 1- and 2-MG. (*Used with permission from Chadha [42].*)

interface. Under low-water conditions, the hydrolysis reaction is reversible, i.e., the ester bond is synthesized rather than hydrolyzed. Scientists are interested in the development of lipase applications to the interesterification reactions of vegetable oils for production of biodiesel.

Nag has reported [43] celite-immobilized commercial *Candida rugosa* lipase and its isoenzyme lipase 4 efficiently catalyzed alcoholysis (dry ethanol) of various TG and soybean oil (see Fig. 6.14). This process has many advantages over chemical processes such as (a) low reaction temperature, (b) no restriction on organic solvents, (c) substrate specificity on enzymatic reactions, (d) efficient reactivity requiring only the mixing of the reactants, and (e) easy separation of the product.

Kaieda et al. have developed [44] a solvent-free method for methanolysis of soybean oil using Rhizopus oryzae lipase in the presence of 4–30 wt% water in the starting materials. Oda et al. [45] have reported methanolysis of the same oil using whole-cell biocatalyst, where *R. oryzae* cells were immobilized within porous biomass support particles (BSP). Köse et al. [46] have reported the lipase-catalyzed synthesis of alkyl esters of fatty acids from refined cottonseed oil using primary and secondary alcohols in the presence of an immobilized enzyme from *C. antarctica*, commercially called Novozym-435 in a solvent-free medium. Under the same conditions, with short-chain primary and secondary alcohols, cottonseed oil was converted into its corresponding esters.

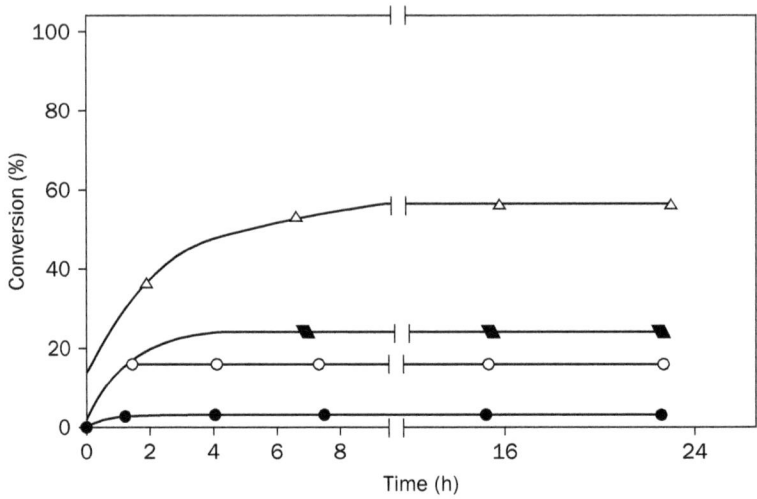

Figure **6.14**   Conversion versus reaction for ethanolysis of soybean oil catalyzed by immobilized lipase 4 at 40°C and 250 rpm. Ethyl oleate (Δ); ethyl palmitate (◆); ethyl stearate (o); ethyl linoleate (·).

Alcoholysis of soybean oil with methanol and ethanol using several lipases has been investigated. The immobilized lipase from *Pseudomonas* cepacia was the most efficient for synthesis of alkyl esters, where 67 and 65 mol% of methyl and ethyl esters, respectively, were obtained by Noureddini et al. [47]. Shimada et al. [48] have reported transesterification of waste oil with stepwise addition of methanol using immobilized *Candida antarctica* lipase, where they have successfully converted more than 90 percent of the oil to fatty acid ME. They have also implemented the same technique for ethanolysis of tuna oil.

Dossat et al. [49] have found that hexane was not a good solvent as the glycerol formed after the reaction was insoluble in n-hexane and adsorbed onto the enzyme, leading to a drastic decrease in enzymatic activity. Enzymatic transesterification of cottonseed oil has been studied using immobilized *Candida antarctica* lipase as catalyst in t-butanol solvent by Royon et al. [50].

Sometimes, gums present in the oils used inhibit alcoholysis reactions due to interference in the interaction of the lipase molecule with substrates by the phospholipids present in the oil gum. Crude soybean oil cannot be transesterified by immobilized *Candida antarctica* lipase. So, Watanabe et al. [51] have used degummed oil as a substrate for a transesterification reaction, in order to minimize this problem, and have effectively achieved conversion of 93.8 percent oil to biodiesel.

Methanol is insoluble in the oil, so it inhibits the lipases, thereby decreasing its catalytic activity toward the transesterification reaction.

Du et al. [52] transesterified soybean oil using methyl acetate in the presence of Novozym-435 (see Fig. 6.15). Further, glycerol was also insoluble in the oil and adsorbed easily onto the surface of the immobilized lipase, leading to a negative effect on lipase activity. They have suggested that methyl acetate was a novel acceptor for biodiesel production and no glycerol was produced in that process, as shown below:

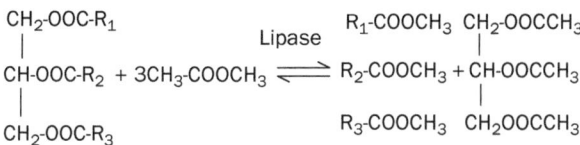

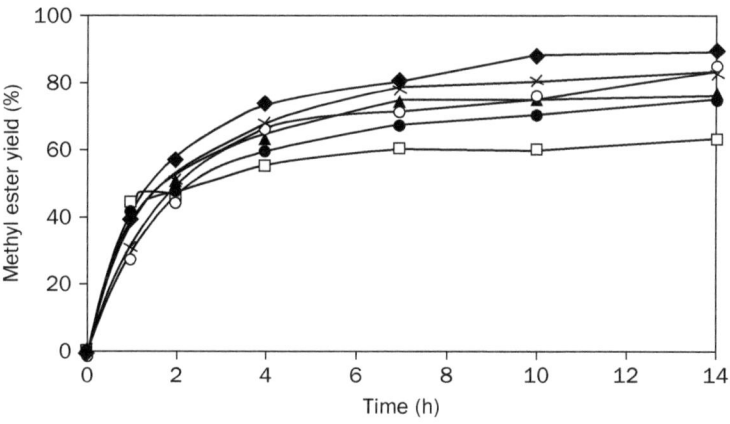

FIGURE 6.15   Effect of the substrate ratio of methyl acetate to oil as biodiesel production. Reaction conditions: 40°C; 150 ppm; 30 percent Novozym-435, based on oil–methyl acetate molar ratio 6:1 (□), 8:1 (·), 10:1 (▲), 12:1 (◆), and 24:1 (s). (*Used with permission from Du et al. [52].*)

They found 92 percent yield with a methyl acetate–oil molar ratio of 12:1, and methyl acetate showed no negative effect on enzyme activity.

The comparison of biodiesel production by acid, alkali, and enzyme is given in Table 6.2.

6.2.4   Engine performance with esters of vegetable oil

Hawkins et al. [53] have conducted combustion studies on methyl and ethyl esters of degummed sunflower oil, maize oil, cottonseed oil, peanut oil, soybean oil, and castor oil. Fuel properties of the esters were very similar to each other, except the esters of castor oil which were much more viscous. The heating values of ethyl esters were also

Acid-catalyzed transesterification	Base-catalyzed transesterification	Enzymatic Transesterification
The glycerides and alcohol need not be anhydrous	The glycerides and alcohol must be substantially anhydrous	The alcohol needs to be anhydrous
Does not make a soap-like product Easy to water wash for product separation	Soap formation taking place during the reaction Water washing is difficult due to soap emulsifier	Does not make a soap-like product Separation of the product is very easy; product is obtained only by filtration
Recommended for any free-fatty acid content of vegetable oil Converts free fatty acid to ester Product yield is high	Recommended for low fatty acid content of vegetable oil Converts free-fatty acid to soap Product yield is comparatively low	Recommended for any free-fatty acid content of vegetable oil Converts free-fatty acid to ester Product yield depends on different types of enzymes used; reaction is selective
Is slower than alkali-catalyzed transesterification Percentage of conversion is low Under low water content in reactants (oil or alcohol), transesterification reaction is not hindered	Is faster comparatively Percentage of conversion is high Water content in the reactant inhibits the reaction rate	Reaction is slower than acid and alkali reactions Percentage of conversion is high Under low water conditions, the hydrolysis reaction is reversible, i.e., the ester bond is synthesized rather than hydrolyzed. Lipases break down the triacylglycerols into free fatty acids and glycerol that exhibit maximum activity at the oil—water interface
Need high temperature	Reacts even at room temperature	Conversion takes place at low temperature

TABLE 6.2 Comparison of Biodiesel Production by Acid, Alkali, and Enzyme

considerably lower. Engine results indicated that the power output for esters varied from 44.4 to 45.5 kW, with diesel delivering 45.1 kW. The brake thermal efficiencies were also slightly higher than diesel. High esterification yields (around 90 percent) must be obtained to avoid choking of injector tips. Further, sticking of injector needles after a shutdown time of 48 h has been reported.

Fort and Blumberg [54] have tested a diesel engine with a mixture of cottonseed oil and ME of this oil. Results indicate that viscosity and density increased whereas the heating value and the cetane number decreased, when the percentage of the cottonseed oil was increased in the blend. The durability test with 50–50 percent cottonseed oil and ME was terminated after 183 h of running the engine, because the engine was noisy. After disassembly, the engine indicated severe wear scouring and heavy carbon deposits. But specific emissions and visible smoke characteristics of diesel fuel and esterified cottonseed oil were comparable.

Ziejewski and Kaufman [55] conducted a long-term test using a 25–75 percent blend of alkali-refined sunflower oil and diesel fuel in a diesel engine, and compared the results with that of a baseline test on diesel fuel. Engine power output over the tested speed range was slightly higher for this blend. At 2300 rpm, the difference was 25 percent. At 1800 rpm, the gain in power was 6 percent. The smoke level increased at a higher engine speed from 1 to 2.2 and decreased at a lower engine speed. Greater exhaust temperature was caused by a higher intake air temperature. The major problems experienced were:

1. Abnormal carbon buildup in the injection nozzle tips.
2. Injector needle sticking.
3. Secondary injection.
4. Carbon buildup in the intake port and exhaust-valve stems.
5. Carbon filling of the compression ring grooves.
6. Abnormal lacquer and varnish buildup.

Tahir [56] has determined the fuel properties of sunflower oil and its ME. The properties were favorable for diesel engine operation, but the problem of high viscosity (14 times higher than diesel at 37°C) of sunflower oil might cause blockage of fuel filters, higher valve-opening pressure, and poor atomization in the combustion chamber. Transesterification of sunflower oil to its ME has been suggested to reduce viscosity of the fuel. The viscosity of ME at 0°C was closer to that of No. 2 diesel fuel, but below 0ᵧC, it was not possible because of the pour point of −4°C.

Pryor et al. [57] have conducted a short-term performance test on a small, test diesel engine using crude soybean oil, crude degummed soybean oil, and soybean ethyl ester. The engine developed about 3 percent more power output with crude legume soybean oil, but the development was insignificant with soybean ethyl ester. The fuel flow of soybean oil was 13–30 percent higher and for the ethyl ester it was 11–15 percent higher, depending upon the load on the engine. The exhaust temperature throughout the test was 2–5 percent higher for soybean oil and 2–3 percent lower for ethyl ester than the diesel fuel.

Clark et al. [58] have tested methyl and ethyl esters of soybean oil as a fuel in a CI engine. Esters of soybean oil with commercial diesel fuel additives revealed fuel properties comparable to diesel fuel, with the exception of gum formation, which manifested itself in problems with the plugging of fuel filters. Engine performance with esters differed little from the diesel fuel performance. Emissions of nitrous oxides for the esters were similar, or slightly higher than diesel fuel. Measurement of engine wear and the fuel injection test showed no abnormal characteristics for any of the fuels after 200 h of testing.

Laforgia et al. [59] has prepared biodiesel from degummed vegetable oil with 99.5 percent methanol and an alkaline catalyst (KOH). On engine performance, pure biodiesel and blends of biodiesel combined with 10 percent methanol had a remarkable reduction in smoke emissions. When the injection timing was advanced, better results were obtained.

Pischinger et al. [60] have conducted engine and vehicle tests with ME of soybean oil (MESO) 75–25 percent gas oil–MESO blend and 68–23–9 percent gas oil–MESO–ethanol blend. The fuel properties of the blend indicated a 6 percent lower volumetric calorific value of the ester, a drastic reduction in kinematics viscosity, and a greater ethane number than that of gas oil. The engine results indicated about 7 percent higher BSFC with a marginal difference in power and torque in comparison with gas oil. The smoke emission was much lower with ME.

Ali et al. [61] have observed that engine performance with diesel fuel–methyl soyate blends did not differ to a great extent up to a 70–30 percent (v/v) from that of diesel-fueled engine performance. There was a slight increase in NOx emissions with increasing methyl soyate content in the blends at higher speeds but at lower speeds there was a quadratic trend with diesel fuel content.

Carbon monoxide emissions were very similar for blends up to 70–30 percent (v/v) diesel fuel–methyl soyate blends at any speed. Visible smoke decreased with increasing speed and methyl soyate content. More smoke was produced with neat diesel fuel at full load.

6.3 Engine Performance with Esters of Tallow and Frying Oil

The estimated amount of good quality and nutritive-value oils and fats used for frying around the world is around 20 million metric tons (MT). In frying, the hot oil serves as a heat exchange medium by which heat is transferred to the material being fried. As a result of frying, the oil darkens from the formation of polar materials such as minor phenolic components; elevated FFA; high total polar materials; compounds having high foaming property, low smoke point, low iodine value, and increased viscosity; and color compounds.

Sims [62] reported has conversion of tallow, a by-product of the meat industry, into esters. The fuel properties of methyl, ethyl, and butyl esters of tallow were similar to diesel fuel, particularly ME, which were remarkably similar except for the higher liquidification temperature of tallow esters. Short-term engine performance tests with methyl, ethyl, and butyl esters gave comparable results as diesel fuel, but at higher BSFC. Blends with diesel in 50–50 percent proportion by volume gave intermediate results between esters and neat diesel fuel.

Richardson et al. [63] have tested an engine with ME of tallow. Preliminary engine tests indicated that the use of 10 percent and 20 percent blends (volume basis) performed similar to diesel fuel. However, lubricant quality aspects were not studied and an endurance test was not conducted. The ignition quality of the blend was significantly better than that of diesel. Overall, it was concluded that tallow ME on 10 percent (volume basis) can be successfully used as diesel fuel where large amounts of tallow are produced and temperatures below 10ʏC are not encountered. The fuel consumption of ME of used frying oil has been measured by Mittelbach and Tritthard [64]. The ester fuel showed slightly lower hydrocarbon and carbon monoxide emissions but increased oxides of nitrogen, compared with that of diesel fuel. The particulate emissions, however, were significantly lower for used frying oil. But, they suggest long-term engine testing to prove the quality of this fuel.

The results discussed contribute to a better understanding of the structure–physical property relationships in different fatty acid esters from different vegetable oils which give the desired biodiesel quality and optimal performance of engines.

6.4   Otto Cycle Engine Performances by Biodiesel

An Otto cycle is a typical spark ignition piston engine commonly found in automobile engines that is subjected to changes of pressure, temperature, volume, addition of heat, and removal of heat. The system functions as the work added to the system, plus the heat added, minus the heat removed, yielding the net mechanical work generated by the system. The four-stroke engine is also referred to as the Otto cycle engine after its inventor N. A. Otto. Most cars use the four-stroke engine. An individual cycle comprises four strokes: 1, intake stroke; 2, compression stroke; 3, power stroke; and 4, exhaust stroke. These four strokes repeat to generate the crankshaft revolution. During one cycle, the piston makes two-round trips and the crankshaft revolves twice. The inlet and exhaust valves open and close only once. The ignition plug also sparks only once. Biodiesel can be used at 100 percent (B100) or in blends with petroleum diesel fuel. Blends are indicated by B##, which correspond to the percentage of biodiesel in the blended fuel. For example, a 20 percent blend of biodiesel with 80 percent diesel fuel is called B20.

Performances of Otto engine by biodiesel have been tested by several scientists. Torres et al. (2006) tested a mono cylinder stationary Agrale engine, model M-85 with 7.36 kW power, and detected no significant differences regarding the use of diesel oil and biodiesel (B100) prepared from olive oil with close results. The hourly engine fuel consumption was determined by an Oval flow meter, model LSN41, together with its display, that was calibrated for reading in liters per hour and 0.84 ($g.cm^{-3}$) fluid specific mass. The specific fuel consumption with B100 was approximately 20 percent greater than with diesel oil, the CO_2 emissions were practically unaltered, but the CO emissions were much greater for the 4 kW power. The tests performed showed the feasibility of operating a diesel cycle engine with olive biofuel (B100). There was gain in torque with the use of biodiesel that was 8.43 percent greater than diesel oil at the work rotation. Olive biodiesel obtained better reduced power results than fossil diesel oil, and 28.57 percent greater than diesel oil at the work rotation. Olive biodiesel presented lower specific and energetic consumption compared to diesel oil that was 12.8 percent and 30 percent, respectively, less at the work rotation. Corrêa et al. (2008) performed an experiment with sunflower biodiesel blends (B5, B10, B20, and B100) and fossil diesel in an IC engine, direct injection, with the engine performance through power take-off (PTO) for each fuel. The lubricating oil was analyzed before and after period of 96 hours with B100. The results showed that the use of blends B5, B10, B20, and B100 decreased the power of PTO maximum 2.2 percent and increased the fuel consumption maximum 7.3 percent. The analysis of lubricating oil showed that the viscosity, water content, and level of iron were the parameters most affected. Barbosa et al. (2008) evaluated the performance of an engine fueled with diesel oil and mineral mixtures with the proportion of biodiesel equivalent B2 (98 percent mineral diesel and 2 percent biodiesel), B5 (95 percent mineral diesel and 5 percent biodiesel), B20 (80 percent diesel mineral and 20 percent biodiesel) and B100 (100 percent biodiesel) and concluded that the increased engine power of the B100 to mineral diesel was in reverse order; the thermal efficiency of the diesel diminished mineral mixtures for growing biodiesel, and 4 percent lower for B100. Volpato et al. (2009) tested the performance of a cycle diesel engine using biodiesel from soy oil (B100) as compared to fossil diesel where they analyzed the effective power and reduced power, rise, specific and energy consumption of fuel, efficiency term-mechanics, and volumetric. The tests showed the viability of operation of a diesel engine with fuels substitute for soy oil (B100). Nietiedt et al. (2011) evaluated the use of different blends of soybean methyl biodiesel (B10, B50, and B100) in comparison to the commercial diesel B5, with 5 percent of biodiesel added to the fossil diesel. The engine performance was analyzed through the tractor power take-off (PTO) for each fuel, and the best performance occurred with the use of B5 and B10 fuel, without significant differences between these blends. The B100 fuel showed significant differences compared to the other fuels.

References

1. D. L. Klass. *Biomass for Renewable Energy, Fuels and Chemicals*, New York: Academic Press, 1998
2. V. V. Kafarov. *Wasteless Chemical Processing*, Moscow, Russia: Mir Publishers, 1985.
3. W. R. Nitseke and C. M. Wilson. *Rudolf Diesel: Pioneer of the Age of Power*, Norman, OK: University of Oklahoma Press, 1965.
4. D. H. Chowdhury. Indian vegetable fuel oils for diesel engines, *Gas and Oil Power*, **37**(5), 80–85, 1942.
5. R. V. Barve and P. V. Amurthe. Ground nut oil for diesel engine, *Current Science* **2**(10), 403, 1942.
6. R. Weibe and J. Nowakowska. The technical literature of agricultural motor fuels, USDA Bibliographical Bulletin No. 10, Superintendent of Documents, US Government Printing Office, Washington, DC, 1949.
7. K. S. Fang. Vegetable Oils and Diesel Fuel for China, Unpublished M.S. Thesis, University of Nebraska, Lincoln, NE.
8. H. W. Engelman, D. A. Guenther, and T. W. Silvis. Vegetable Oil as a Diesel Fuel, ASME, New York, ASME Paper No. 78-DG-P-19, November 5, 1978.
9. Z. M. Cruz, A. S. Ogunlowo, W. J. Chancellor, and J. R. Goss. Vegetable oils as fuels for diesel engines, *Resources and Conservation* **6**, 69–74, 1981.
10. E. H. Pryde. Vegetable oils as fuel alternatives, *American Society of Agricultural Engineers* **4**, 101, 1984.
11. J. J. Bruwer, B. D. Boshoff, F. J. C Hugo, L. M. DuPlessis, J. Fulsand, and C. Hawkins. Using unmodified vegetable oils as a diesel fuel extender, *American Society of Agricultural Engineers* **32**(5), 1503, 2002.
12. R. A. Baranescu and J. J. Lusco. Performance, durability and low temperature evaluation of sunflower oil as a diesel fuel extender, *American Society of Agricultural Engineers* **4**, 312, 1982.
13. L. E. Wagner and C. L. Peterson. Performance of winter rape based fuel mixtures in diesel engines, *American Society of Agricultural Engineers* **4**, 329–336, 1982.
14. S. C. Borgelt and F. D. Harris. Endurance tests using soybean oil–diesel fuel mixture to fuel small pre-combustion chamber engines, *American Society of Agricultural Engineers* **82**(4), 364, 1982.
15. N. J. Barsic and A. L. Humke. Vegetable oils: Diesel fuel supplements? *Automotive Engineering* **89**(4), 37–41, 1981.

16. M. Ziejewski and K. R. Kaufman. Endurance tests of a sunflower oil/diesel fuel blends, *Journal of American Oil Chemists' Society* **61**(10), 1620, 1984.

17. T. K. Bhattacharya, T. N. Mishra, B. Singh, and D. P Darmora. Potential of some alternative fuels for I.C. engines, In: *20th Annual Convention of Indian Society of Agricultural Engineers*, 1983, Pantnagar, India, Paper No. 83–4723.

18. W. D. Samson, C. G. Vidrine, and J. W. D. Robbins. Chinese tallow seed oil as a diesel fuel extender, *Transactions of the American Society of Agricultural Engineers* **28**(5), 1406, 1983.

19. P. D. Dunn and E. D. I. H. Perera. Rubber seed oil for diesel oil in Sri Lanka, In: *Proceedings of the International Conference on Biomass*, March 25–29, 1985, Venice, Italy, p. 1172.

20. B. S. Samga. Use of Honni oil as an alternative fuel for C.I. engine, In: *Proceedings of the 9th National Conference on IC Engines and Combustion*, 1986, Dehradun, India, Paper No. A-20, 1.

21. D. L. Auld, B. L. Bettis, and C. L. Peterson. Production and fuel characteristics of vegetable oil from oilseed crops in the Pacific Northwest, *American Society of Agricultural Engineers* **4**(82), 92–100, 1982.

22. T. W. Ryan, L. G. Dodge, and T. J. Callahan. The effects of vegetable oil properties on injection and combustion in two different diesel engines, *Journal of American Oil Chemists' Society* **61**(10), 1610, 1984.

23. H. B. Mathur and L. M. Das. Utilization of nonedible wild oil as diesel engine fuel, In: *Proceedings of Bioenergy Society*, 1985, Hyderabad, India.

24. C. E. Goering, A. W. Schwab, R. M. Campion, and E. H. Pryde. Evaluation of soybean oil–aqueous ethanol microemulsions for diesel engines, *Journal of American Society of Agricultural Engineers* **4**(82), 279–286, 1982.

25. A. Nag, S. Haldar, and B. B. Ghosh. Studies on the Comparison of Performance and Emission Characteristics of a Diesel Engine Using Three Degummed Non-Edible Vegetable Oils, *Biomass and Bioenergy*, 33, 1013–1013.

26. F. Ma and M. A. Hanna. Biodiesel production: A review, *Bioresource Technology* **70**, 1, 1999.

27. D. Noureddini and D. Zhu. Kinetics of transesterification of soybean oil, *Journal of American Oil Chemists' Society* **74**, 1457–1463, 1997.

28. R. O. Feuge and T. Gros. Modification of vegetable oils. VII. Alkali-catalyzed interesterification of peanut oil with ethanol, *Journal of American Oil Chemists' Society* **26**, 97–102, 1949.

29. Y. Kawahara and T. Ono. US Patent No. 4,164,506, 1979.

30. K. J. Harrington and C. D'Arcy-Evans. Transesterification in situ of sunflower seed oil, *Industrial & Engineering Chemistry Product Research and Development* **24**, 314, 1985.

31. R. Stern, G. Hillion, P. Gateau, and J. C. Gulbet. Energy from biomass, In: *Proceedings of the International Conference on Biomass,* March 25–29, 1985, Venice, Italy.
32. A. W. Schwab, M. O. Bagby, and B. Freedman. Preparation and properties of diesel fuels from vegetable oils, *Fuel* **66**, 1372–1378, 1987.
33. K. Liu. Preparation of fatty acid methyl esters for gas-chromatographic analysis of lipids in biological materials, *Journal of American Oil Chemists' Society* **71**, 1179–1187, 1994.
34. J. M. Goff, S. N. Bauer, S. Lopes, W. R. Sutterlin, and J. G. Suppes. Acid-catalyzed alcoholysis of soybean oil, *Journal of American Oil Chemists' Society* **81**, 415–420, 2004.
35. L. C. Meher, S. N. Naik, and L. M. Das. Methonolysis of *Pongamia Pinnata* (Karanja) oil for production of biodiesel, *Journal of Scientific & Industrial Research* **63**, 913, 2004.
36. A. Isigigur, F. Karaosmanoglu, and H. A. Aksoy. Methyl ester from safflower seed oil of Turkish origin as a biofuel for diesel engines, *Applied Biochemistry and Biotechnology* **45**, 103, 1994.
37. D. Darnoko and M. Cheryan. Continuous production of palm methyl esters, *Journal of American Oil Chemists' Society* **77**(12), 1269–1272, 2000.
38. G. Knothe, R. O. Dunn, and M. O. Bagby. The use of vegetable oils and their derivatives as alternative diesel fuels, In: *Fuels and Chemicals from Biomass,* Saha, B. C. and Woodward, J. (Eds.), Washington, DC: American Chemical Society, pp.172–208, 1977.
39. B. Freedman, E. M. Pryde, and T. L. Mounts. Variables affecting the yields of fatty esters from transesterified vegetable oils, *Journal of American Oil Chemists' Society* **61**, 1638–1643, 1984.
40. H. P. Kruclen, H. C. A. VanBeek, E. Vander Drift, and G. Spruiji. *Proceedings of the International Conference on Biomass,* March 25–29, 1985, Venice, Italy, p. 1069.
41. G. Gelbard, O. Bres, R. M. Vargas, E. Vielfaure, and U. F. Schuchardt. Optimization of alkali-catalyzed transesterification of *Pongamia pinnata* oil for production of biodiesel, Journal of American Oil Chemists' Society 72, 1239, 1995.
42. A. Chadha, K. S. Karmee, P. Mahesh, and R. Ravi. Study of the kinetics of base catalyzed transesterification of monoglycerides from Pongamia oil, *Journal of American Oil Chemists' Society* **81**, 425–430, 2004.
43. A. Nag. Alcoholysis of vegetable oil catalyzed by an isozyme of Candida rugosa lipase for production of fatty acid esters, *Indian Journal of Biotechnology* **5**, 175–178, 2006.
44. M. Kaieda, T. Samukawa, T. Matsumoto, K. Ban, and A. Kondo. Biodiesel fuel production from plant oil catalyzed by Rhizopus oryzae lipase in a water-containing system without an organic solvent, *Journal of Bioscience and Bioengineering* **88**(6), 627, 1999.

45. M. Oda, M. Kaieda, S. Hama, H. Yamaji, A. Kondo, E. Izumoto, and H. Fukuda. Facilitatory effect of immobilized lipase-producing Rhizopus oryzae cells on acyl migration in biodiesel-fuel production, *Journal of Biochemical Engineering* **23**, 45, 2005.

46. O. Köse, M. Tüter, and A. H. Aksoy. Immobilized Candida antarctica lipase-catalyzed alcoholysis of cotton seed oil in a solvent-free medium, *Bioresource Technology* **83**(2), 125, 2002.

47. H. Noureddini, X. Gao, and R. S. Philkana. Immobilized Pseudomonas cepacia lipase for biodiesel fuel production from soybean oil, *Bioresource Technology* **96**, 769, 2005.

48. Y. Shimada, Y. Watanabe, A. Sugihara, and Y. Tominaga. Enzymatic alcoholysis for biodiesel fuel production and application of the reaction to oil processing, *Journal of Molecular Catalysis B: Enzymatic* **17**, 133, 2002.

49. V. Dossat, D. Combes, and A. Marty. Continuous enzymatic transesterification of high oleic sunflower oil in a parked bed reactor: Influence of the glycerol production, *Enzyme and Microbial Technology* **25**, 194, 1999.

50. D. Royon, M. Daza, G. Ellenriedera, and S. Locatellia. Enzymatic production of biodiesel from cotton seed oil using t-butanol as a solve, *Bioresource Technology* **98**, 648, 2007.

51. Y. Watanabe, Y. Shimada, A. Sugihara, and Y. Tominaga. Conversion of degummed soybean oil to biodiesel fuel with immobilized Candida antarctica lipase, *Journal of Molecular Catalysis B: Enzymatic* **17**, 151, 2002.

52. W. Du, Y. Xu, D. Liu, and J. Zeng. *Journal of Molecular Catalysis B: Enzymatic* **30**, 125, 2004.

53. C. S. Hawkins, J. Fuls, and F. J. C. Hugo. Society of Automotive Engineers Trans (Index Abstracts), Paper No. 831356, 92, 191.

54. E. F. Fort, and P. N. Blumberg. Society of Automotive Engineers Trans (Index Abstracts), Paper No. 820317, 9l, 63, 1982.

55. M. Ziejewski and K. R. Kaufman. Laboratory endurance test of a sunflower oil blend in a diesel engine, *Journal of American Oil Chemists' Society* **60**, 1567, 1983.

56. A. R. Tahir. Sunflower oil: An anticipated diesel fuel alternative, *Agricultural Mechanization in Asia, Africa, and Latin America* **16**(3), 59, 1985.

57. R. W. Pryor, M. A. Hanna, J. L. Schinstock, and L. L. Bashford. Soybean oil fuel in a small diesel engine, Transactions of the *American Society of Agricultural Engineers* **26**(2), 333, 1983.

58. S. J. Clark, L. Wagner, M. D. Schrock, and P. G. Piennaer. Methyl and ethyl soybean esters as renewable fuels for diesel engines, *Journal of American Oil Chemists' Society* **61**(10), 1632, 1984.

59. F. Laforgia, and V. Ardito. Biodiesel fueled IDI engines: Performances, emissions and heat release investigation, *Bioresource Technology* **51**, 53, 1995.

60. G. H. Pischinger, R. W. Siekmann, A. M. Falcon, and F. R. Fernandes. Methyl Esters of Plant Oils as Diesels Fuels Either

Straight or in Blends. Vegetable Oil Fuels, 1982, American Society of Agricultural Engineers, St. Joseph, MI, Publication No. 4–82, p. 101.

61. Y. Ali, M. A. Hanna, and L. L. Leviticus. Emissions and power characteristics of diesel engines on methyl soyate and diesel fuel blends, *Bioresource Technology* **52**, 185, 1995.

62. R. E. H. Sims. Tallow esters as an alternative diesel fuel, *Transactions of the American Society of Agricultural Engineers* **28**(3), 716, 1985.

63. D. W. Richardson, R. J. Joyee, T. A. Lister, and D. F. S. Natusch. In: *Proceedings of the International Conference on Biomass, March* 25–29, 1985, Venice, Italy, p. 735.

64. M. Mittelbach and P. Tritthard. Diesel fuel derived from vegetable oils: Emission tests using methyl esters of used frying oil, *Journal of American Oil Chemists' Society* **65**, 707, 1988.

65. E. A. Torres, D. C. Santos, L. B. Peixoto, and T. Franca. Performance of cycle diesel engine using biodiesel of olive oil (b100), *Encontro de Energia No Meio Rural*, **6**, 2006.

66. R. Leite Barbosa, F. Moreira da Silva, N. Salvador, and C. Eduardo Silva Volpato. Comparative performance of a cycle diesel engine using diesel and biodiesel mixtures, *Ciência e Agrotecnologia, Lavras*, **32**, 5, 1588–1593.

67. I. M. Corrêa. Desempenho de motor diesel com misturas de biodiesel de óleo de girassol. *Ciência e Agrotecnologia, Lavras*, **32**, 3, 923–928, 2008.

68. C. Eduardo Silva Volpato, A. do Prado Conde, J. Antonio Barbosa, and N. Salvado. Performance of four stroke diesel cycle engine supplied with soybean oil biodiesel (B 100). *Ciênc. agrotec., Lavras*, **33**, 4, 1125–1130, 2009.

69. U. Nieteiedt, J. F. Schlosser, A. Russini, U. G. Frantz, and L. Ribas. Performance evaluation of a direct injection engine using different blends of soybean (Glycine max) methyl biodiesel, *Engenharia Agrícola*, **31**, 916–922, 2011.

Questions

1. Why is the degumming of vegetable oils performed in the preparation of biodiesel?

2. Why is the transesterification process preferred in the preparation of biodiesel from vegetable oil?

3. Compare acid, alkali, and enzymes in the preparation of biodiesel.

4. What abnormalities and efficiencies are observed when biodiesel is properly produced?

5. How is enzymatic transesterification of vegetable oil used to prepare biodiesel?

CHAPTER 7

Ethanol and Methanol as Fuels in Internal Combustion Engines

B. B. Ghosh and Ahindra Nag

7.1 Introduction

The increasing industrialization and motorization of the world has led to a steep rise in the demand of petroleum products. Petroleum-based fuels are stored fuels in the earth. There are limited reserves of these stored fuels, and they are irreplaceable. Figure 7.1 shows the difference in demand and supply of petroleum products, and how this depletion will create a problem for the world within a decade or two.

Geologists throughout the world have been searching for further deposits. Although the present reserves seem vast, the accelerating consumption is challenging the world to create new types of fuels to replace the conventional ones. New oil reserves appear to grow arithmetically while consumption is growing geometrically. Under this situation, when consumption overtakes discovery, the world will be heading toward an industrial disaster.

Apart from the problems of fast-vanishing reserves and the irreplaceable nature of petroleum fuels, another important aspect of their use is the extent and nature of environmental pollution caused by combustion in vehicular engines. Petroleum-fueled vehicles discharge significant amounts of pollutants like CO, HC, NO_x, soot, lead compounds, and aldehydes.

A light-vehicular engine (car engine) discharges 1–2 kg of pollutants a day, and a heavy automobile discharges 660 kg of CO a year. CO is highly toxic, and exposure for a couple of hours to concentrations of 30 ppm can cause measurable impairments to physiological functions.

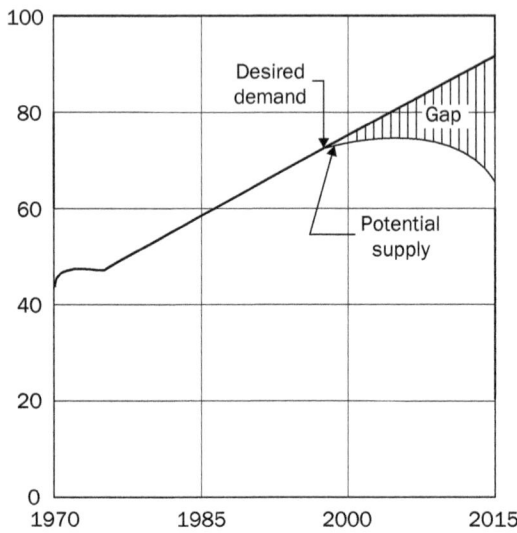

FIGURE 7.1   Difference in demand and supply of petroleum products.

Oxides of nitrogen and unburned hydrocarbons from exhausts cause environmental fouling by forming photochemical smog. Their interaction involves the formation of certain formaldehydes, peroxides, and peroxyacylnitrate, which cause eye and skin irritation, plant damage, and reduced visibility. Present day leaded gasoline contains lead compounds. Lead coming out with the exhaust finds its way into the human body and causes brain damage in infants and children.

Vehicular exhaust fouling of the environment has already become a serious problem in Western countries and is a growing menace in developing countries like India [1]. They exhaust huge quantities of harmful pollutants in urban areas. Every day, vehicles running in Delhi discharge about 240 tons of CO, 30 tons of HC, 20 tons of NO_x, and 2 tons of SO_2. The disastrous effect of these pollutants on human health, animal and plant life, and property are well known.

In view of these problems, attempts must be made to develop technology to produce alternative, clean-burning synthetic fuels. These fuels should be renewable, should perform well in the engine, and their potential for environmental pollution should be quite low.

Various fuels have been considered as substitutes for petroleum fuels used in automobiles. The most prominent of these include ethanol, methanol, NH_3, H_2, and natural gases [2]. The suitability of each of these fuels for internal combustion (IC) engines used in automobiles has been under investigation throughout the world. A few of them are already in use in different countries. This chapter introduces different types of unconventional fuel such as ethanol and methanol, their burning properties when used in IC engines, their performance characteristics compared with conventional engines, the modifications required in the engine if used in practice, and their environmental pollution characteristics.

7.2   Alcohols as Substitute Fuels for IC Engines

Due to the global energy crisis and continuous increase in petroleum prices, scientists have been in search of new fuels to replace conventional fuels that are used in IC engines. Among all the fuels, alcohols, which can be produced from sugarcane waste and many other agricultural products, are considered the most promising fuels for the future. There are two types of alcohols: ethanol (C_2H_5OH) and methanol (CH_3OH). Many other agricultural products (renewable sources) also have a vast potential for alcohol production, and it is necessary to tap this source to the maximum level in national interest. The use of alcohol as a motor fuel is itself not a new idea. Nicolas Otto, the pioneering German engine designer, suggested it as early as 1895. But, as long as crude oil was plentiful and inexpensive, petroleum gasoline was the most economical fuel for the IC engine.

Due to the global energy crisis, many countries that used to export molasses to be used as cattle feed are now setting up distilleries to manufacture ethanol.

7.2.1   Ethanol as an alternative fuel

Ethanol (ethyl alcohol) as a transport fuel has attracted a lot of attention because it is seen as a relatively cheap nonpetroleum-based fuel. It is produced to a large extent from biomass, which aids agricultural economies by creating a stable market. Ethanol, being a pure compound, has a fixed set of physical as well as chemical properties. This is in contrast to petrol and diesel, which are mixtures of hydrocarbons [3].

The use of alcohol in spark ignition (SI) engines began in 1954 in countries like the United States, Germany, and France. During World Wars I and II, gasoline shortages occurred in France and Germany, and alcohol was used in all types of vehicles, including military planes. Nowadays, it is used with gasoline (a mixture) in the United States and has become a major fuel in Brazil.

Ethyl alcohol can be produced by fermentation of vegetables and plant materials. But in countries like India, ethanol is a strong candidate since they possess the agricultural resources for the production of ethyl alcohol. It is a more attractive fuel for India because the productive capacity from sugarcane crops is high, of the order 1345 L/ha. Earlier, this fuel was not used in automobiles due to low energy density, high production cost, and corrosion. The current shortage of gasoline has made it necessary to substitute ethanol as fuel in SI engines.

Any new fuel that is going to be introduced should be evaluated from the aspect of availability, renewability, safety, and cost adaptability

Sr no.	Property	Petrol	Diesel	Ethanol
1.	Specific gravity (at 15°C)	0.73	0.82	0.79
2.	Boiling point (°C)	30–225	190–280	78.3
3.	Specific heat (MJ/kg)	43.5	43.0	27.0
4.	Heat of vaporization (kJ/kg)	400	600	900
5.	Octane number (Research)	91–100	NA	NA
6.	Cetane number	Below 15	40–60	Below 15

TABLE 7.1 Comparative Properties of Ethanol with Petrol and Diesel

to the existing engines' performance, economy, and finally emission. A massive research effort has been put into the study and analysis of all these aspects for ethanol, which is now an established, viable alternative fuel for IC engines. The comparative properties of ethanol with petrol and diesel are shown in Table 7.1.

7.2.2 Production of ethanol

Ethanol is the most appropriate fuel for India to replace petrol, and the utmost of efforts have been made to increase alcohol production in the country. India is in an extremely happy position in this regard as it is the world's largest producer of sugarcane, a major source of alcohol. India topped the world in sugar production with 181 Mton (in 1978), followed by Brazil (130 Mton) and Cuba (67 Mton).

Alcohol is derived not directly from sugarcane but molasses–sugarcane by-products. All starch-rich plants like maize, tapioca, and potato can be used to produce alcohol; cellulosic waste materials can also be used. Production of ethanol from biomass involves fermentation and distillation of crops. India has a vast potential to produce ethanol, and only 2.5 percent of the country's irrigated land is used to produce sugarcane. This can be raised to a much higher level without adversely affecting the production of food-bearing crops.

At present, Brazil is the only country that produces fuel alcohol on a large scale from agricultural products (mainly sugarcane). Other countries, especially those with substantial agricultural surpluses, such as the United States and Canada, are bound to enter into this field of so-called energy forming. The area of land required is substantial. A medium-sized car with an annual run of 15,000 km needs 2000 L of ethanol. To produce this amount, the crop areas required are given in Table 7.2. To provide enough sugar beet alcohol to fuel 20 million cars in Germany requires half the area of the entire country.

Crop	Sugarcane	Sweet sorghum	Sugar beet	Cassava	Potatoes	Wheat
Area (ha)	0.49	0.38	0.5	1.43	1.2	2.52

TABLE 7.2 Crops Area Required for Growth

Sugarcane.   The present method adopted to obtain alcohol for energy purposes requires three stages: (1) extracting the juice from sugarcane, (2) fermentation of the juice, and (3) distillation into 90–95 percent alcohol.

Molasses.   The black residue remaining after the sugar is extracted from sugarcane is called molasses. It contains mostly invert sugars and some sucrose. This sucrose also undergoes hydrolysis to produce invert sugar by a catalytic action of acids in molasses.

$$C_{12}H_{22}O_{11} + H_2O \rightarrow C_6H_{12}O_{11} \text{ (D-Glucose)} + C_6H_{12}O_6 \text{ (D-Fructose)}$$

This mixture product is not crystallizable. Yeast organisms in the presence of oxygen oxidize sugars into CO_2 and H_2O and convert sucrose mostly into ethyl alcohol.

$$C_6H_{12}O_6 \rightarrow 2C_2H_5OH + 2CO_2$$

Process adopted.   Molasses is mixed with water so that the concentration of sugar in it is 10–18 percent (optimum is 12 percent). If the concentration is high, more alcohol may be produced and may kill the yeast. Then, a selected strain of yeast is added (it should not contain any wild yeast). For some nutrient substances like ammonium and phosphates, the pH value is kept between 4 and 5, which favors the growth of yeast organisms. H_2SO_4 is used for lowering pH. The temperature of the mixture is kept at 15–25°C. The fermentation takes place as follows:

1. First, the yeast cells multiply at an optimum temperature (30°C).
2. Rapid fermentation takes place at the boiling temperature, and oxygen is given off. The optimum temperature (50°C) is maintained, and the process is continued for 20–30 h.
3. The fermentation rate is reduced, and alcohol is produced slowly. Total time for fermentation is 36–48 h, depending upon the temperature and sugar content. Last, the formed ethanol is distilled.

Starch.   In this process, starchy materials are first converted into fermentable sugars. This is done by enzymatic conversion (by means of malt process) or by acid hydrolysis.

$$\text{Starch} \rightarrow C_{12}H_{22}O_{11} \text{ (Maltose)} + C_6H_{12}O_6 \text{ (Dextrose)}$$

Malt process.   Malt is prepared by germination of barley grains to produce required enzymes. The grain is ground and steam cooked at 100–150°C to break the cell wall of starch. For every 25 kg of grain, 100 L of water is added. Then the formed mass is cooled to 60–70°C and taken to large vessels where malt is added within 2 h and 60–70 percent of the stock is converted into maltose. Converted mash is cooled to a fermenting temperature of 20–25°C. pH is adjusted and fermentation is affected, producing ethanol.

Acid hydrolysis.   This process involves treatment with concentrated sulfuric or hydrochloric acid at pH 2–3 and 10–20 kg pressure in an autoclave to make sugar and then conversion of sugar to alcohol by yeast.

Cellulose material
Wood.   Cellulose from wood is hydrolyzed into simple sugars by using diluted acid at a high temperature or concentrated acid at a low temperature. Similarly, cellulosic agricultural waste and straws can be used in place of wood.

Sulfite waste liquor from paper manufacture.   Waste liquor contains 2–3.5 percent of sugar, out of which 65 percent is fermentable into alcohol. Before fermentation, all acids in the liquor are removed by adding calcium. Then fermentation is carried out by special yeasts. Generally, 1 percent of liquor is converted into alcohol.

Hydrocarbon gases.
   Hydration of ethylene.   Conversion of ethylene to ethyl alcohol can be carried out with high yield by first treating ethylene with H_2SO_4, forming ethyl hydrogen sulphate and diethyl sulfate, as given by the following reactions:

$$C_2H_5HSO_4 \rightarrow (C_2H_5)_2SO_4$$

$$2C_2H_4 + H_2SO_4 \rightarrow (C_2H_5)_2SO_4$$

These products, ethyl sulfuric acid and diethyl sulfate, when treated with water give ethanol as per the following reactions:

$$C_2H_5HSO_4 + H_2O \rightarrow C_2H_5OH + H_2SO_4$$

$$(C_2H_5)_2SO_4 + 2H_2O \rightarrow 2C_2H_5OH + H_2SO_4$$

Direct hydration.   Ethanol is also formed as per the following chemical reaction:

$$C_2H_4 + H_2O \rightarrow C_2H_5OH$$

This type of conversion is very small as the reaction is exothermic; it is not a suitable method for mass production. The corn is first ground, then mixed with water and enzymes, and cooked at 150°C to convert starch to sugar. The mixture is then cooled and sent to fermentation tanks, where yeast is added and the sugar is allowed to ferment into ethanol. After 60 h in the tanks, the mixture is sent to distillation columns, where ethanol is evaporated out, condensed, and mixed with unleaded gasoline to form gasohol, which contains 90 percent gasoline and 10 percent ethanol.

Tapioca materials. Tapioca is available in plenty in Asia, the United States, central Europe, and Africa. Its production can be increased through modern cultivation techniques. The process consists of converting the tapioca flour into fermentation sugars with enzymes prior to fermentation with yeast. Modern technology uses α-amyl glycosidase, one of two enzymes required in the process, and then saccharification of the material into alcohol by using yeast.

Anhydrous alcohol from vegetable wastes. The Philippines has embarked on an "alcogas program" to produce its own anhydrous alcohol from local vegetable wastes for blending with petrol. The program is currently based on sugarcane juice and molasses, but it plans to diversify by using other raw materials. In the basic process, cellulose conversion begins with the pretreatment of the raw materials, which may include coffee hulls, rice straw, grass—even sawmill wastes. Enzymes then take over by converting the feedstock into a sugary liquid that is fermented and finally distilled into anhydrous alcohol. After distillation, waste residues can be evaporated into syrup to feed animals, while unconverted cellulose is used as the primary fuel for the plant. If the Philippines could engineer a breakthrough in this area, its agricultural and forestry wastes could supply energy equivalent to 9720 mL of oil annually. In the years to come, this new energy source could make a significant economic impact on a country that depends on imports of crude oil for 95 percent of its energy.

Manioc. As oil prices continue to rise, more and more work is being done on alternatives. Manioc is one such staple crop in many tropical lands. Brazil has planned to use manioc in its ethanol production plants, aiming to make 35,000 bbl a day from 400×10^3 ha of manioc plantation. Conversion of manioc to ethanol is somewhat more complex than is the case with sugarcane. The raw material has to be turned into sugar by fermentation. This first step requires the use of enzymes. Danish Co. has developed the necessary heat-resistant enzymes in a pilot plant in Brazil.

Manioc does not grow in higher temperature zones, so scientists have turned to other plants, and there is work being done in Sweden that is in an advanced stage. They have developed fast-growing poplars and willows. Their yield is 30 ton/ha, which is equal to 12 tons of fuel oil.

It is estimated that 1000×10^3 ha planted with such trees can provide 10 percent of Sweden's electricity. Also in Sweden, work has been carried out on the common reed, and the estimated yield is 10 ton/(yr · ha), which is equal to 4.5 tons of oil. Sweden has plans to have 100×10^3 ha of reeds. Brazil's program of ethanol from sugarcane and manioc may employ 200×10^3 people and save $1600 million each year in foreign exchange.

7.3   Distillation of Alcohol

If a mixture of water and alcohol is boiled, the percentage of alcohol to water is greater in vapor than in liquid. Therefore, by repeated distillation and condensation, the alcoholic strength of the distillate can be increased until it contains 97.6 percent alcohol. There are different methods of distillation, but they are not discussed here, as ethanol production is our prime concern.

7.4   Properties of Ethanol and Methanol

Both ethanol and methanol, as listed in Table 7.3, have high knock resistance (as the octane numbers are 89 and 92, against 85 for gasoline), wide ignition limit, high latent heat of vaporization, and nearly the same specific gravity. All those properties are of great advantage if used in SI

Sr no.	Property	Gasoline C_8H_{18} isooctane	Ethyl alcohol	Methyl alcohol
1.	Molecular weight (g)	114.2	46	32
2.	Boiling point at 1 bar (°C)	43–170	78	66
3.	Freezing point (°C)	–107.4	117.2	–161.8
4.	Specific gravity (150°C)	0.72–0.75	0.79	0.79
5.	Latent heat (kJ/kg)	400	900	1110
6.	Viscosity (centipoise)	0.503	0.60	0.596
7.	Stoichiometric A:F (ratio)	14.6	9	6.45
8.	Mixture heating value (kJ/kg) (for stoicmixture)	2930	2970	3070
9.	Ignition limits (A/F)	8–19	3.5–17	2.15–2.8
10.	Self-ignition temperature	335	557	574
11.	Octane number			
	a. Research	80–90	111	112
	b. Motor	85	92	91
12.	Cetane number	15	8	3
13.	Lower CV (kJ/kg)	44,100	26,880	19,740
14.	Vapor pressure at 38°G (bar)	0.48–1	0.17	0.313
15.	Flame speed (m/sec)	0.43	—	0.76
16.	Autoignition temperature (°C)	222	—	467

TABLE 7.3 Important Alcohol Properties

engines. Some important advantages of alcohol-fueled engines compared with gasoline engines are listed below:

1. The alcohols (both) have higher heat of vaporization. As the liquid fuel evaporates into the air stream being charged to the engine, a higher heat of vaporization cools the air, allowing more mass to be drawn into the cylinder. This increases the power produced from the given engine size. High latent heat of vaporization leads to higher volumetric efficiency and provides good internal cooling.

2. The high octane number of alcohols compared to petrol means higher compression ratios can be used, which results in higher engine efficiency and higher power from the engine.

3. Ethanol burns faster than petrol, allowing more uniform and efficient torque development. Both alcohols have wider flammability limits, which results into a rich air–fuel (A:F) ratio being used when needed to maximize power by injecting more fuel per cycle.

4. Alcohols also have lower exhaust emissions than gasoline engines except for aldehydes. Both alcohols have lower carbon–hydrogen ratio than petrol and diesel, and produce less CO_2. For the same power output, CO_2 produced by an ethanol-fired engine is about 80 percent of the petrol engine. Because of high heat of vaporization, the fuels burn at lower flame temperatures than petrol, forming less NO_x. The CO percentage in both cases (alcohol and petrol) remains more or less the same.

5. Contamination of water in alcohols is less dangerous than petrol or diesel because alcohols are less toxic to humans and have a recognizable taste.

6. The alcohols can also be blended with gasoline to form the so-called gasohol (80 percent petrol and 20 percent alcohol), which is widely used in the United States.

7. Ethyl alcohol as a fuel offers great safety due to its low degree of volatility and higher flash point (17°C).

8. The heating value of alcohol is 60 percent of that of petrol (60 percent only), and it shows equally good thermal efficiency and lower fuel consumption, because the air required for petrol and alcohol is in the ratio of 15:9 by weight, which is the same as their calorific value, i.e., the same heat is developed per cylinder charge in petrol and alcohol engines. The power per unit volume of cylinder for petrol, ethanol, and methanol are closely similar.

9. In many hot-climate countries, more precautions are often taken for the use of more volatile spirit-based fuels, while alcohol is perfectly safe in the hottest climate.

10. The major problem faced with ethanol is corrosion; special metals should be used for the engine parts to avoid corrosion.

Alcohols are clean-burning, renewable alternative fuels that can come to our rescue to meet the duel challenge of vehicular fuel oil scarcity and fouling of the environment by exhaust emissions.

Alcohols inherently make very poor diesel engine fuels as their cetane number is considerably lower. They can be used in dual-fuel engines or with assisted ignition in diesel engine. In the dual-fuel mode, alcohol is inducted along with air, compressed, and then ignited by a pilot spray of diesel oil.

7.5   Use of Blends

Alcohol can be used as a blend with gasoline as this has the advantage that the existing engines need not be modified and tetra-ethyl lead (TEL) can be eliminated from gasoline, due to the octane-enhancing quality of alcohol. If the engine is to be operated using only pure alcohol, then some major modifications are required in the engine and fuel system, as listed below:

1. Both alcohols and blends with gasoline are corrosive to many of the engine materials. These materials have to be changed.

2. Adjustment of the carburetor and fuel injection need to be made to compensate for the leaning effect.

3. Change in the fuel pump and circulation system need to be made to avoid vapor lock, as the methanol vaporization rate is very high.

4. Introduction of high energy ignition system with lean mixture.

5. Increase in compression ratio to make better antiknock properties of the fuel.

6. Addition of detergent and volatile primers to reduce engine deposits and assist in cold starting.

7. Use of cooler-running spark plugs to avoid preignition.

General properties of the blends are listed in Table 7.4. The volatility shown by the American Standard Testing Method (ASTM) distillation characteristics of petrol is a compromise between opposing factors to ensure good performance in petrol engines. This requires petrol to have a sufficiently lighter reaction and a 10 percent distillation temperature in order to start the engine as well as warm up, but the temperature should not be so low that vapor-locking takes place and stops the engine due to the nonsupply of fuel. As far as volatility is concerned, ethanol–petrol blends are as

Characteristics	BIS specification for petrol	petrol	Ethyl alcohol	Petrol and Ethyl Alcohol (Gasohols)			
				95% + 5%	90% + 10%	85% + 15%	80% + 20%
ASTM distillation							
Initial boiling point, °C	. . . .	55	78	55	50	48	46
10% volume	70 (max)	64	. . . .	59	56	57	57
50% volume	125 (max)	92	. . . .	95	73	70	70
90% volume	180 (max)	128	. . . .	145	127	130	125
Final boiling, °C	215 (max)	143	. . . .	. . . .	147	256	156
Gum residue, mg/100	4	22	55	51	91	131	180
Aniline point, °C	. . . .	44	30	40	35	32	30
Specific gravity	. . . .	0.720	0.7966	0.7230	0.722	0.7289	0.733

TABLE 7.4 General Properties of the Blends (Petrol and Ethyl Alcohol)

good as petrol, if not better. Also gum resistance is greater than that of petrol. Aniline points for blends are lower, which indicates more aromatic content than petrol, due to the adding of ethanol to petrol, which helps to improve the octane number marginally. If a small quantity of water is introduced into a gasoline–alcohol blend, phase separation takes place, with gasoline–content in the upper phase and alcohol in the lower. This separation produces some undesirable effects. The alcohol–water mixture tends to pick up sediment and stall the engine on reaching the carburetor [4]. To improve the water tolerance of the blend, benzene and heptanes are added.

Since 1979, gasohol has been sold at 500 filling stations in the midwestern United States, where the corn from which alcohol is commonly made is abundant. This blend yields about the same mileage as unleaded gasoline and even offers an ever renewable source of energy. Moreover, if this blend were to replace gasoline, it could cut as much as 10 percent of the nation's oil imports, which totalled $40 billion in 1979. This fuel has a good future in wealthy countries. The blends have some important advantages over pure ethanol, as listed below:

1. The starting difficulty can be removed.
2. There is no abnormal corrosion compared with pure ethanol.

3. Lubrication in a petrol–alcohol blend is more or less the same.

4. Some benzene is added to prevent separation of the layers of petrol and alcohol.

If blends are used, some minor modifications in the engine are required, as listed below:

1. The carburetor jet should be increased to increase the flow 1.56 times that of petrol.

2. The float has to be weighted down to correct levels due to higher specific gravity.

3. The air inlet should be modified to get less air as blends require less air for complete combustion than petrol.

4. Specific arrangement of heating the carburetor and intake manifold should be provided as lower vapor pressure of alcohol makes the starting difficult below 70°C.

7.6 Performance of Engine Using Ethanol

The effect of speed on power output, brake specific fuel consumption (BSFC), and thermal efficiency of an engine using ethanol is compared with gasoline engine, is shown in Figures 7.2 through 7.5.

The observations are listed below:

1. The power output of the ethanol engine is higher, compared to a gasoline engine at all speeds.

2. The BSFC is improved with an ethanol engine, compared to a petrol engine.

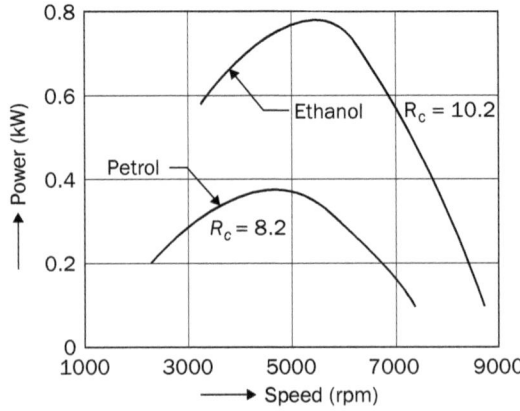

FIGURE 7.2 Effect of speed on power at different compression ratios.

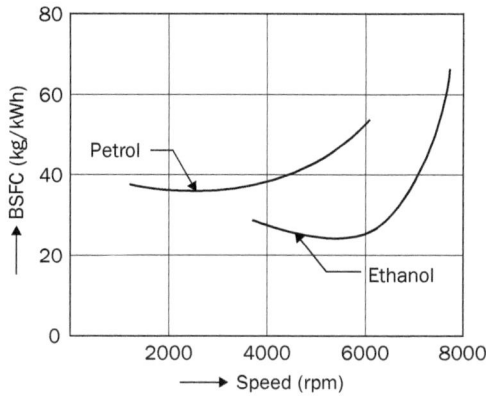

FIGURE 7.3   Effect of speed on BSFC (brake specific fuel consumption).

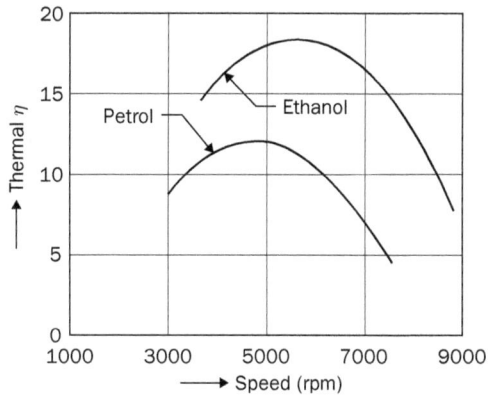

FIGURE 7.4   Effect of speed on thermal efficiencies.

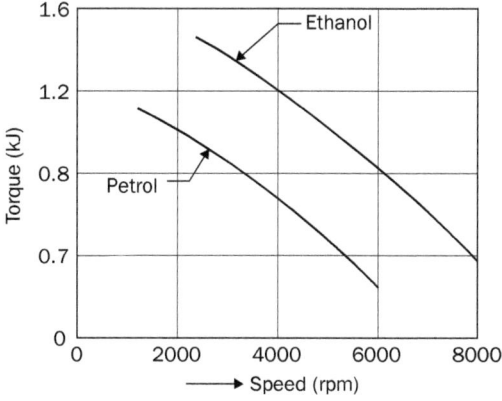

FIGURE 7.5   Effect of speed on the torque.

3. The maximum thermal efficiency of an ethanol engine is higher than that of a petrol engine. The efficiency curve of an ethanol engine is flat for a wide range of speeds, which indicates that the partial-load efficiency is much better, compared with a petrol engine.

4. The engine torque is considerably higher for ethanol as compared to a petrol engine.

7.7 Alcohols in CI Engine

Although the physical and thermodynamic characteristics of alcohols do not make them particularly suitable for compression ignition (CI) engines, with certain modifications, however, they can also be used in CI engines. In heavy vehicles powered by CI engines, ethanol carburetion can be employed for bi-fuel operation of the engine with proportional savings in diesel oil. The various methods for using alcohols with diesel are fumigation, dual injection, and alcohol–diesel emulsions.

In a fumigation system the engine is fitted with a suitable carburetor and auxiliary ethanol tank. An ethanol-air mixture is carbureted during the induction stroke to provide 50 percent of the total energy of the cycle and the remaining energy is provided by diesel oil being injected in the conventional manner near the end of the compression stroke. The materials of a fuel tank and fuel system must be compatible with alcohol. The entire system can be used as a retrofit kit, as shown in Figure 7.6.

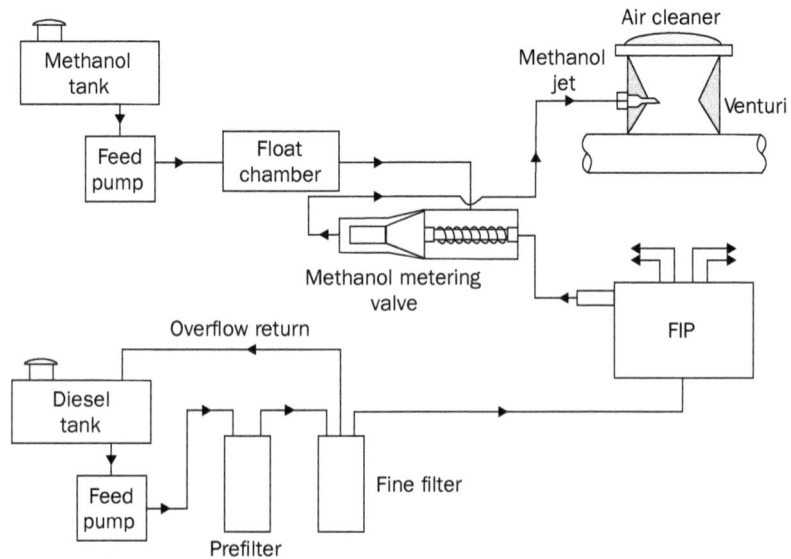

Figure 7.6 Fuel circuit for fumigation.

FIGURE 7.7   Experimental setup of ethanol fumigation.

FIGURE 7.8   Ethanol fumigation nozzle.

Ghosh et al. [4] carried out an investigation on the performance of a tractor diesel engine with ethanol fumigation (see Figures 7.7 and 7.8). The following observations were recorded:

1. The brake thermal efficiency decreases with an increase in ethanol fumigation rate at a constant engine speed.

2. The BSFC decreases with an increase in ethanol fumigation rate at a constant engine speed.

3. The diesel substitution and the energy replacement increases with an increase in an ethanol fumigation rate at a constant engine speed.

4. The NO_x emission level and the exhaust gas temperature decreases with an increase in a ethanol fumigation rate at a constant engine speed.

5. The CO emission level increases with an increase in an ethanol fumigation rate at a constant engine speed.

6. The smoke level decreases with an increase in an ethanol fumigation rate at a constant engine speed.

7. The fumigation rate of 1.06 kg/h (40 percent diesel substitution) is optimal for good engine performance.

Ethanol fumigation in diesel engines can play a major role in environmental air pollution control, and ethanol is a viable alternative fuel for diesel engines.

Ethanol is a very good SI engine fuel and a rather poor CI engine fuel. Ethanol has a high octane rating of 90 and a low cetane rating of 8, and will not self-ignite reasonably in most CI engines. Dehydrated ethanol is fumigated into the air stream in the intake manifold of a 42-hp tractor diesel engine to improve its self-ignition quality. The performance of the engine under dual-fuel (diesel and fumigated ethanol) operation is compared with diesel fuel operation at various speeds (800, 900, 1000, 1100 rpm), loads (0, 4, 8, 12, 16 kgf), and fumigation rates (0.00, 1.06, 1.45, 2.06 kg/h). Analysis of the results shows that ethanol fumigation has the advantages of reduction in BSFC, NO_x emission, and smoke level and the disadvantage of slight reduction in brake thermal efficiency. The fumigation rate of 1.06 kg/h (40 percent diesel substitution) is optimal for good engine performance.

It has been concluded that ethanol is a viable alternative fuel for diesel engines. A dramatic reduction in the NO_x and the smoke level suggests that fumigation, as an emission control technique in diesel engines, can play a vital role in environmental air pollution control on a farm.

In the dual-injection method, two injection systems are used, one for diesel and the other for alcohol. This method can replace a large percentage of diesel fuel. In this method, air is sucked and compressed, and then methanol is injected through a primary injector. To ignite this, a small amount of diesel is injected through a pilot injector. The relative injection timing of alcohol and diesel is an important aspect of the system.

As two injection systems are required, two injectors are required on the cylinder head, which limits the application of this method to large-bore engines. An additional pump, fuel tank, and fuel line are also required, making the system more complicated. But this method replaces 60 percent of diesel at a partial load and 90 percent at a full load, and provides higher thermal efficiency.

7.7.1 Alcohol–diesel fuel solution

This method is the easiest but requires anhydrous ethanol, because methanol has limited solubility. A maximum of 10 percent diesel can be substituted due to the lower solubility of methanol in diesel. No component changes; only adjustments of injection timing and fuel volume delivery are required to restore full power. Dodecanol is an effective surfactant for methanol–diesel fuel blends. Straight-run gasoline is an economical additive for ethanol–diesel blends.

Solubility of alcohols in diesel fuels is a function of (a) fuel temperature, (b) alcohol content, (c) water content, (d) specific gravity of diesel, (e) wax content, and (f) hydrocarbon composition. Methanol solubility in diesel increases as the aromatic content goes up.

7.7.2   Alcohol–diesel fuel emulsions

Here, an emulsifier extends the water tolerance of alcohol–diesel blends. In general, equal volumes of alcohols and emulsifiers are required for suitable emulsions. No component changes, but injection volume and timing are adjusted for diesel fuel with alcohol then solutions, i.e., up to 35 percent diesel substitution is possible. Addition of ignition improvers, e.g., cyclohexanol nitrate, up to 1 percent helps increase the alcohol percentage up to 35 percent while maintaining a cetane rating at permissible levels. Cost of emulsifiers and poor low-temperature physical properties of emulsions limit the use of this technique. Stable emulsion requires the use of costly surfactants. Using higher-order alcohols improves the stability of blends at temperatures as low as −20°C.

7.7.3   Spark ignition

This technique replaces 100 percent diesel. The injection system can be retained as is or replaced by carburetion or port-type fuel injection. A spark plug is introduced in the combustion chamber, and the associated ignition system is added. High compression ratio and positive ignition result in smooth combustion, thereby improving thermal efficiency.

This approach is quite attractive as it uses the high latent heat of the vaporization of alcohols and their octane rating to good advantage. Power output is reduced due to lower heat content of alcohols. Changes in engine operability are not noticeable with alcohol-fired SI engines, relative to the same engines using diesel fuel due to their similar torque. The engines are as efficient as their diesel-fueled counterparts. In fact, huge torque is available at engine speeds below 1400 rpm, which increases engine flexibility and response in use. Converting an existing diesel fleet to an SI technique involves engine modification. Space at the appropriate place must be available for spark plugs in the cylinder head. Lubricants need to be added to alcohols to increase lubricity and prevent wear. Small amounts of cetane improvers may be added, but they are not required. It is not easy to switch between fuels after conversion to the SI technique.

7.7.4   Ignition improvers

Neat alcohols are used in diesel engines by increasing the cetane number sufficiently using ignition improvers. This technique saves the expense and complexity of engine component changes but adds the cost of ignition improvers. The cost of 10–20 percent ignition improvers is quite prohibitive.

The most effective ignition improvers are nitrogen-based compounds, which can aggravate exhaust emissions of NO_x. Ethylene glycol nitrates have shown promising trends at 5 percent concentration.

Engines operating on cetane-enhanced alcohol need a few changes, e.g.,

- Injection volume and timing must be adjusted to obtain optimum performance.
- A large pump, fuel lines, and injectors are required to satisfy total fuel requirements of the engine for the desired output.
- A lubricant (generally castor oil used so far) is required to be added to alcohols using improvers.

7.8   Methanol as an Alternate Fuel

Methanol behaves much like petroleum, so it can be stored and shifted in the same manner. It is a more flexible fuel than hydrocarbon fuels, permitting wider variation from the ideal A:F ratio. It has relatively good lean combustion characteristics compared to hydrocarbon fuels. Its wider inflammability limits and higher flame speeds have shown higher thermal efficiency and less exhaust emissions, compared with petrol engines.

Methanol can be used directly or mixed with gasoline. Tests have shown improvements in fuel economy by 5–13 percent, decreases in CO emission by 14–70 percent, and reductions in exhaust temperature by 1–9 percent, with varying methanol in petrol from 5 to 30 percent. Depending on the gasoline–methanol mixture, some changes in fuel supply are essential. Simple modifications to the carburetor or fuel injection can allow methanol to replace petrol easily. Some important features of methanol as fuel are listed below:

1. The specific fuel consumption with methanol as fuel is 50 percent less than a petrol engine.
2. Exhaust CO and HC are decreased continuously with blends containing higher percentage of methanol. But exhaust aldehyde concentration shows the opposite trend.
3. Like ethanol, methanol can also be used as a supplementary fuel in heavy vehicles powered by CI engines with consequent savings in diesel oil and reduced exhaust pollution. No undue wear of engine components is encountered with methanol as a fuel, while engine peak power improves and smoke density and NO_x concentration in exhaust is reduced.

Phase separation, vapor lock, and low-temperature starting difficulties are the problems associated with the use of methanol or its blends as IC engine fuels. Availability from indigenous sources, ease of handling, low emission, and high thermal efficiency obtainable with its use make methanol a logical alternative fuel for vehicular engines.

7.8.1   Production of methanol

Methanol can be produced from resources such as coal, natural gas, oil shell, and farm waste, which are abundant worldwide. But methanol from natural gas is unlikely to provide a large greenhouse benefit, not more than a 10 percent reduction in emissions with quite optimistic assumptions. It is not considered as a main raw material to produce methanol. For countries having vast reserves of coal but small oil deposits, methanol from coal can provide an indigenous substitute to oil. But this method has an adverse effect on greenhouse gases and is very expensive, requiring capital investments that can increase the price by 50 percent.

In India, there is an abundant production of sugarcane. The government can divert this feedstock to produce methanol. The production of methanol by using water and methane is shown in Figure 7.9, and by using methane and a catalyst in Figure 7.10.

Producing methanol from methane with the technology available today generally involves a two-step process. Methane is fuel reacted with water and heat to form carbon monoxide and hydrogen—together called synthesis gas. Synthesis gas is then catalytically converted to methanol. The second reaction unleashes a lot of heat, which must be removed from the reactor to preserve the activity of the temperature-sensitive catalyst. Efforts to improve methanol synthesis technology focus on sustaining the

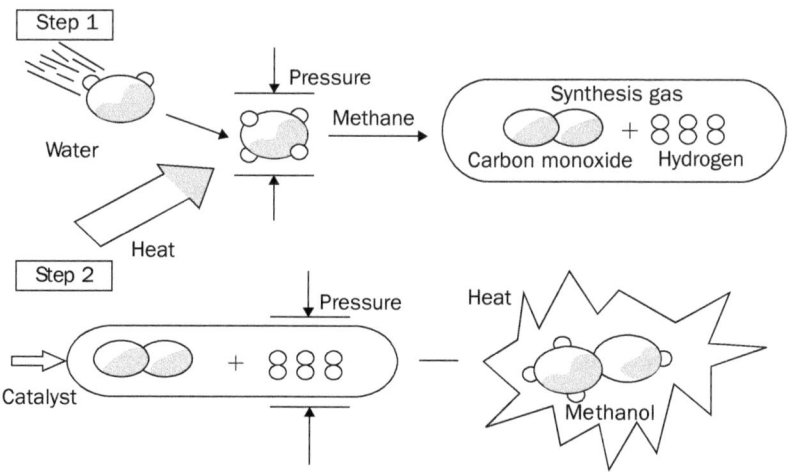

FIGURE 7.9   Conversion of methane to ethanol.

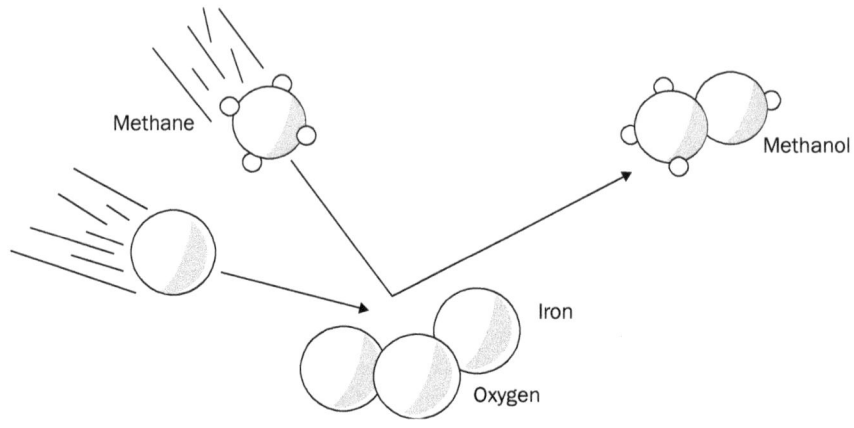

Figure 7.10 Production of methanol by using methane and a catalyst.

catalyst life and increasing reactor productivity. As a novel alternative to the two-step method, a chemical catalysis that mimics biological conversion of methane by enzymes is being developed. The iron-based catalyst captures a methane molecule, adds oxygen to it, and ejects it as a molecule of methanol. If this type of conversion could be performed on a commercial scale, it would eliminate the need to first reform methane into a synthesis gas, which is a costly, energy-intensive step. Conversion of coal to methanol is simpler and cheaper as compared to its liquefactions to gasoline.

Advantages of methanol

1. One percent methanol in petrol is used to prevent freezing of fuel in winter.
2. Tertiary-butyl alcohol is used as an octane improving agent.
3. Because of the excellent antiknock characteristics of the fuel, it is very suitable for SI engines.
4. Isopropyl alcohol is used as an anti-icing agent in carburetors.
5. Addition of methanol causes a methanol–gasoline blend to evaporate at a much faster rate than pure gasoline below its boiling point (bp).
6. Due to an increase in emission levels of conventional fuels, the percentage of O_3 in the atmosphere is increasing. This increase in the O_3 in the atmosphere might cause biomedical and structural changes in the lungs, which might cause chronic diseases. O_3 content of even between 0.14 and 0.16 ppm temporarily affects lung function if the person is exposed to it for 1–2 h. An annual crop yield is also reduced if exposed to O_3; some trees suffer injury to needles or leaves, and lower

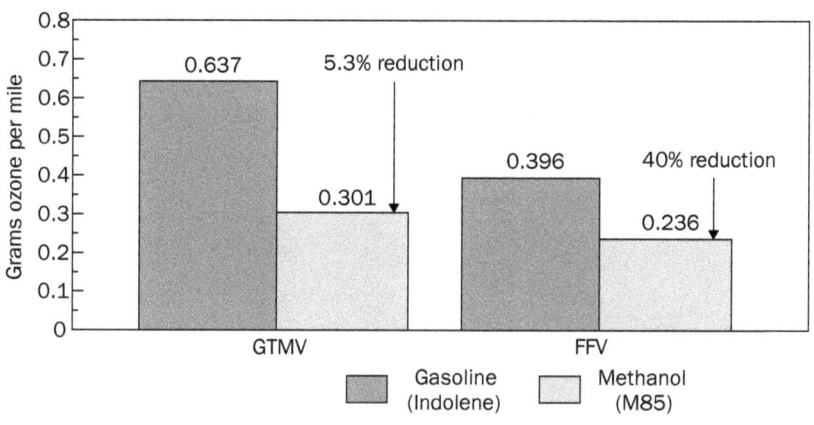

FIGURE 7.11   Effects of methanol on O_3 emission compared with petrol.

productivity or even die. High content of O_3 has disturbed the natural ecological balance of species in national forests in California. The effects of methanol on O_3 emission as compared with petrol are shown in Figure 7.11.

7.8.2   Emission

Methanol-fueled vehicles emit less CO_2 and other polluting gases compared to gasoline-fueled vehicles. Therefore, methanol use maintains good air quality. For a higher compression ratio compared to gasoline, a higher level of NO_x can be achieved. But low flame temperature and latent heat of vaporization tend to decrease NO_x emissions. The overall effect is a lower level of NO_x emissions.

7.8.3   Fuel system and cold starting

Methanol has high latent heat; therefore, some provision must be provided to heat the intake manifold, because cold starting problems are often caused by A:F vapor mixture being outside the flammability range. Specially, methanol in its pure form is much more inferior to petrol for cold starting. Cold starting more or less becomes impossible with methanol when the ambient temperature falls below system on chip (SOC). Figure 7.12 shows the modification that is provided to avoid the difficulty of cold starting. By preheating, methanol dissociates into $CO + 2H_2$ to obtain gaseous H_2, which gives a broad flammability limit. While cranking the engine, a rich gaseous A:F mixture of methanol is collected near the spark plug, which enables good starting of the engine.

7.8.4   Corrosion

Corrosion of the engine parts has been one of the main reasons for not using alcohols as fuels. The problem of corrosion is severe during starting and

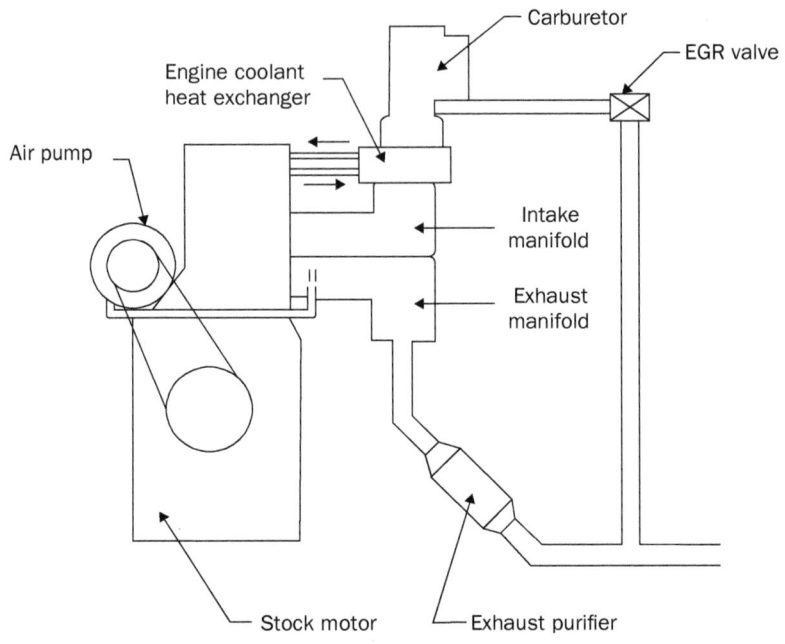

FIGURE 7.12   Modification provided to avoid the difficulty of cold starting.

idling; but once the engine starts and gets heated, corrosion does not take place. Severe corrosion is noticed with Zn, Pb, Cu, Mg, and Al. This problem has been solved by using a methanol-resistant filter before the carburetor. Corrosion by methanol has been prevented by using the corrosion inhibitor LZ541 manufactured by M/S Lubrizol India. Being solvent, it swells or softens many parts of plastic or rubber commonly used for gaskets or floats in the carburetor. This is solved by using elastomers instead of rubber or plastic. American Motors' Gremlin model of 1970 has been used continuously for 9 years using pure methanol without facing any difficulty of corrosion. Two 1972 Plymouth Valiants have been used for 7 years: one using pure methanol and the other using a methanol blend without any difficulty. None of these vehicles has had a failure of engine components or fuel system components.

7.8.5 Toxicity of methanol

Methanol is more toxic as compared to petrol, which creates difficulty in its handling. The toxicity of methanol is reduced by adding chemical emetics.

7.8.6 Formaldehyde emission

The major problem with methanol is high levels of formaldehyde emission, which is negligible with conventional fuels. Formaldehyde

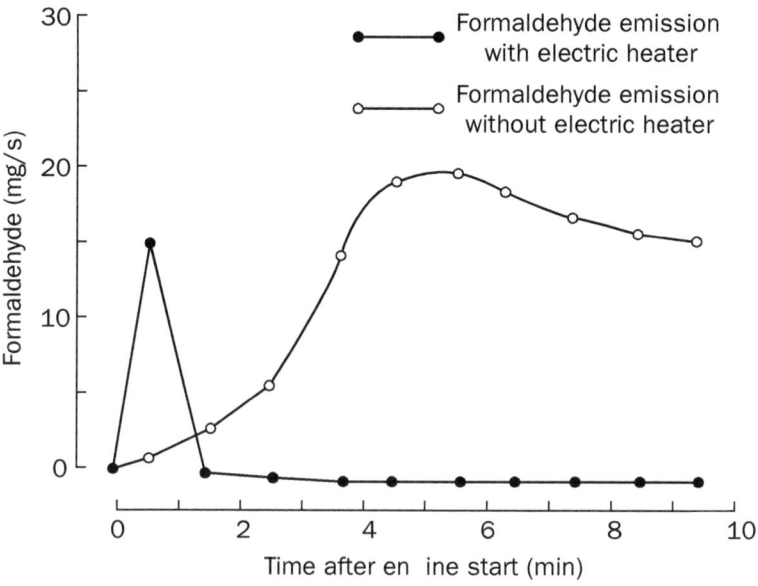

FIGURE 7.13   Performance of methanol as an IC engine fuel.

emission levels with and without an electric heater are shown in Figure 7.13. The level with an electric heater is considerably lower compared with its absence.

The performance characteristics compared with petrol engine are considered as brake thermal efficiency versus air fuel (A:F) ratio, the effect of speed power output and specific heat consumption. In addition, the performance characteristics also include the effect of A:F ratio on exhaust emission. The effects of A:F ratio and speed on brake power are shown in Figures 7.14a and 14b. Another important characteristic is the effect of speed on volumetric efficiency, which is shown in Figures 7.15 and 7.16.

Both alcohols, as well as their blends, are studied as alternative fuels for IC engines. The power can be increased from 6 to 10 percent with alcohols or their blends. The use of a leaner mixture provides more O_2, which reduces the emission. Because of the high heat of vaporization of these fuels compared to petrol, greater cooling of the inlet mixture occurs, which gives higher thermal efficiency, less specific heat consumption, and smooth operation. At higher speeds, the specific heat consumption is lower than that of petrol. Methanol dissociates in the engine cylinder forming H_2. This H_2 gas helps the mixture to burn quickly and increases the burning velocity, which brings about complete combustion and makes a leaner mixture more combustible. In a petrol engine, misfiring occurs while operating at a lean

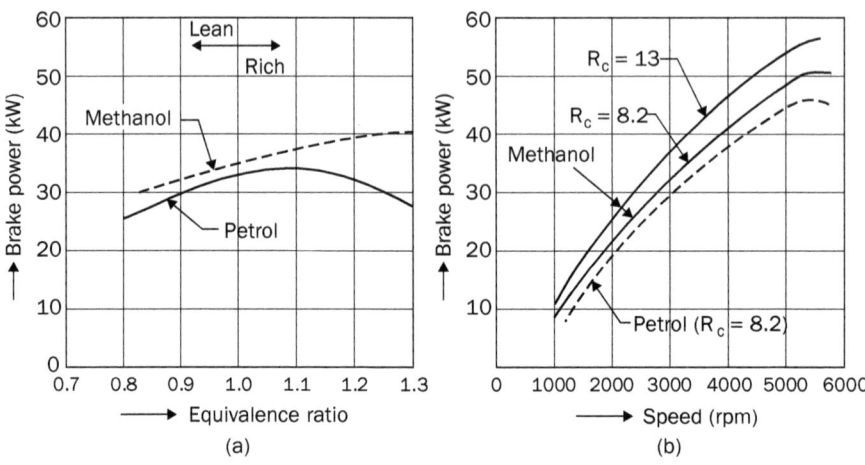

Figure 7.14 Effect of (a) equivalence ratio and (b) speed on brake power.

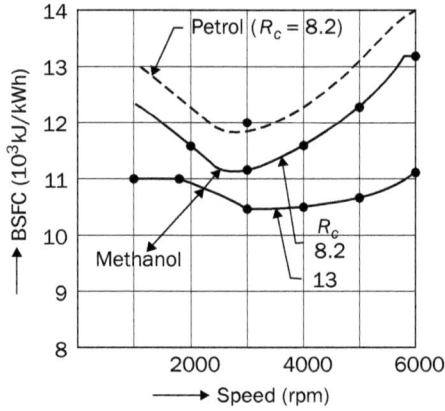

Figure 7.15 Effect of speed on BSFC.

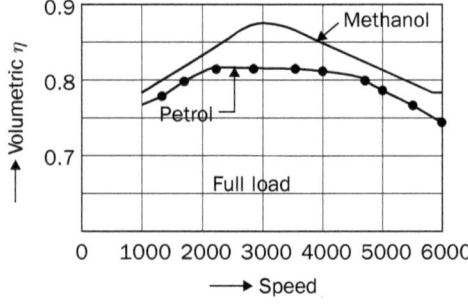

Figure 7.16 Effect of speed on volumetric efficiency.

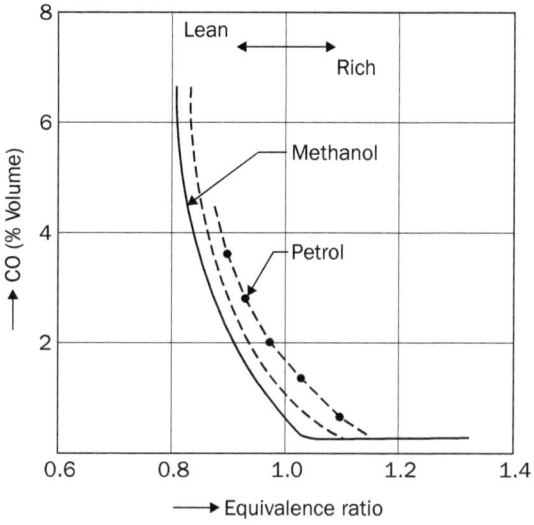

FIGURE 7.17   Effect of equivalence ratio on CO.

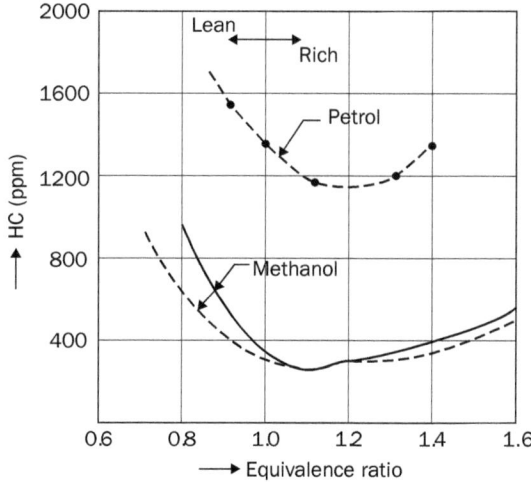

FIGURE 7.18   Effect of equivalence ratio on HC.

A:F ratio, whereas in an engine using alcohol, the engine can manage to handle leaner mixtures without any misfire. Important objectionable emissions are CO, HC, NO_x, and aldehydes. The effect of equivalence ratio on all these emissions for petrol and methanol is shown in Figures 7.17 through 7.20.

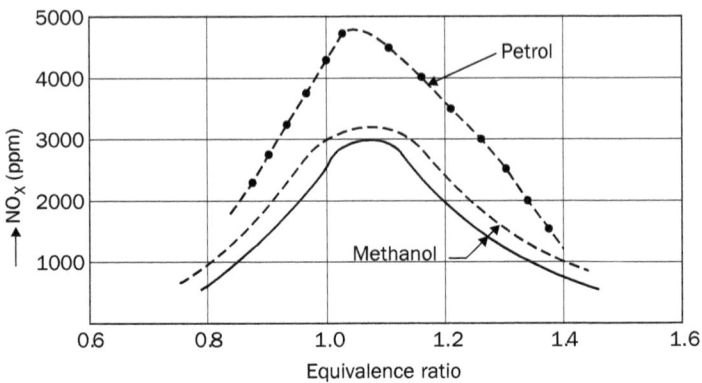

FIGURE 7.19 Effect of equivalence ratio on NO$_x$.

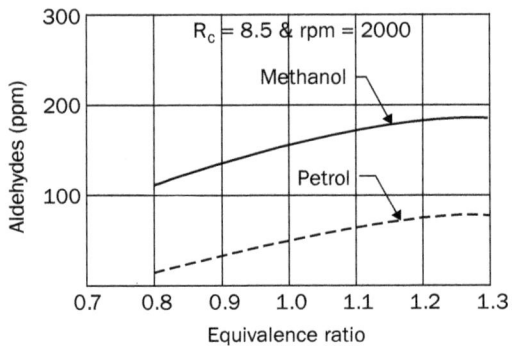

FIGURE 7.20 Effect of equivalence ratio on aldehyde.

For all the above graphs, the engine details and compression ratio are as follows:

1. Full throttle rpm = 2500
2. Compression ratio

   Methanol **Petrol**

   $R_c = 9$ $R_c = 9$
   $R_c = 12.6$

Regarding emission, ethanol and methanol are considered as clean fuels, as emissions of CO, HC, and NO$_x$ are reduced by nearly 10–15 percent compared with a petrol engine. The flame speed of alcohol mixtures is higher than a petrol A:F mixture, and this helps in making the combustion more complete without misfiring.

Regarding the production of formaldehyde, its percentage in exhaust is much higher, which is a great problem to extract methanol in pure form as a replaceable fuel. To avoid this, blends (15–25 percent) of both alcohols are preferred over pure ethanol or methanol. The properties of blends and their effect lie in between pure alcohol and petrol. As we know, methanol blends have lower stoichiometric air requirements compared to petrol. Therefore, if we use a methanol–petrol blend without any modification in the carburetor, we get more air for combustion, which will reduce the emission of CO and HC as well as NO_x as the engine works cooler with the blend compared with a petrol engine.

Oxidizing catalytic devices can control aldehyde emissions. Platinum–rhodium and platinum–palladium catalysts are considered the most effective in tackling aldehyde emissions of methanol fueling.

Concerning the alcohol fuels, the following conclusions can be drawn:

1. Alcohol is potentially a better fuel than gasoline for SI engines.
2. Its use improves the thermal efficiency as a higher compression ratio (12:16) can be used.
3. It can avoid knocking even at a higher compression ratio because of the high octane number.
4. It provides better fuel economy and less exhaust emissions.
5. High latent heat of alcohol reduces the working temperature of the engine.
6. It gives more power, specially when used as a blend.
7. Easy availability of raw materials.
8. Cost of production is low because of the price hike in crude petroleum.

In agricultural countries like India, we can get ethyl alcohol easily from vegetables, agricultural material, and sugarcane waste at a much lower cost compared with the cost of petrol today. Therefore, replacing petrol with alcohol in an SI engine has a good future.

7.9   Comparison of Ethanol and Methanol

Most of the properties are similar, with differences of only 5–10 percent. Ethanol is superior to methanol as it has a wider ignition limit (3.5–17) than methanol (2.15–12.8). Its calorific value (CV) (26,880 kJ/kg) is considerably higher than methanol (19,740 kJ/kg).

Ethanol is a much more superior fuel for diesel engines as its cetane number is 8 compared to the cetane number of 3 for methanol. There are wide resources for manufacturing ethanol compared with methanol. Therefore, ethanol is widely used as SI engine fuel in many countries. Methanol is superior to ethanol in one respect: Its vaporization rate is much higher than ethanol. Therefore, mixing with air rapidly forms a uniformly vaporized mixture and also burns uniformly. One major drawback of methanol is that it creates vapor locks because of the higher vaporization rate. Properties of ethanol and methanol as compared with petrol are listed in Table 7.3.

7.10   Ecosystem Impacts Using Alcohol Fuels

7.10.1   Aquatic system impacts

The biological consequences of alcohol spills or leaks into marine water are sensitive to many factors such as scale and duration of the spill, tidal patterns, water currents, flow rate, temperature, and available oxygen. Marine life can tolerate low concentrations of alcohol.

In general, methanol and ethanol are significantly less toxic than gasoline or crude oil. Because alcohols are miscible, volatile, and degradable, they are dispersed readily, and diluted and neutralized in aquatic environments. The aquatic environment recovers more rapidly and completely from an alcohol spill than from a gasoline or crude oil spill of the same volume.

7.10.2   Terrestrial system impacts

The direct exposure of soils to methanol spills results in immediate damage of surface vegetation. The miscibility, volatility, and degradability of alcohols reduce the alcohol residence time in soil and minimizes the environmental impact. Fungal and bacterial populations, which are important agents of nutrient cycling, exhibit 80–90 percent recovery with 3 weeks of exposure. Total recovery of the site occurs within a period of weeks or months. In comparison, recovery of biodegradation by crude oil and petroleum products takes months or years.

7.10.3   Occupational health impacts

Occupational health risks associated with using alcohol fuels are lower than those associated with conventional fuels. The relative toxicity of alcohol fuels depends on the means of exposure, inhalation, and ingestion. Gasoline poses a greater occupational health risk than either methanol or ethanol as carcinogens in gasoline can be readily absorbed by the skin or inhaled.

7.10.4   Occupational safety impacts

Two major safety hazards of all fuels are fire and explosion, which can occur because of improper fuel storage, spills, or vehicle accidents. The properties of alcohols and gasoline that pertain to fire and explosion risks include the flash point, auto-ignition temperature, flammability limits, and saturated vapor concentrations. While ethanol and methanol have broader flammability limits than gasoline, gasoline poses a greater risk of fire in open air. Because of the low flash point and auto-ignition temperature of gasoline, gasoline is more likely to ignite and burn rapidly; therefore, the fire hazard is greater for gasoline.

Alcohol-fueled fire can be more readily contained than a gasoline-fueled fire of equivalent volume because alcohols have a lower heat of combustion than gasoline and less of the energy released is converted to radiant heat. Therefore, energy release and potential damage from an explosion caused by alcohol would be less than that of an explosion caused by gasoline.

7.10.5 Socioeconomic impacts

Substitution of alcohol fuels for conventional fuels will increase the number of jobs in fuel production, distribution, and handling industries. Alcohol fuels are expected to cost more than gasoline over the next 10 years.

As a result, vehicle-operating costs will be somewhat higher if alcohol blends are used. The price of alcohol blends varies significantly, depending upon the type of alcohol and feedstock used. Blends containing methanol derived from coal are the least expensive. The most expensive are alcohol blends containing ethanol produced from corn.

7.10.6 Transportation and infrastructure impacts

The existing fuel distribution system must be modified and expanded to accommodate the increasing use of alcohol fuels in the long run. The changes required will include construction of new pipelines, storage facilities, and retrofitting of existing facilities with alcohol-compatible pumps, hoses, valves, and other components.

The vehicle support services such as refueling, maintenance, repairs, and vehicle sales will be unaffected by the use of alcohol fuels. The use of alcohol fuels is not expected to have a significant impact on the existing transportation system infrastructure.

References

1. A. Nag. *Analytical Techniques in Agriculture, Biotechnology and Environmental Engineering*, New Delhi, India: Prentice-Hall of India, 2006.
2. A. Nag, Bio system Engineering, McGraw Hill Publishers, USA, 2009.
3. E. S. Lipinsky. Chemicals from biomass: Petrochemical substitution options, *Science* 212, 1465–1471, 1992.
4. B. B. Ghosh, E. V. Thomas, and S. Natarajan. The Performance of a Tractor Diesel Engine with Ethanol Fumigation, Ph.D. Thesis, Mechanical Engineering Department, Indian Institute of Technology, Kharagpur, India, 1992.

Questions

1. How is methanol prepared?
2. Compare the properties of methanol and ethanol.
3. Why are blends (alcohols and gasoline) necessary for optimal engine performance?
4. What are the problems associated with engine performance when applying with alcohols?
5. What is ignition improvers? What is formaldehyde emission?

CHAPTER 8

Cracking of Lipids for Fuels and Chemicals

Ernst A. Stadlbauer and Sebastian Bojanowski

8.1   Introduction

Lipids [1] in the form of fat and edible oils are important energy sources for humans due to the high calorific value of triacylglycerols (~37 kJ/g, or ~9 kcal/g) and the nutritional benefits of both essential fatty acids and phosphate. In addition, energy stored in lipids may be technically realized by either direct use in combustion or by upgrading into a more versatile fuel. In this respect, lipids play an important role for providing lighting and warmth.

Historically, whale oil lamps and tallow candles were gradually displaced by kerosene lamps and electric bulbs [2]. Nowadays, lipids are attracting interest as a renewable source of fuels and chemical feedstock. Therefore, segmentation in the marketplace for lipids is noticeable [3]. In emerging economies of eastern Asia, there is a demand for cheap, edible commodity oils, such as soybean or palm oils. In developed economies, a nutritionally led demand for niche oils, such as low-trans-fat oils, high-omega-3 oils, and enhanced lipophilic vitamins (especially A and E), prevails. More recently, nonedible uses of lipids arise from the proliferating demand for alternative fuels [4] to substitute liquid hydrocarbons derived from mineral oil [5]. Such strategies fall into four broad categories. One is aimed at fueling diesel engines with pure vegetable oils [6] or vegetable oil–fossil fuel blends [7]. The other focuses on biodiesel (alkyl esters of fatty acids), which is mainly sourced from rapeseed and palm oils [8–10]. Problems [11] associated with the more polar characteristics of vegetable oil and biodiesel in comparison to conventional diesel has given rise to studies for cracking of lipids (vegetable oils/animal fat) into nonpolar hydrocarbons [12] to be used as a base for fuels or chemical commodities. Decomposition studies with and without catalysts (metallic salts, metal oxides) have been performed. Finally, lipids (and proteins) in dead cellular matter such as sewage sludge or meat and bonemeal may be converted by natural catalysts present in the substrate to oil having properties similar to diesel fuel [13].

In the following sections, basic processes of converting lipids into nonpolar hydrocarbons with alkanes, alkenes, and arenes as main constituents are discussed. Details of pure vegetable oils or biodiesel are outlined elsewhere (see Chaps. 4, 5, 6).

8.2 Thermal Degradation Process

Thermal decomposition of vegetable oil was performed to prove the theory of the origin of mineral oil from organic matter [14] as early as 1888. Literature up to 1983 has been reviewed by Schwab et al. [15]. In many cases, inadequate characterization of products formed in pyrolysis of vegetable oils was found. Therefore, analytical data obtained by gas chromatography–mass spectrometry (GC-MS) from thermally decomposed soybean oil and high oleic safflower oil in the presence of air or nitrogen were reported [15].

The ASTM standard method for distillation of petroleum products D86-82 has been used for decomposition experiments. Catalytic systems were excluded in this destructive distillation. The actual temperature of the oil in the feeder flask was about 100°C higher than the vapor temperature throughout the distillation. Under these conditions, GC-MS analysis showed that approximately 75 percent of the products were made up of alkanes, alkenes, aromatics, and carboxylic acids with carbon numbers ranging from 4 to more than 20 (see Table 8.1).

A comparison of fuel properties is given in Table 8.2. The carbon-hydrogen ratio shows 79 percent C and 11.88 percent H for the pyrolyzate of soybean oil. This indicates considerable amounts of oxygenated compounds in the distillate. Consequently, methylation of these oils

	Percent by mass high oleic safflower		Soy	
Class of compounds	N_2 sparge	Air	N_2 sparge	Air
Alkanes	37.5	40.9	31.3	29.9
Alkenes	22.2	22.0	28.3	24.9
Alkadienes	8.1	13.0	9.4	10.9
Aromatics	2.3	2.2	2.3	1.9
Unresolved unsaturates	9.7	10.1	5.5	5.1
Carboxylic acids	11.5	16.1	12.2	9.5
Unidentified	8.7	12.7	10.9	12.6

TABLE 8.1 Composition Data of Pyrolyzed Oil

ASTM test no.	Specification	Distilled soybean oil (N_2 sparge)	No. 2 diesel fuel	Soybean oil	High oleic safflower oil
D613	Cetane rating	43[*]	40 (min.)	37.9[*]	49.1
	Higher heating value, BTU/lb	17,333	19,572	17,035	17,030
D129	Sulfur, %	<0.005	<0.5	0.01	0.02
D130	Copper corrosion, 3 h at 50°C standard strip	1[*]	<3	1[*]	1[*]
D524	Carbon residue at 10% residium	0.45%	<0.35%	0.27%	0.24
D1796	Water and sediment, % by volume	0.05	<0.05	Trace	Trace
D482	Ash, % by weight	0.015	<0.01	<0.01	<0.01
D97	Pour point, °C	+7	−7C (max.)	+12	−21
D445	Viscosity, mm^2/s at 38°C	10.21	1.9–4.1	32.6	38.2
DE191	Carbon, %	79.00	86.61	—	—
	Hydrogen, %	11.88	13.20	—	—

* ASTM test D613 with ignition delays observed visually.

TABLE 8.2 Comparison of Fuel Properties

has revealed 9.6–12.2 percent of carboxylic acids ranging from C-3 to over C-18. This is reflected in the higher viscosity compared to diesel.

Mass-spectral fingerprints of the entire pyrolysis product slate from tripalmitin, different vegetable oils, and extracted oils from microalgae confirm that the decomposition of ester bonds in the absence of external catalysts is extensive [16–18]. However, a great variability in primary pyrolysis/vaporization product slates was observed [18].

Thermodynamic calculation in the degradation process shows that the cleavage of C-O bond takes place at 288°C and fatty acids are the main product [19]. The actual pyrolysis temperature should be higher than 400°C to obtain maximum diesel yield [20]. The mechanism of pyrolysis of vegetable oil has been discussed by various authors [9, 15, 19]. Generally, thermal decomposition proceeds through either a free-radical or carbonium ion mechanism. The primary R-COO splits off carbon dioxide. The alkyl radicals (R), upon disproportionation and elimination of ethene, give rise to alkanes and alkenes. The formation of aromatics is facilitated by a Diels-Alder addition of ethene to a conjugated diene formed in the pyrolysis reactions. However, the product mix and product quality are influenced by many factors such as feed pretreatment, heating rate, and temperature. As vegetable oils may contain trace elements, catalytic effects cannot be completely excluded from any thermal degradation process [21].

8.2.1 Catalytic cracking (CC)

In 1979, a paper [22] from the petrochemical industry reported for the first time that high-molecular-weight triglycerides such as corn oil ($C_{57}H_{104}O_6$) and castor oil ($C_{57}H_{104}O_9$) were convertible to a high-grade gasoline when passed over H-ZSM-5, a catalyst. The latter is a synthetic, medium-pore, shape-selective acid catalyst. Lipids were fed with a piston displacement pump at a rate of 2 mL/h with flowing hydrogen (300 mL/h) over 2 mL of H-ZSM-5 catalyst (0.77 g, 14–30 mesh) contained in a vertical Pyrex reactor at atmospheric pressure and $T = 400$–$450°C$. Paraffins, olefins, aromatics, and nonaromatics could be detected in the product mixture. The distribution of hydrocarbons is similar to selective conversion of methanol into hydrocarbon units with up to 10 carbon atoms per molecule. In all cases, a high degree of BTX aromatics (benzene, toluene, and xylene) was achieved. The precondition for the catalytic conversion is that the molecule penetrate the cavities of microporous zeolite.

This new catalytic approach has paved the way for a variety of applications. A schematic diagram of experimental arrangements for pyrolysis and catalytic conversion is given in Figure 8.1.

Conversion of different kinds of vegetable oils over medium-pore H-ZSM-5 have been investigated in detail [23–26]. Catalytic cracking of by-products from palm oil mills with a selectivity of 51wt.% toward aromatic hydrocarbon formation has been reported [27]. To achieve higher yields, this type of work was extended to pyrolysis and zeolite conversion of both whole algae and their major components as well as whole seeds and selected vegetable oils [18, 28–31]. Hot vapors from

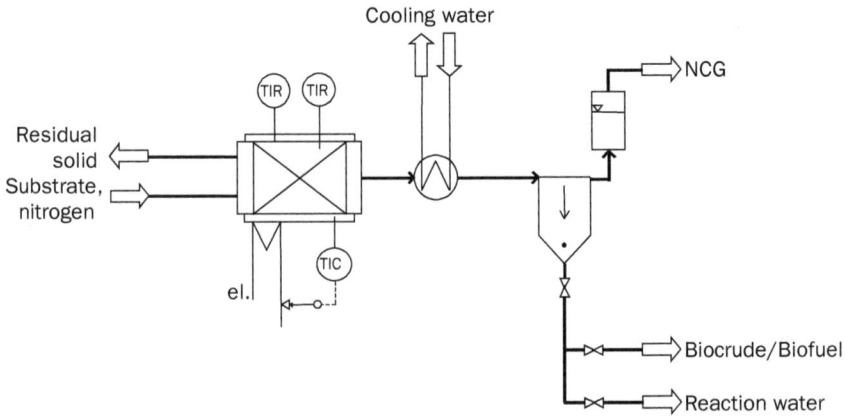

FIGURE 8.1 General scheme of pyrolysis and catalytic conversion reactor.

solid organic material (microalgae, seeds, etc.) or vaporized vegetable oils were passed directly over the H-ZSM-5 catalyst. Products of different algae, seeds, or vegetable oils emerging from the passage showed a uniform, high-octane, aromatic gasoline product. Obviously, the molecular pattern of products is insensitive to the nature of lipids used. This is in contrast to pyrolysis without a catalyst [18].

Upgrading of crude tall oil to fuels and chemicals has been studied at atmospheric pressure and in the temperature range of 370–440°C, in a fixed-bed microreactor containing H-ZSM-5 [32]. The oil was co-fed with diluents such as tetralin, methanol, and steam. High oil conversions, in the range of 80–90 wt.%, were obtained using tetralin and methanol as diluents. Conversions under steam were reduced to 36–70 wt.%. The maximum concentration of gasoline-range aromatic hydrocarbons was 52–57 wt.% with tetralin and steam, but only 39 percent with methanol. The amount of gas product in most runs was 1–4 wt.% [32].

8.3   Vegetable Oil Fuels/Hydrocarbon Blends

At first glance vegetable oil offers a favorable CO_2 balance. However, when the extra N_2O emission from biofuel production is calculated in "CO_2-equivalent" global warming terms, and compared with the quasi-cooling effect of "saving" emissions of fossil fuel derived CO_2, the outcome is that production of commonly used biofuels can contribute as much or more to global warming by N_2O emissions than cooling by fossil fuel savings [33]. In addition, widespread use of vegetable oil fuels is limited by high viscosity, low volatility, poor cold flow behavior, and lack of oxidation stability during storage [6, 7]. Partial conversion of vegetable oil to hydrocarbons offers the possibility to preserve the favorable environmental characteristics of vegetable oil–based fuels while improving viscosity and cold flow behavior [34, 35]. Figure 8.2 depicts thermogravimetry of vegetable oil without pure oil (dashed line) and in the presence of a Y-zeolite (Koestrolith). The dotted line represents the first derivative from the catalyzed conversion reaction.

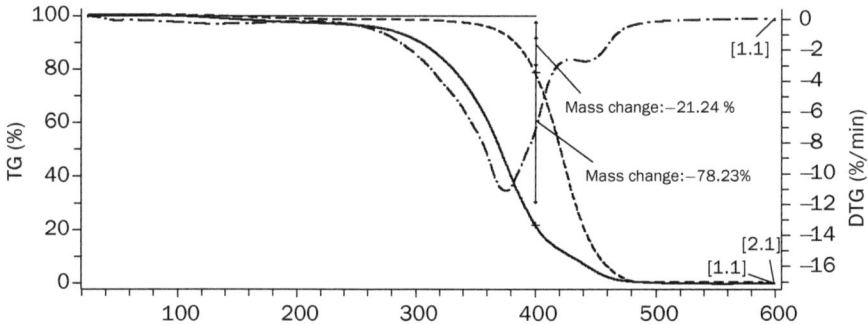

FIGURE 8.2   Thermogravimetry of commercial vegetable oil fuel without pure oil (dashed line) and in the presence of a Y-zeolite.

The efficiency of the decarboxylation effect of Y-zeolite activity on pure vegetable oil at $T = 450°C$ may be seen by comparing the IR spectrum of pure vegetable oil fuel in Figure 8.3 with the corresponding spectrum of the conversion product in Figure 8.4. The carbonyl band at around 1700 cm^{-1} is an indicator for conversion efficiency.

Table 8.3 summarizes physical and chemical parameters of vegetable oil fuel and conversion products at different temperatures. The change

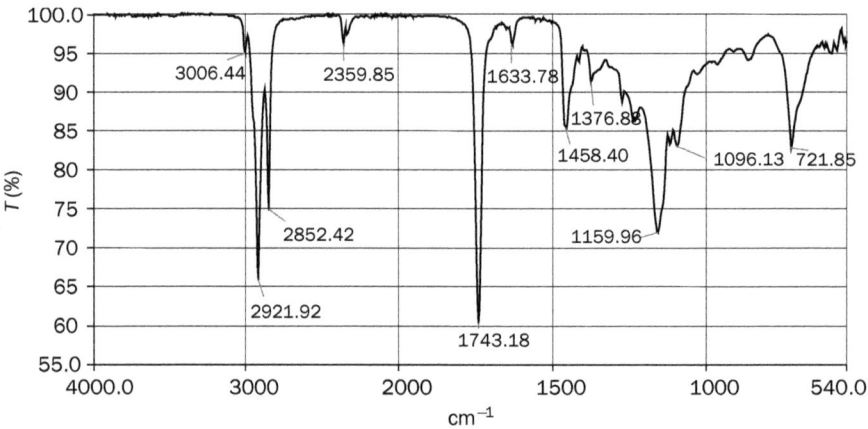

FIGURE 8.3 IR spectrum of commercial vegetable oil fuel. The bands observed at around 2900 and 1740 cm^{-1} are due to absorption of IR radiation, the absorbed energy causing transitions between energy levels for the stretching vibrations of C-H (hydrocarbons) and C=O bonds (ester function R_1COOR_2), respectively.

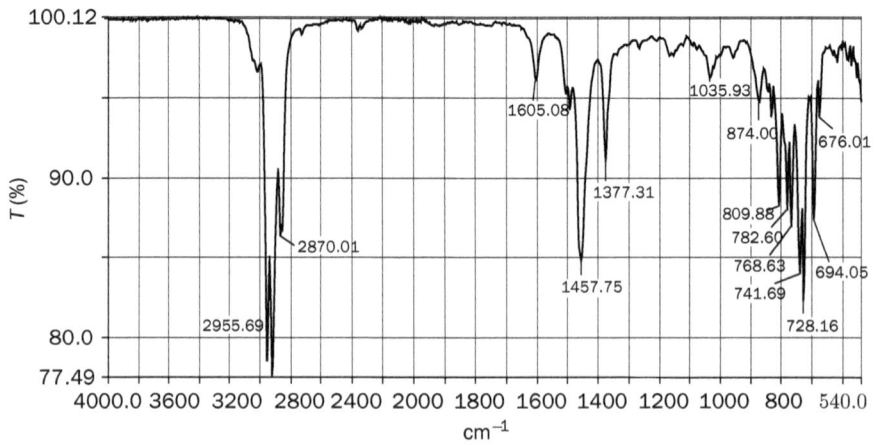

FIGURE 8.4 IR spectrum of conversion product. The carbonyl function at 1720–1740 cm^{-1} is missing. Absorption bands between 1300 and 650 cm^{-1} are generally associated with complex vibrational and rotational energy changes (fingerprint region) of the molecules.

Parameter	Commercial vegetable oil fuel	Y-zeolite (koestrolith), $T = 430°C$	Y-zeolite (koestrolith), $T = 450°C$
Yield, %	—	34.3	43.5
NCV, MJ/kg	37.0	42.4	42.2
Density, g/mL	0.91	0.79	0.81
Viscosity, mm^2/s	32.87	0.73	0.79
C, %	77.04	87.8	88.32
H, %	12.0	9.59	9.67
N, %	0.29	<0.14	<0.14
S, %	<0.34	<0.34	<0.34

TABLE 8.3 Characteristics of Commercial Vegetable Oil Fuel and Its Y-Zeolite Conversion Product

in viscosity is quite remarkable. In accordance with Figure 8.2, a reaction temperature of $T = 450°C$ is preferred.

8.3.1 Refitting engines

Presently, vegetable oil is regarded as a niche application. One liter of rapeseed oil substitutes for approximately 0.96 L of diesel. The annual yield is 1480 L/ha. CO_2 reduction in relation to the diesel equivalent is about 80 percent [36]. However, this is questioned in newer literature [33] in terms of global warming reduction considering the effects of extra N_2O entering the atmosphere as a result of using nitrogen-based fertilizers to produce crops for biofuels. Before unmodified vegetable oil is used as a fuel, the engine must be refitted for the fuel to correspond to the viscosity and combustion properties of vegetable oil. Refitting concepts include preheating either the fuel and the injection system or the equipment with a two-tank system. The engine is started with diesel and changes to vegetable oil only when the operating temperature has been reached. Blends of pure vegetable oils and a conversion product together with additives (antioxidants) increase oxidation stability, reduce viscosity, and give a better perspective for vegetable oil fuel markets.

8.3.2 Tailored conversion products

The chemical nature of conversion products depends both on the structure or type of the zeolite used and the reaction temperatures, because restructuring occurs at the inner surface, which acts as a reaction vessel at the molecular scale. Specific reactions depend on the diameters of pores, the resident time of molecules within the pores or channels and voids of the microporous zeolite, and the temperature. The penetration of lipids into a zeolite is depicted in Figure 8.5. The scheme is based on [22].

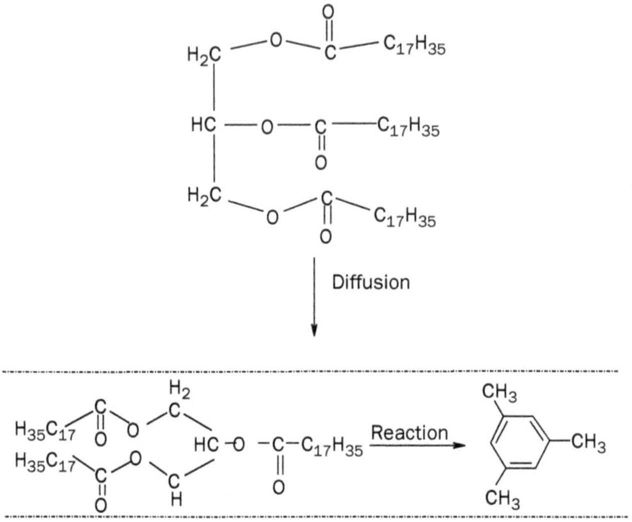

FIGURE 8.5 Scheme of restructuring triglycerides with shape-selective H-ZSM-5 to aromatic hydrocarbons.

Parameter	H-ZSM-5 PZ-2/50H, $T = 550°C$	H-ZSM-5 PZ-2/50H, $T = 550°C$	DAY-Wessalith, $T = 400°C$
Yield, %	31.48	56.74	72.9
NCV, MJ/kg	40.1	40	41.3
Density, g/mL	0.83	0.85	0.81
Viscosity, mm^2/s	0.92	1.01	2.29
C, %	88.6	84.5	83.4
H, %	10.7	12.5	13.5
N, %	<0.14	<0.14	<0.14
S, %	<0.34	<0.34	<0.34

TABLE 8.4 Yields and Physical Characteristics of Hydrocarbons from Catalytic Conversion of Animal Fat Using Zeolite Types H-ZSM-5 (Pentasil, PZ-2/50H) and DAY-Wessalith at Different Temperatures

To demonstrate this influence of catalysts and reaction temperature on yields and products, Table 8.4 considers a shape-selective zeolite type H-ZSM-5, commercially available as Pentasil, PZ-2/50H, and Y-zeolite (DAY-Wessalith). The physical characteristics of oils formed from the conversion of animal fat (rendering plant) are depicted [56]. Yields are between 30 percent and 70 percent, depending on the type of zeolite and temperature. Net calorific values are in the range of 40 MJ/kg compared to 35 MJ/kg of animal fat. All reaction products show relatively low viscosity and densities.

Products at $T = 400°C$.   Again, the chemical nature of products formed from animal fat was analyzed by spectroscopic methods (see Fig. 8.6). The IR spectrum reveals the hydrocarbon nature of products. The strong C-H stretching vibrations (frequencies) at 2900 cm^{-1} is characteristic for alkanes. Functional groups are widely missing. The comparison to diesel from a commercial gas filling station (imprinted spectrum) shows a similar pattern [37].

Proton resonance spectroscopy depicts the chemical environment of protons in the product formed from the conversion of animal fat. Figure 8.7 shows the dominance of aliphatic protons at chemical shifts of 0.9–2.25 ppm. Aromatic protons absorb at 6.5–8 ppm. The inspection of the ratio of the integral of absorptions reveals 5 percent aromatics for catalysis at $T = 450°C$. This is also reflected in the ^{13}C-NMR spectrogram (see Fig. 8.8). However, with increasing temperature in the catalytic bed, the content aromatic alkylbenzenes increase.

Using ^{13}C-NMR spectroscopy in-depth mode (see Fig. 8.9), negative signals at 30–20 ppm are characteristic for CH$_2$-groups. The intensity indicates the presence of long-chain hydrocarbons. Peaks between 140 and 120 ppm denote carbon atoms of aromatic systems. The low intensity reflects the low content. Obviously, catalytic cracking over a Y-zeolite widely preserves hydrocarbon moiety in vegetable oil.

These spectroscopic findings are confirmed by gas chromatography (GC) [56]. Pyrolyzates (see Fig. 8.10) and commercial diesel (see Fig. 8.11) have a similar GC pattern. However, crude conversion products contain more volatile hydrocarbons.

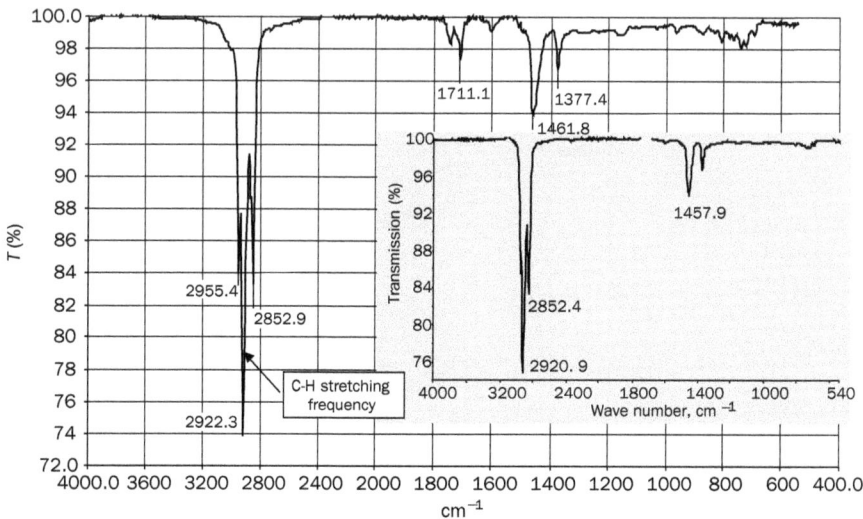

FIGURE 8.6   IR spectrum of hydrocarbons derived from animal fat at 400°C (Y-zeolite catalyst, DAY-Wessalith).

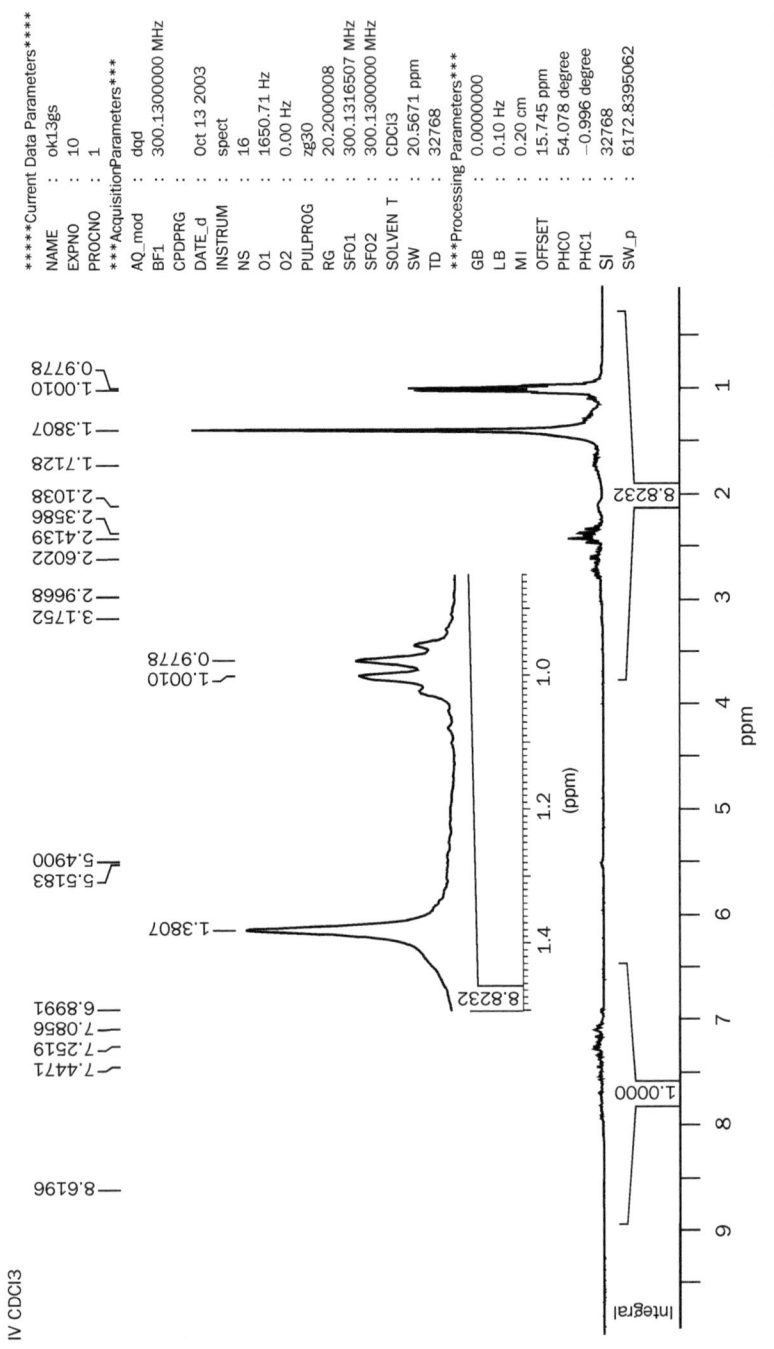

FIGURE 8.7 ¹H-NMR spectrogram of hydrocarbons from animal fat at $T = 400°C$ (Y-zeolite catalyst, DAY-Wessalith).

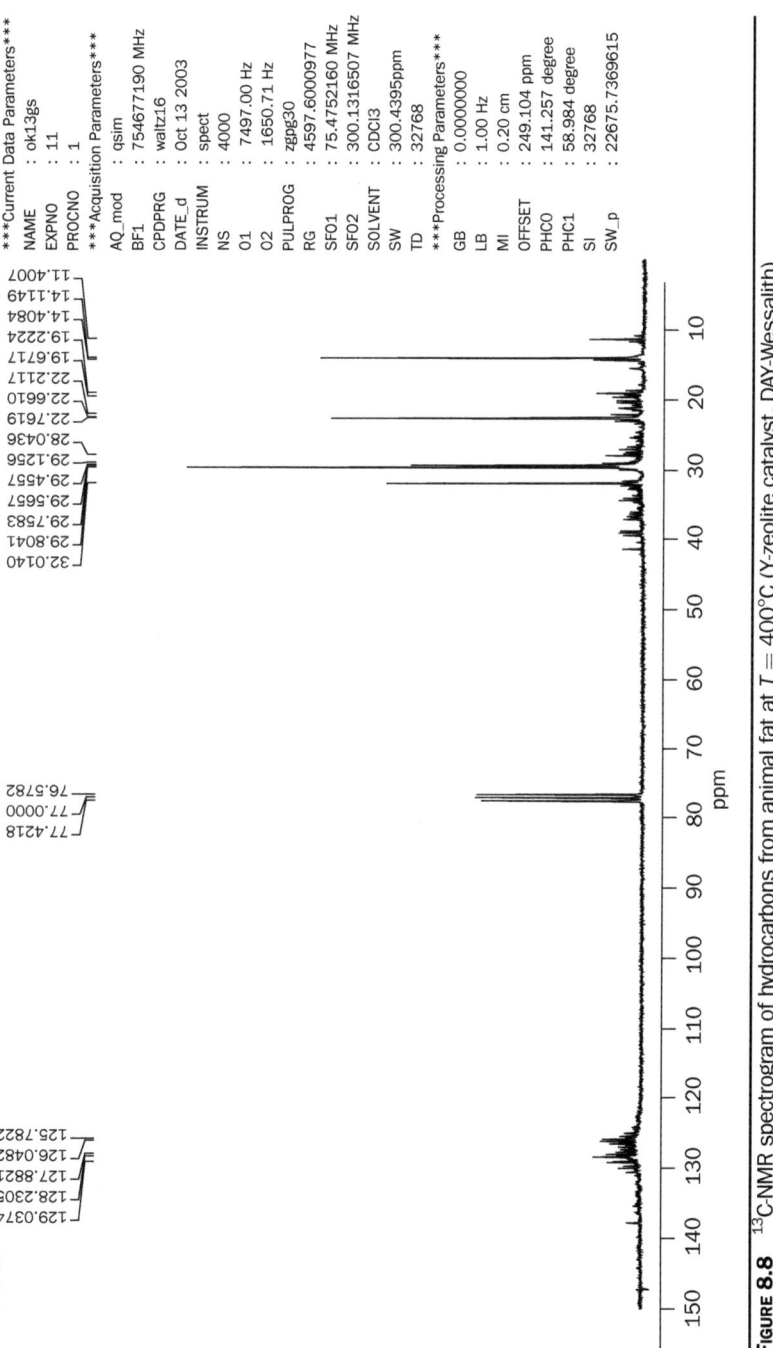

Figure 8.8 ^{13}C-NMR spectrogram of hydrocarbons from animal fat at $T = 400°C$ (Y-zeolite catalyst, DAY-Wessalith).

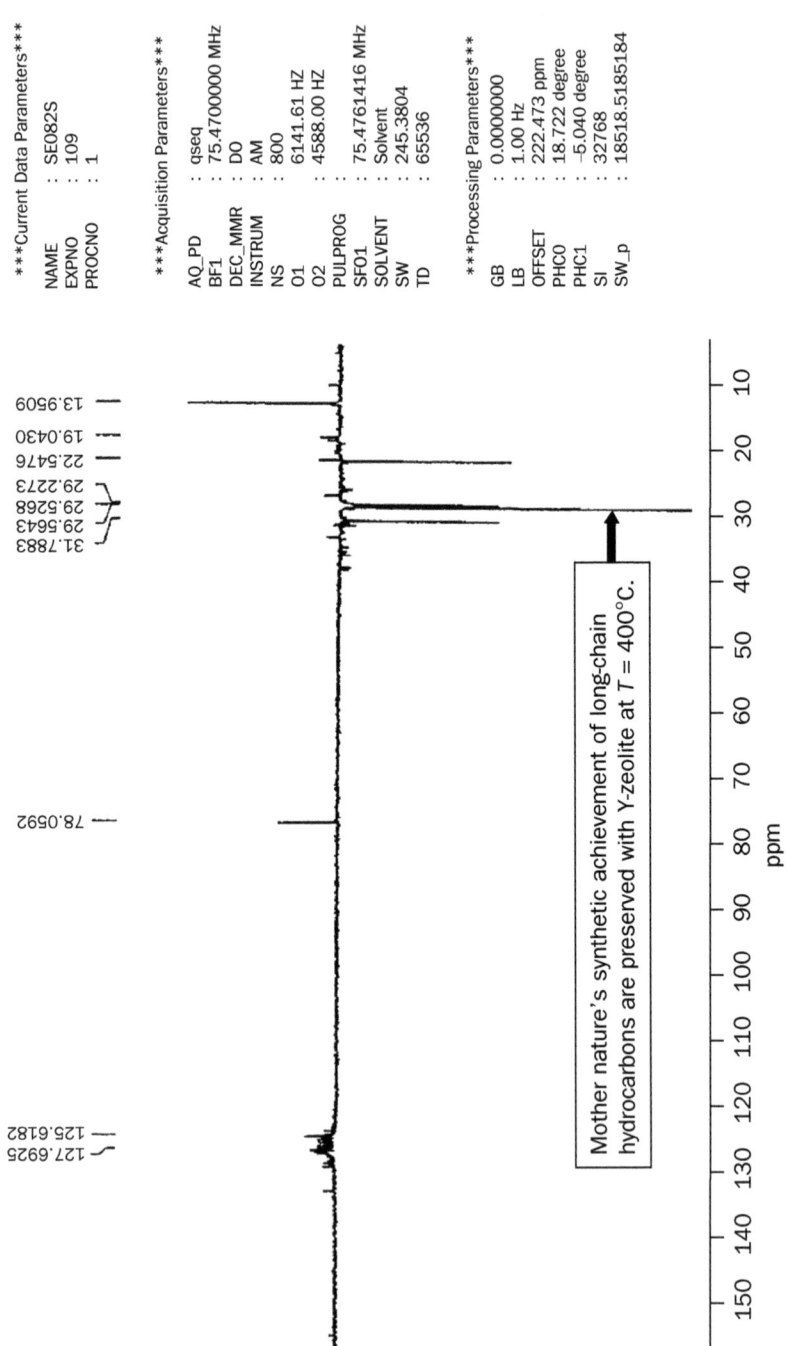

Figure 8.9 DEPT-135 ^{13}C-NMR spectrogram of biofuel from animal fat at 400°C (Y-zeolite catalyst, DAY-Wessalith).

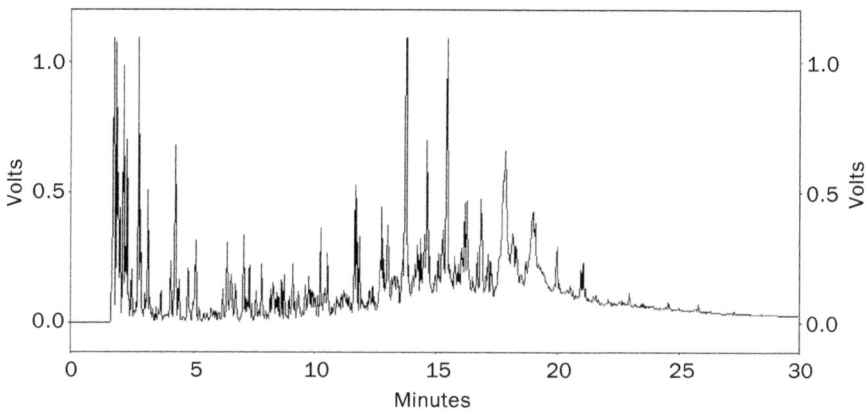

FIGURE 8.10   GC pattern of Y-zeolite conversion product of animal fat at reaction temperature $T = 400°C$. GC-14A Shimadzu, column: FS-Supreme-5/H53, 30 m; temperature program: 50°C (5 min); 158C/min to 320°C (10 min); FID detector at 320°C.

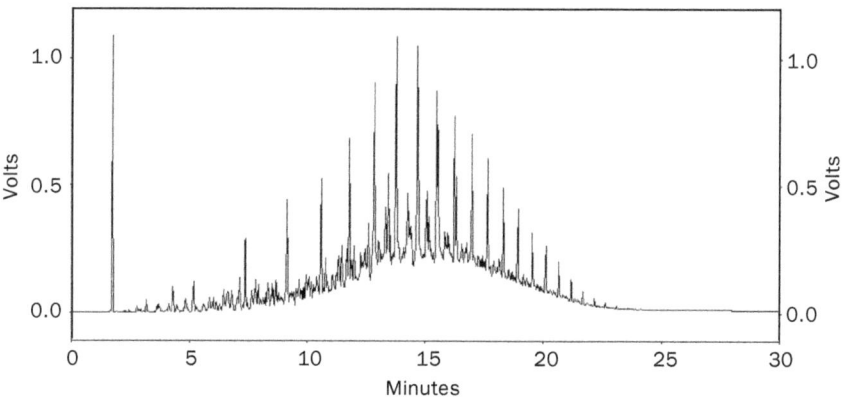

FIGURE 8.11   GC pattern of commercial diesel. GC-14A Shimadzu, column: FS-Supreme-5/ H53, 30 m; temperature program: 50°C (5 min); 15°C/min to 320°C (10 min); FID detector at 320°C.

GC separation on an OV101 capillary [column: 20 m × 0.3 mm, split 1:25; temperature program: 25°C (2 min), 4°C/min to 320°C] reveals double peaks in more detail (see Fig. 8.12). The first peak is for the alkene with a double bond of a given C number. The second peak is for the alkane having the same C number.

You may use these hydrocarbons as a base for biofuels. However, there are markets for certain fractions of this hydrocarbon mixture. For example, the C-12 to C-18 fraction is a raw material widely used for bulk commodities. As mineral oil prices increase, it is becoming more financially viable to produce chemical feedstock for commodities and specialities from wastes. Wastes are an energy and carbon source of the future.

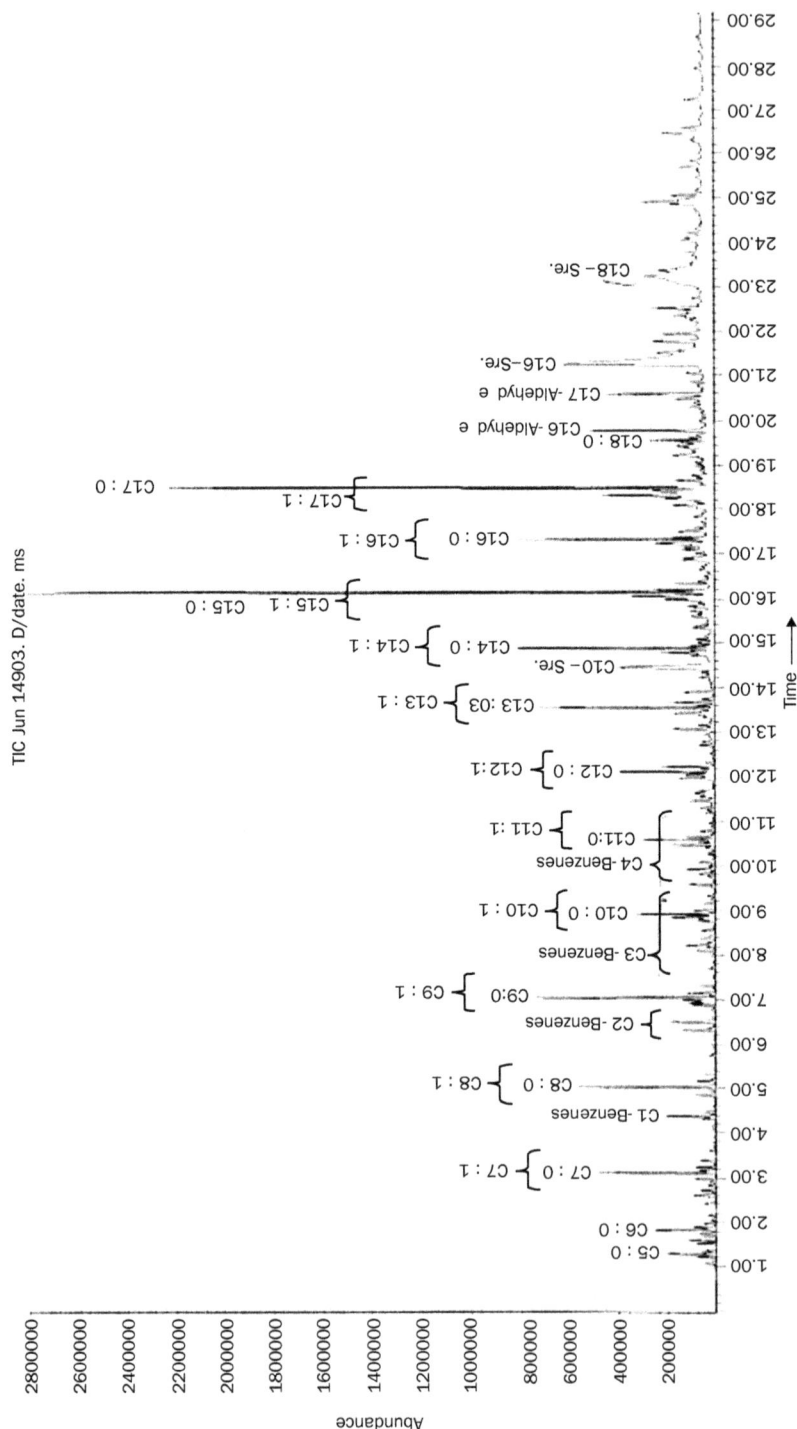

FIGURE 8.12 GC separation of products from degradation fuel from animal fat at $T = 400°C$ (Y-zeolite catalyst, DAY-Wessalith) with characteristic double peaks.

Products at $T = 550°C$. For a given H-ZSM-5-zeolite, the nature of conversion products of lipids (animal fat) shifts to more aromatic compounds as the temperature increases. This is demonstrated by different NMR findings [56] for animal fat as a substrate at a reaction temperature of $T = 550°C$ (see Figs. 8.13 through 8.15). Especially, DEPT-135 ^{13}C-NMR pattern of oil from catalytic conversion of animal fat at 550°C shows the dominance of aromatic protons and a very low amount of CH_2 groups. Chromatographic separation revealed alkylbenzenes (especially 1,3,5-trimethybenzene) as main products [38].

Heating oil and a conversion product from animal fat have been used in a commercial burner (Buderus, Germany). Both oils resulted in emissions within legal limits (see Table 8.5).

A straightforward approach to apply vegetable oil in the most-talked-about biomass-to-liquid-fuel scheme is to use it as a co-substrate in mineral oil refineries. Advantages are low investments for peripheral facilities such as loading and storage and use of an existing infrastructure for distribution and marketing. The processing of rapeseed

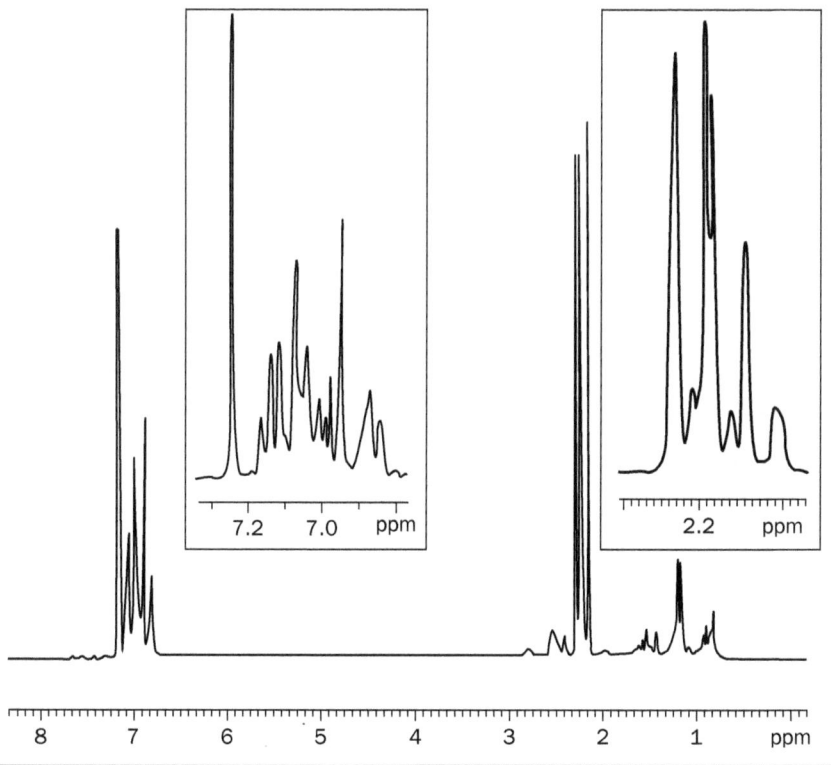

FIGURE 8.13 ^{1}H-NMR spectrogram of hydrocarbons from animal fat at $T = 550°C$ with the commercial catalyst H-ZSM-5 (Pentasil, PZ-2/50H).

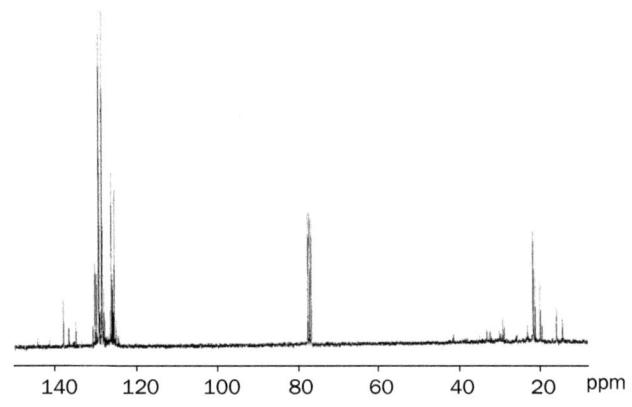

FIGURE 8.14 ^{13}C-NMR spectrogram of hydrocarbons from animal fat at $T = 550°C$ with the commercial catalyst H-ZSM-5 (Pentasil, PZ-2/50H).

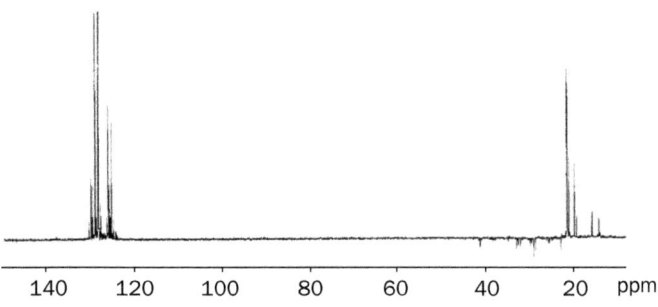

FIGURE 8.15   DEPT-135 ^{13}C-NMR spectrogram of hydrocarbons from animal fat at $T = 550°C$ with the commercial catalyst H-ZSM-5 (Pentasil, PZ-2/50H).

Parameter	Heating oil	Heating oil & oil from AF 1:1	Oil derived from AF	Limiting value
NCV, MJ/kg	42.0	41.5	41.3	>42*
Kinetic viscosity, mm²/s	3.25	2.74	2.51	<6.0*
C, %	86.5	85.7	83.4	—
H, %	14.0	13.8	13.5	—
N, %	<0.14 (d/l)	<0.14 (d/l)	<0.14 (d/l)	—
S, %	<0.34 (d/l)	<0.34 (d/l)	<0.34 (d/l)	<0.20*
NO$_X$, mg/m³	162	186	233	250†
SO$_2$, mg/m³	87	26	0	350†
Smoke pot no.	0.0	0.4	0.4	1‡

*DIN 51 603; †TA Luft; ‡1. BImSchV; d/l: detection limit

TABLE 8.5   Comparison of Combustion Parameters: Heating Oil versus Oil Derived from Y-Catalytic Conversion of Animal Fat (AF) at $T = 400°C$

Fraction	Total oil		Gasoline		Middle distillate		VGO*	
Rapeseed oil, %	0	20	0	20	0	20	0	20
Density (15°C), g/mL	0.815	0.815	0.753	0.759	0.830	0.817	0.852	0.847
Carbon, mass %	86.04	85.33	85.39	85.31	86.06	85.27		
Hydrogen, mass %	14.01	14.42	14.48	14.64	13.82	14.66		
Sulphur, ppm	284	114	29	39	103	18	38	11
Nitrogen, ppm	<1	2	<1	0.5	<1	<1	0.7	<1
Oxygen, mass %	0.1	0.1	0.05	0.1	<0.1	0.06		
NCV, MJ/kg			43.9	44.0	43.4	44.0		
Octane number (MOZ)			63.2	61.4				
Cetane number					48	64		
Pour point,°C					−35	+3		

*VGO, vacuum gas oil.

TABLE 8.6 Product Quality of the Hydrocracker with 20 Percent and without Rapeseed Oil as a Feed Component

oil as a feed component in a hydrocracker was described in 1990 [39]. The results are summarized in Table 8.6.

It is worth mentioning that rapeseed oil is converted in the hydrotreatment step to paraffins. The oxygen content of the vegetable oil causes an increased consumption of hydrogen to form water. Changes in quality occur in the middle distillate. A lower density and a higher cetane number are a quality-enhancing advantage. A drawback is the susceptibility to freezing point of the fuel. This kind of cold flow behavior would make its use in winter impossible unless special additives are supplemented [40].

8.3.3 Feed component in FCC

In 1993, the influence of 3–30 percent rapeseed oil in vacuum distillate FCC feed on product slate and quality both at laboratory and at a continuously operated bench-scale apparatus was reported for the first time [41]. On the one hand, results showed decreasing yields of liquid hydrocarbons with increasing rapeseed oil concentrations. On the other hand, the gasoline portion in the liquid product increased. Considering propenes, butanes, and *i*-butenes as gasoline potentials, low rapeseed oil portions in the FCC feed seem to result in an optimum yield of gasoline plus gasoline potentials. Most interestingly, the gasoline fraction recovered from a 500-h bench scale run using a feed with 30 percent rapeseed oil proved suitable for standardized gasoline blending. Calcium concentration $c(Ca) > 2$ ppm gradually decreases FCC catalyst activity. Oxygen contained in the vegetable oil was mainly converted to water. Moreover, traces of phenols and carboxylic acid were detected in the liquid reaction product.

MAT with animal fat. In a laboratory scale, mixtures of vacuum gas oil and up to 15 percent of animal fat were converted in a Micro-Activity Test (MAT) unit [37]. Results are given in Figures 8.16 and 8.17. Two aspects

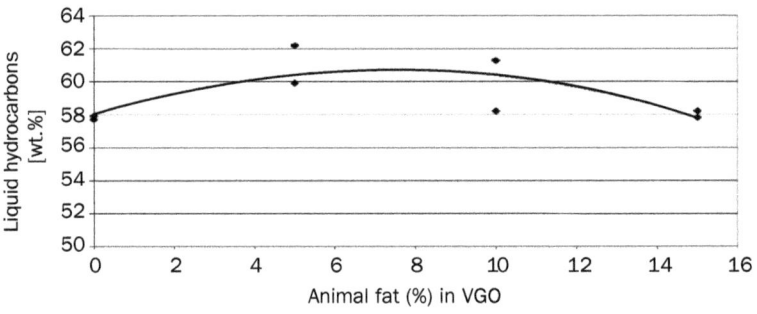

FIGURE 8.16 Co-catalytic cracking of animal fat and vacuum gas oil (VGO) in MAT experiments. At around 7 percent feed component, the maximum yield of liquid hydrocarbons is found; weight–hourly space velocity (WHSV) = 2 h^{-1}.

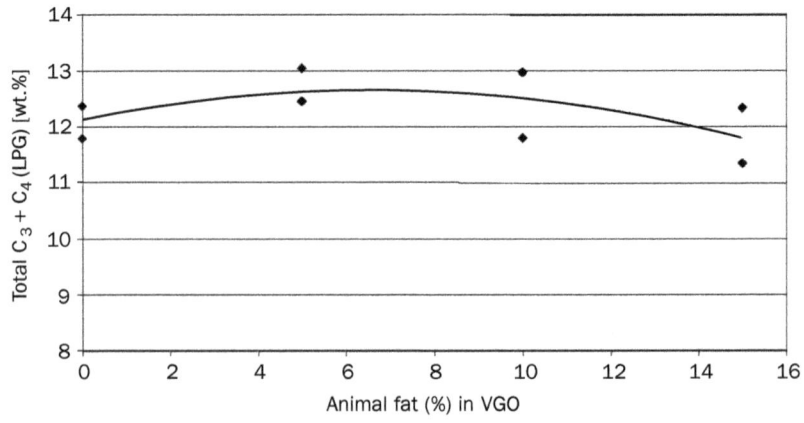

FIGURE 8.17 Co-catalytic cracking of animal fat and vacuum gas oil (VGO) in MAT experiments. Gasoline potentials show a maximum at low rapeseed portions around 7 percent; weight–hourly space velocity (WHSV) = 2 h^{-1}.

are of special interest. First, yields of propene and butene increase with animal fat as a co-substrate. This is an advantageous finding as C-3 and C-4 are gasoline potentials. C-3 and C-4 liquefied petroleum gas can be used for the production of isoparaffins for motor gasoline through alkylation and polymerization processes.

Second, a higher yield of gasoline fraction is observed. This is a consequence of the high hydrogen:carbon ratio of about 2 and the low heteroatom content. For this reason, biomaterials with a hydrocarbon-like structure are particularly interesting candidates for conversion to low-molecular-weight fuels or chemical raw materials. Problems to be investigated are possible calcium and phosphate deposits on the catalyst particles, which may impair catalyst activity and process stability of the riser. Therefore, the process must include a regeneration step. The market will decide whether or not animal fat can substitute a bit of nonrenewable resources in petroleum refining.

8.4   Other Metal Oxide Catalysts

Cottonseed oil has been thermally decomposed at 450°C using 1 percent Na_2CO_3 as a catalyst [42]. Pyrolysis produced a yellowish-brown oil with 70°C yield. The fuel properties of original and pyrolyzed cottonseed oil are summarized in Table 8.7. Results of ASTM distillation compared to diesel are given in Table 8.8 showing a higher volatility for the conversion product.

Rapeseed oil was pyrolyzed in the presence of about 2 percent calcium oxide up to a temperature of 450°C [43]. An oil was obtained with a heating value of 41.3 MJ/kg, a kinematic viscosity of 5.96 mm^2/s, a cetane number of 53, and a flash point of 80°C. When tested on a diesel

Property	Diesel fuel	Pyrolyzed oil	Original oil
API gravity	35	35	21.5
Specific gravity (at 15.6°C)	0.8504	0.8500	0.9246
Kinetic viscosity (mm/s^2) at 40°C	0.0213	0.0178	0.0357
Cetane index	33	28	20
Flash point, °C	96	53	268
Sulfur content, wt.%	0.04	Nil	0.02
Pour point, °C	0.0	>15	23
Sediment content, wt.%	Nil	0.04	5.0
Calorific value, kJ/g	45.57	45.57	41.80
Water content, vol.%	Nil	2.98	1.20
Ash content, wt.%	Nil	Nil	Nil
Carbon residue, wt.%	0.01	0.16	1.06

TABLE 8.7 Fuel Properties of Original and Pyrolyzed Cottonseed Oil and No. 2 Diesel Fuel

	Temperature °C	
Parameter	**Diesel oil**	**Pyrolyzed oil**
Distillate, %		
0	63	55
10	105	79
20	174	116
30	192	131
40	200	157
50	210	178
60	235	186
70	245	220
80	250	247
90	255	269
98	260	–
Recovery, %	98	90
Residue, %	1	9
Loss, %	1	1

TABLE 8.8   Results of ASTM Distillation of No. 2 Diesel Oil and Pyrolyzed Cottonseed Oil as Volume Percent

engine, the thermal efficiency (η_{th}) and brake specific fuel consumption were improved. The concentration of nitrogen oxide in the exhaust gas was less than diesel. The absence of sulfur in the pyrolytic oil was seen as an advantage to avoid corrosion problems and the emission of polluting sulfur compounds from combustion.

Triolein, canola oil, trilaurin, and coconut oil were pyrolyzed over activated alumina at 450°C and atmospheric pressure [44]. The products were characterized by IR spectrometry and decoupled ^{13}C-NMR spectroscopy. The hydrocarbon mixture contained both alkanes and alkenes. These results are significant for the pyrolysis of lipid fraction in sewage sludge as well as for wastes from food-processing industries [44].

Pyrolysis of rapeseeds, linseeds, and safflowers results in bio-oil containing oxygenated polar components. Hydropyrolysis at medium pressure in the presence of 1 percent ammonium dioxydithiomolybdenate $(NH_4)_2MoO_2S_2$ can remove two-thirds to nine-tenths of the oxygen present in the seeds to generate bio-oils in yields up to 75 percent [45]. In addition, extraction with organic solvents including diesel oil gave yields up to 40 percent.

The potential of liquid fuels from *Mesua ferrea* seed oil [46], *Euphorbia lathyris* [47, 48], and underutilized tropical biomass [49] has been investigated in the search for "energy farms" involving the purposeful cultivation of selected plants to obtain renewable energy sources.

8.5   Cracking by In Situ Catalysts

This method is applicable for cellular biomass containing lipids, e.g., sewage sludge or organic residues from rendering plants. The European Union is looking for new markets for both materials. On the one hand, treatment of municipal and industrial wastewaters generates huge quantities of sludge, which is the unavoidable by-product especially if biological processes are used. Management of this residue poses an urgent problem. The residue contains about 60 percent of bacterial biomass and up to 40 percent of inorganic materials such as alumina, silicates, alkaline and alkaline earth elements, phosphates, and varying amounts of heavy metals [56]. On the other hand, returning animal meal (AM) or meat and bone meal (MBM) from the rendering plant into the food cycle is forbidden by law since the BSE crisis [50, 51]. Besides burning, low-temperature conversion (LTC) of these organic materials offers an alternative disposal method [52–54]. LTC is a thermocatalytic process whereby organics react to hydrocarbons as the main product [12].

The conversion of bacterial biomass or organic residues from rendering plants to oil may be formally defined by considering the starting materials and the end products. The principal components of these substrates are proteins and lipids. They make up about 60–80 percent of this biomass. The average elemental composition of neutral lipids is $C_{50}H_{92}O_6$. An empirical formula for proteins is $(C_{70}H_{135}N_{18}O_{38}S)_x$. From these compounds, nonpolar hydrocarbons of the general elemental composition C_nH_m have to be produced [13, 55].

Obviously, LTC removes the heteroatoms from both principal components. In general, it splits off functional groups from complex biomass. The process operates at moderate temperatures (380–450°C), essential atmospheric pressure, and the exclusion of oxygen. Under these conditions, heteroatoms from organics are removed as ammonia (NH_3), dihydrogensulfide (H_2S), water (H_2O), and carbon dioxide (CO_2). This decomposition scheme may serve as a model for the formation of coal from primarily plant sources. Carbohydrates $(C_6H_{10}O_5)n$ are the principal components in plants. The elimination of water from carbohydrates produces elemental carbon, according to the following reaction:

$$(C_6H_{10}O_5)_n - 5H_2O \rightarrow C_m$$

Consequently, carbohydrates of bacterial mass will be converted to carbon, mainly in the form of graphite [56, 57]. Therefore, the formation of oil from complex biomass will always be accompanied by the formation of carbon. Figure 8.18 depicts the mechanism for the production of oil from lipids by LTC [58].

It is worth mentioning that the ash content (Table 8.9) includes natural catalysts (e.g., alumina and silicates) that substantially influence the yield and composition of LTC products. Table 8.9 shows results of

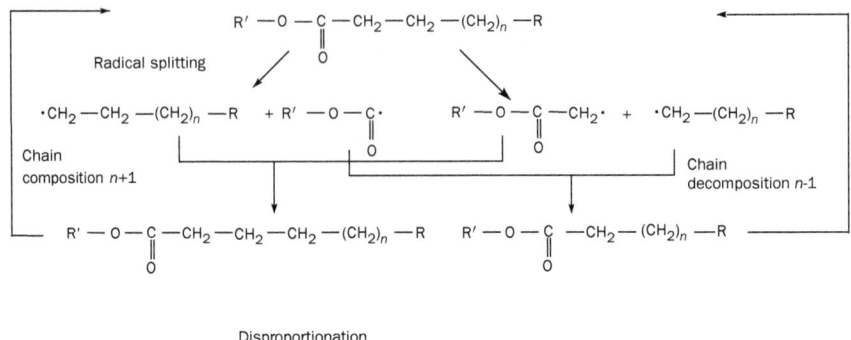

Figure 8.18 Mechanistic aspects of the formation of hydrocarbons by cracking of lipids [58].

Parameter	AS	DS	AM	MBM
Dry solids, %	95.0	79.6	94.3	95.0
Ash content, %	35.1	40.7	23.2	38.2
Protein, %	32.9	26.6	52.3	49.6
Fat, %	—	—	14.4	8.9
Calcium as Ca, %	—	9.6	9.7	20.1
Phosphorus as P_2O_5, %	—	6.3	8.7	16.0
NCV, MJ/kg	14.2	9.9	18.8	15.4
C, %	31.6	23.0	42.5	30.5
H, %	4.4	5.0	6.6	4.8
N, %	5.0	3.3	8.3	7.6
S, %	0.6	1.0	0.5	0.3

AS: aerobically stabilized sewage sludge; DS: digested sewage sludge;
AM: animal meal; MBM: meat and bone meal.

Table 8.9 Chemical and Physical Characteristic Substrates for LTC

the conversion of these organic residues. Yields of oil, solid product, water, volatile salts (NH_4Cl, $NaHCO_3$), and noncondensable gases (NCG: CO_2, H_2, C-1–C-4 alkanes and different alkenes) are given in Figure 8.19. Digested sludge produces less oil than aerobically stabilized sludge. This correlates with the carbon content in Table 8.9. The food chain of anaerobic bacteria efficiently removes organic carbons as biogas (CH_4/CO_2). Thus it is no longer available for the production of oil in subsequent LTC. AM shows higher yields of oil due to its higher content of fat and proteins (Table 8.9). The viscosities of untreated oils at 40°C are as follows: DS, 14 mm^2/s; AS, 35 mm^2/s; AM, 27 mm^2/s; and MBM,

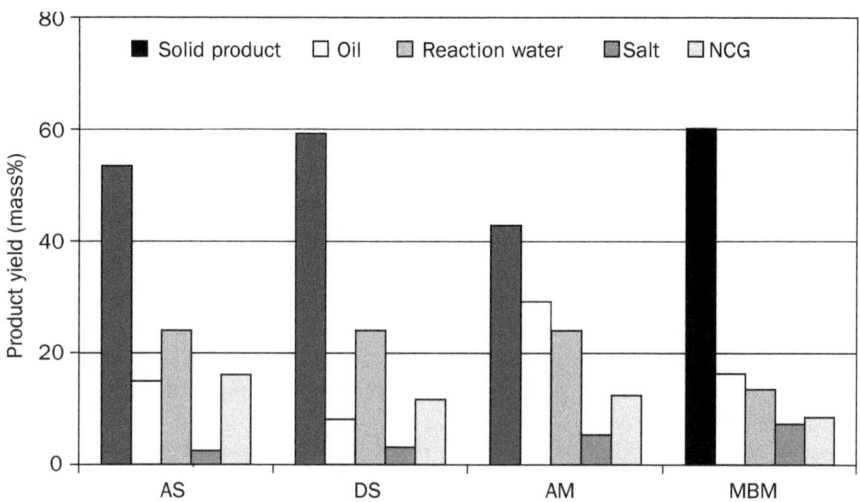

FIGURE 8.19   Mass yield of LTC products from different substrates. AS: Aerobically stabilized sewage sludge; DS: digested sewage sludge; AM: animal meal; MBM: meat and bone meal [60].

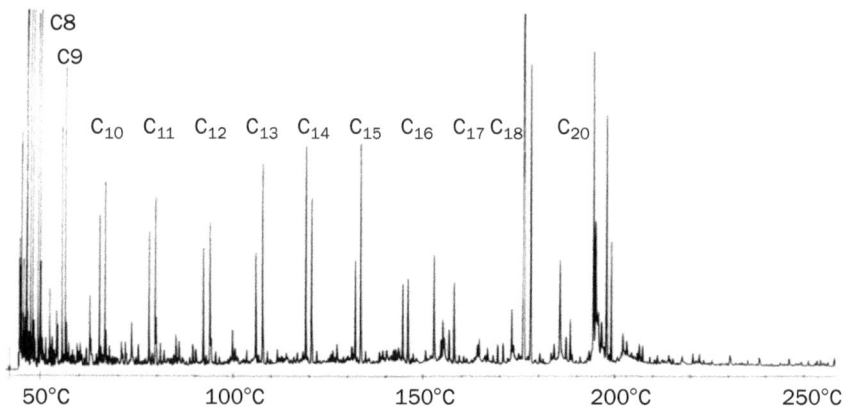

FIGURE 8.20   Chromatographic separation of oil derived from sewage sludge; separation was performed on OV101 capillary column 20 m × 0.3 mm, split 1:25; temperature program 25°C (2 min), 4°C/min to 320°C.

21 mm^2/s. In comparison, diesel from a filling station has a viscosity of about 4 mm^2/s. The solid products consist of carbon, nonvolatile salts (e.g., $CaKPO_4$), and metal oxides or sulfides. Especially in the case of AM and MBM, the solid product is of commercial interest due to its high content of phosphate. It is free of proteins [59].

As with natural crude oils, the hydrocarbon mixtures obtained by LTC of lipids containing biomass are of a highly complex composition. For example, Figure 8.20 shows the gas chromatogram of oil derived from sewage sludge AS [61]. Peaks assigned by numbers correspond to the aliphatic, unbranched saturated hydrocarbons. The peak appearing before the *n*-alkane corresponds to the *n*-alkenes.

The predominant aliphatic nature of oils produced is readily ascertained by NMR spectroscopy. Figure 8.21 depicts the ^{1}H-NMR spectrogram of oil from DS with about 5 percent of aromatic protons.

Infrared spectroscopy (see Fig. 8.22) reveals the presence of C-H-stretching frequencies at 2850–3000 cm^{-1}. In addition, the spectrum provides clear evidence of hydrogen bonding due to a broad absorption band of 3350 cm^{-1}. Thus, decarboxylation of lipids in the presence of in situ catalysts is not complete. This is consistent with the higher viscosities in comparison to diesel. A special loop reactor for recycling catalytic activity to overcome these problems has been designed [62].

Hydrocarbons are derived from both lipids and proteins in the sewage sludge in the presence of in situ catalysts. However, oil produced from proteins under anaerobic LTC conditions is high in nitrogen and sulfur: Amines, purins, and mercaptanes are trace contaminants that are formed. Consequently, this oil smells and is a nuisance, and upgrading (e.g., over H-ZSM-5 as catalyst) is essential [64]. The useful oil is produced from lipids. When sewage sludge was spiked with triolein, representative of unsaturated triglycerides, the compound did not survive the LTC [65]. As a result, sludge was extracted with toluene

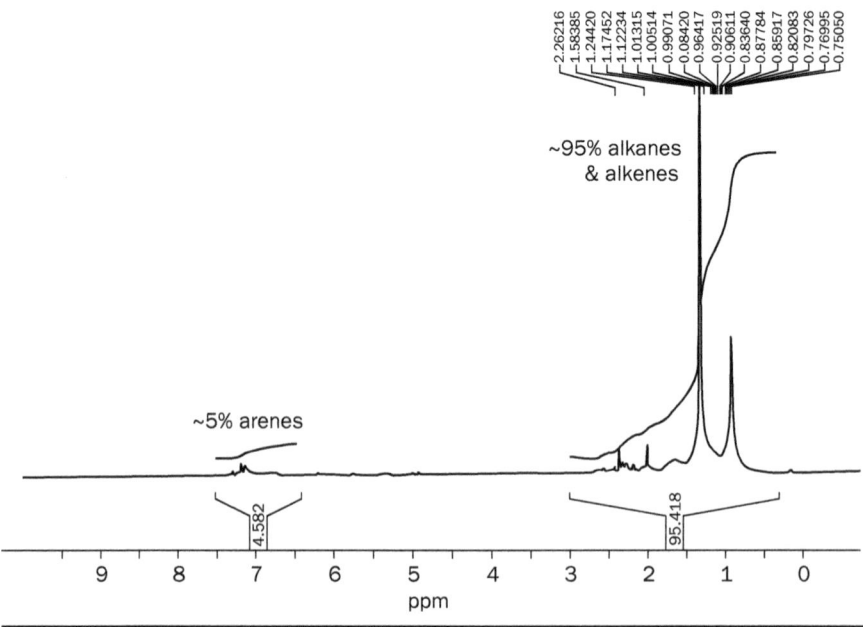

~5% arenes

~95% alkanes & alkenes

FIGURE 8.21 ^{1}H-NMR of oil from LTC of DS at $T = 400°C$.

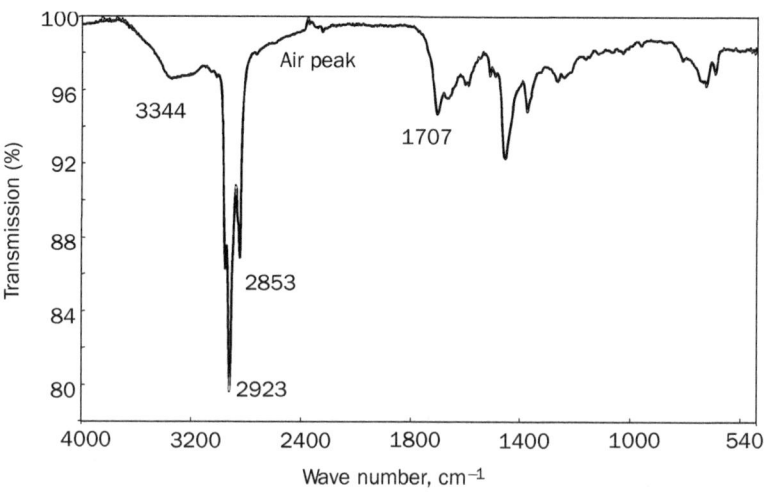

FIGURE 8.22 Infrared spectrum of oil from DS shows associated –OH and –NH bonds (3350 cm^{-1}) from the remaining carboxylic acids R-COOH or amides R-CONH$_2$ [63].

Component	C, %	H, %	O, %	N, %	S, %	Ash, %
Original sludge	39.5	6.0	26.1	6.00	0.7	20.5
Extracted lipids	72.2	10.7	14.2	0.35	0.85	0.6
Liquid product	86.6	13.5	0.0*	0.08	0.22	0.0*

*By difference.

TABLE 8.10 Elemental Composition of Original Dried Sludge, Extracted Lipids, and Pyrolyzed Liquid Product

using a Soxleth extraction method to yield 12 wt.% lipids. Pyrolysis of sewage sludge lipids over activated alumina produced liquid hydrocarbons containing mostly alkanes [65]. Even the carboxylic acid fractions of the lipids that were separated were completely converted. This is in contrast to direct sewage sludge LTC, where long-chain carboxylic acids are detectable in the IR spectrum (see Fig. 8.22). The reason is the lower content of catalytically active in situ material. Pyrolyzed liquid products from sewage sludge lipids contain virtually no nitrogen or sulfur (see Table 8.10). Only this liquid has a potential for use as a base for commercial fuels [65].

8.6 Conclusion

The potential offered by lipids for alternative fuel and chemicals is widely recognized. Various sources from plant seeds to animal fat are commercially available. Cracking converts polar esters into nonpolar hydrocarbons. Highly efficient conversion technology should include use of catalysts, e.g., zeolites such as H-ZSM-5 or Y-type representatives. At 380–450°C, alkanes and alkenes are predominantly found in the liquid product. With increasing temperatures up to 550°C, the product spectrum shifts to alkylbenzenes with 1,3,5-trimethylbenzene as the main product. For commercial fuel production based on lipids, assessment of oxidation stability and deposit formation are essential. Influences on regulated and nonregulated emissions have to be analyzed. Attention should be paid both to the NO_x content of exhaust gas and to the particle size distribution with special focus on ultrafine particles. In addition, mutagenic tests for potency of particulate matter extracts are recommended. Finally, it has to be kept in mind that the replacement of fossil fuels by biofuels may not bring the intended climate cooling due to the accompanying emissions of N_2O from the use of N-fertilizers in crop production. Much more research on the sources of N_2O and the nitrogen circle in connection with biofuels from lipids is needed.

References

1. A. Lehninger, M. M. Cox, and O. Lehninger. *Principles of Biochemistry*, New York: W. H. Freeman & Company, 2004.
2. E. S. Lipinsky, D. Anson, J. R. Longanbach, and M. Murphy. Thermochemical applications for fat and oils, *Journal of American Oil Chemists' Society* **62**, 940–942, 1985.
3. D. J. Murphy. Production of healthy vegetable oils, In: *Book of Abstracts, 4th European Federation Lipid Congress: Oils, Fats and Lipids for a Healthier Future*, October 1–4, 2006, University of Madrid, Spain, p. 27.
4. Directive 2003/30/EG of European Parliament and Council from May 8, 2003, for Promotion of the Use of Biofuels or Other Renewable Fuels in the Traffic Sector.
5. J. Krahl, A. Minack, N. Grope, Y. Tuschel, and O. Schroeder. Emissions from biodiesel and vegetable oil, In: *Fuels of the Future 2007*, Conference Papers, Domnik Rutz and Rainer Janson (Eds.) UFOP & BBE, Berlin Germany November 27–28, 2006.
6. G. Vellguth. Performances of Vegetable Oils and Their Monoesters as Fuels for Diesel Engines, SAE Paper No. 831358, Society of Automotive Engineers, Inc. Warrendale, PA, 2000
7. K. R. Kaufman, T. J. German, G. L. Pratt, and J. Derry. Field evaluation of sunflower oil/diesel fuel blends in diesel engines, *Transactions of the ADSE* **29**(1), 2–9, 1986.
8. K. L. Harrington. Chemical and physical properties of vegetable oil esters and their effect on diesel fuel performance, *Biomass* **9**, 1–17, 1986.

9. A. Srivasrava and R. Prasad. Triglycerides-based diesel fuels, *Renewable & Sustainable Energy Reviews* **4**, 111–133, 2000.

10. M. B. Wahid. Biofuel production and market in Malaysia—Current developments and outlook, In: *Fuels of the Future 2007*, Conference Papers, UFOP & BBE, Berlin, Germany, November 27–28, 2006.

11. O. Loest, J. Ullmann, and J. Winter. Investigations on the Addition of FAME to Diesel Fuels, DGMK-Research Report 639, Hamburg, Germany, July 2006.

12. E. A. Stadlbauer, S. Bojanowski, M. S. Hossain, and A. Fiedler. Thermocatalytic production of biofuels from animal fat and solid residues from extraction of rapeseeds, In: *Book of Abstracts, 4th European Federation of Lipids Congress: Oils, Fats and Lipids for a Healthier Future*, October 1–4, 2006, University of Madrid, Spain, p. 115.

13. E. Bayer and M. Kutubuddin. Thermocatalytic conversion of lipid-rich biomass to oleochemicals and fuel, In: *Research in Thermochemical Biomass Conversion*, Bridgwater, A. V. and Kuester, J. I. (Eds.), London: Elsevier Applied Science, 1988, pp. 518–30.

14. Engler, E., Ber. **22** [1888], 1816.

15. A. W. Schwab, G. J. Dykstra, E. Selke, S. C. Sorenson, and E. H. Pryde. Diesel fuel from thermal decomposition of soybean oil, *Journal of American Oil Chemists' Society* **65**, 1781–1786, 1988.

16. R. J. Evans and T. A. Milne. Molecular characterization of the pyrolysis of biomass. I. Fundamentals, *Energy and Fuels* **1**, 123–137, 1987.

17. R. J. Evans and T. A. Milne. Molecular characterization of the pyrolysis of biomass. II. Applications, *Energy and Fuels* **1**, 311–319, 1987.

18. R. J. Evans, T. A. Milne, and N. Nagle. Catalytic conversion of microalgae and vegetable oils to premium gasoline, with shape-selective zeolites, *Biomass* **21**, 219–232, 1990.

19. C. Zhenyi, J. Xing, and L. Li. Thermodynamics calculation of the pyrolysis of vegetable oils, *Energy Sources* **26**, 849–856, 2004.

20. R. O. Idem, S. P. Katikaneni, and N. N. Bakhshi. Thermal cracking of canola oil: Reaction products in the presence and absence of steam, *Energy and Fuel* **10**, 1150–1162, 1996.

21. Bridgwater, A. V. Catalysis in thermal biomass conversion, *Applied Catalysis A: General* **116**, 5–47, 1994.

22. P. B. Weisz, W. O. Haag, and P. G. Rodewald. Catalytic production of high-grade fuel (gasoline) from biomass compounds by shape-selective catalysis, *Science* **1079**, 57–58, 2005.

23. Y. S. Prasad and N. N. Bakhshi. Effect of pretreatment of HZSM-5 catalyst on its performance in canola oil upgrading, *Applied Catalysis A: General* **18**, 71–85, 1984.

24. Y. S. Prasad and N. N. Bakhshi. Catalytic conversion of canola oil to fuels and chemical feedstock. Part I: Effect of process conditions on

the performance of HZSM-5 catalysts, *Canadian Journal of Chemical Engineering* **64**, 278–284, 1986.

25. E. C. Novella, G. O. Escudero, J. A. Alonso, F. P. M. Andres, and P. Canizares. Conversion of vegetable oils and extracts to hydrocarbons, In: *Actas Simp. Iberoam. Catal., 9th*, Vol. 2., Lisbon, Portugal: Soc. Iberoam. Catal., 1984, pp. 1623–1624.

26. Y. S. Prasad, H. Yaoliang, and N. N. Bakhshi. Effect of hydrothermal treatment of HZSM-5 catalyst on its performance for the conversion of canola and mustard oil to hydrocarbons, *Industrial & Engineering Chemistry Product Research Review* **25**, 257–267, 1986.

27. J. Graille, L. P. Geneste, A. Guida, and O. Norina. Production of hydrocarbons by catalytic cracking of the by-product of the palm oil mills, *Revue Francaise des Corps Gras* **28**, 421–426, 1981.

28. O. Onay, S. H. Beis, and O. M. Kockar. Fast pyrolysis of rapeseed in a well-swept fixed bed reactor, *Journal of Analytical and Applied Pyrolysis* **58/59**, 995–1007, 2001.

29. O. Onay and O. M. Kockar. Slow, fast and flash pyrolysis of rapeseed, *Renewable Energy* **28**, 2417–2433, 2003.

30. A. E. Putun, O. M. Kockar, S. Yorgun, H. F. Gercel, J. Andresen, and C. E. Snape. Fixed bed pyrolysis and hydropyrolysis of sunflower bagasse: Product yield and compositions, *Fuel Process Technology* **46**, 49–62, 1996.

31. E. A. Stadlbauer, S. Bojanowski, S. Hossain, S. Stengl, B. Weber, A. Bone, B. Jehle, E. Ruenagel, *Stoffstrommanagement und Energieeffizienz, η[energie]*, **1**, 48–52, 2007.

32. R. K. Sharma and N. N. Bakhshi. Upgrading tall oil to fuels and chemicals over HZSM-5 catalyst using various diluents, *Canadian Journal of Chemical Engineering* **69**, 1082–1086, 1991.

33. P. J. Crutzen, A. R. Mosier, K. A. Smith, and W. Winiwarter. N_2O release from agro-biofuel production negates global warming reduction by replacing fossil fuels, *Atmos. Chem-Phys. Discuss.*, **7**, 11191–11205, 2007 (www.atmos-chem-phys-discuss.net/7/11191/2007/).

34. E. A. Stadlbauer, S. Bojanowski, S. M. Hossein. Unpublished Results.

35. W. Funktion and E. A. Stadlbauer. Verfahren und Vorrichtung zur Herstellung von Kohlenwasserstoffen aus biologischem Fett DE 10 2004 012 583 A1 2005.12.01, Patent Office, Munich, 2004.

36. Fachagentur Nachwachsende Rohstoffe E.V. (FNR). Biofuels, Guelzow, Germany, 2006, p. 14.

37. E. A. Stadlbauer, et al. Herstellung von Kohlenwasserstoffen aus Tierfett durch thermokatalytisches Spalten, *Erdöl Erdgas Kohle* **122**, 64–69, 2006.

38. S. Bojanowski, A. Fiedler, A. Frank, J. Rossmanith, G. Schilling, and E. A. Stadlbauer. Catalytic production of liquid fuels from organic residues of rendering plants. In: *Proceedings of Environmental Science and Technology*, Vol. II, pp. 106–107, American Science Press, New Orleans, 2005.

39. M. Rupp. Verarbeitung von Rapsöl in Mineralölraffinerien, VDI Berichte Nr. 704, Germany, pp. 97–111, 1990.
40. D. Schliephake and C.-M. Hacker. Joint project for the assessment of the agricultural, process and chemical engineering framework for utilization of rapeseed oil and its conversion products as fuels, Project 0310026 A, BMBF/BML, Germany, 1994.
41. K. Bormann, H. Tilgner, and H.-J. Moll. Rapeseed oil as a feed component for the catalytic cracking process, *Erdöl Erdgas Kohle* **109**, 172–176, 1993.
42. F. A. Zaher and A. R. Taman. Thermally decomposed cottonseed oil as a diesel engine fuel, *Energy Sources* **15**, 400–504, 1993.
43. O. A. Megahed, N. M. Abdelmonem, and D. M. Nabil. Thermal cracking of rapeseed oil as alternative fuel, *Energy Sources* **26**, 1033–1042, 2004.
44. D. G. Boocock, S. K. Konar, A. Mackay, P. T. Cheung, and J. Liu. Fuels and chemicals from sewage sludge. 2. The production of alkanes and alkenes by pyrolysis of triglycerides over activated alumina, *Fuel* **71**, 191–197, 1992.
45. O. Onay, A. F. Gaines, O. M. Kockar, M. Adams, T. R. Tyagi, and C. E. Snape. Comparison of the generation of oil by extraction and the hydropyrolysis of biomass, *Fuel* **85**, 382–392, 2006.
46. D. Konwer, S. E. Taylor, B. E. Gordon, J. W. Otvos, and M. Calvin. Liquid fuels from Mesua ferrea L. seed oil, *Journal of American Oil Chemists' Society* **66**, 223–226, 1989.
47. E. K. Nemethy, J. W. Otvos, and M. Calvin. Hydrocarbons from *Euphorbia lathyris, Pure and Applied Chemistry* **53**, 1101–1108, 1981.
48. M. Oberdörfer. Niedertemperaturkonvertierung von Ölsaaten, Doctoral Thesis, Faculty for Chemistry and Pharmacy, University of Tuebingen, Germany, 1990.
49. K. Esuoso, H. Lutz, M. Kutubuddin, and V. Bayer. Chemical composition and potential of some underutilized tropical biomass. I: Fluted pumpkin (*Telfairia occidentalis*), *Food Chemistry* **61**, 487–492, 1998.
50. Amtsblatt der Europäischen Gemeinschaften. Über Maßnahmen zum Schutz gegen die transmissiblen spongiformen Enzephalopathien bei der Verarbeitung bestimmter tierischer Abfälle und zur Änderung der Entscheidung 97/735/EG der Kommission, 1999.
51. E. A. Stadlbauer, S. Bojanowski, A. Frank, S. Skrypsky-Mäntele, and C. Zettel. Treatment of bovine carcasses from veterinary clinics in times of BSE, In: *1st International Symposium on Residue Management in Universities*, November 6–8, 2002, Santa Maria, Brasilien, pp. 31–32.
52. E. A. Stadlbauer, S. Bojanowski, A. Fiedler, A. Frank, and J. Rossmanith. Niedertemperaturkonvertierung von Klärschlamm zu Kohlenwasserstoffen (Treibstoff für Motoren). Fachtagung: Wasser- und Abwassertechnologie—Neue Märkte erschließen, Justus-Liebig-University of Giessen, Germany, IFZ, March 17, 2004.

53. S. Bojanowski. Untersuchungen zur Niedertemperaturkonvertierung von Tiermehl, Diploma Thesis, University of Applied Sciences Giessen-Friedberg, Giessen, Germany, August 2002.

54. J. Piskorz, D. S. Scott, and I. B. Westerberg. Flash pyrolysis of sewage sludge, *Industrial & Engineering Chemistry Process Design and Development* **25**, 265–270, 1986.

55. S. Skrypski-Mäntele. Untersuchung über die Eigenschaften der Konvertierungskohle der Niedertemperaturkonvertierung, Doctoral Thesis, University of Tuebingen, Germany, 1992.

56. E. A. Stadlbauer. Thermokatalytische Niedertemperaturkonvertierung (NTK) von tierischer und mikrobieller Biomasse unter Gewinnung von Wertstoffen und Energieträgern im Pilotmaßstab, Interim Report DBU, Germany Grant No. 18153, Giessen, Germany, November 2003.

57. E. A. Stadlbauer. Thermokatalytische Niedertemperaturkonvertierung (NTK) von tierischer und mikrobieller Biomasse unter Gewinnung von Wertstoffen und Energieträgern im Pilotmaßstab, Final Report DBU, Germany Grant No. 18153, Giessen, Germany, July 2005.

58. L. Rupp. Mechanismus des Thermischen Abbaus von Fetten und Fettsäuren, Doctoral Thesis, University of Tuebingen, Germany, 1986.

59. M. S. Hossain. Neuwertschöpfung durch Treibstoffproduktion aus tierischen Substraten unter Berücksichtigung arbeitshygienischer Aspekte, Diploma Thesis, University of Applied Sciences, Giessen-Friedberg, Giessen, Germany, July 2005.

60. E. A. Stadlbauer, S. Bojanowski, A. Frank, R. Lausmann, and W. Grimmel. Untersuchungen zur thermokatalytischen Umwandlung von Klärschlamm und Tiermehl. KA – Abwasser, Abfall 50, 1558–1562, 2003.

61. M. Kutubuddin. Niedertemperaturkonvertierung von Biomasse zu Öl und Kohle, Doctoral Thesis, University of Tuebingen, Germany, 1982.

62. W. Funktion and E. A. Stadlbauer and Catalytic Reactor, PCT DE 2004/000329, WO 2004/074181 A2, 2004.

63. E. A. Stadlbauer, S. Bojanowski, A. Fiedler, J. Hoogveldt, J. Rossmanith, G. Schilling, A. Frank, and R. Lausmann. Thermocatalytic reactor to convert organic residues from wastewater treatment and rendering plants to bio-fuels and chemicals, In: *Proceedings: Anaerobic Digestion 2004, 10th World Congress*, August 29–September 3, 2004, Montreal, Canada, pp. 914–920.

64. R. K. Sharma and N. N. Bahkashi. Catalytic upgrading of pyrolysis oil, *Energy and Fuel* **7**, 306–314, 1993.

65. S. K. Konar, D. G. Boocock, V. Mao, and J. Liu. Fuels and chemicals from sewage sludge: 3. Hydrocarbon liquids from the catalytic pyrolysis of sewage sludge lipids over activated alumina, *Fuel* **73**, 642–646, 1994.

Questions

1. What is the mechanism of vegetable oil pyrolysis?
2. What is the catalyst or co-catalyst cracking of fat?
3. What are tailored conversion products of vegetable oil?
4. Where are cracking in situ catalysts used?
5. What is refitting? Why is it essential before engine starting?

CHAPTER 9

Bioethanol: Current Status and Future Perspectives

Behzad Satari, Keikhosro Karimi, and Mohammad J. Taherzadeh

9.1 Introduction

Ethanol or ethyl alcohol is a clear, colorless, flammable chemical with formula CH_3CH_2OH, molecular weight of 46.07 g/mol, and density of 789 kg/m^3 at 294 K. It has been produced and used as an alcoholic beverage for thousands of years. Ethanol also has several industrial applications in, e.g., detergents, toiletries, coatings, and pharmaceuticals, and has been used as a transportation fuel for more than a century. Ethanol is mainly produced biologically from sugar, starch, and lignocellulosic materials. The biologically produced ethanol is sometimes called "fermentative ethanol" or "bioethanol." Ethanol produced from sugar- and starch-based substrates is considered a first-generation ethanol. Second-generation ethanol is produced from lignocellulosic substrates derived from industrial, forestry, agricultural, domestic, and municipal solid wastes. Microalgae-derived carbohydrates are used as substrates for third-generation ethanol production [1, 2]. Application of bioethanol as a fuel has no or limited net emission of CO_2 [3] and is able to limit warming to 2°C (or less), as envisaged in the Paris Agreement (2015) [4]. In this chapter, the current status and future perspective of bioethanol production for fuel applications are briefly reviewed.

9.2   Global Market for Ethanol and Its Future Prospects

Nicholas Otto used ethanol and vegetable oil in an internal combustion engine invented in 1897 [5]. In World War II, processes for production of ethanol from isolated hemicellulose and celluloses from plants were built in Germany, Russia, China, and the United States. However, most of them were shut down after the war because they were not able to economically compete with petroleum-based fuels and the latter dominated the fuel market in the 20th century because of its abundance and low price.

Ethanol did not have a major role in the transportation energy supply until the 1970s when two oil crises occurred: in 1973, the OPEC embargo, and the 1979 Iranian revolution. Brazil started producing sugarcane ethanol in the 1980s for a domestic fuel market, while other countries did not show any interest in the production since the price of oil was stable at below 20 $/barrel (Fig. 9.1). After the U.S. attacks of September 11, war in Iraq resulted in an abrupt growth in oil price (Fig. 9.1) and efforts have developed for commercialization of biofuels. It was not only the cost derives but also some policy initiatives, e.g., replacement of MTBE in gasoline with ethanol as an oxygenator, Renewable Fuel Standard (RFS) introduction in the United States in 2005, and launching Europe's Emissions Trading System (ETS) in 2005, made the development of ethanol production technologies burgeon. The market for ethanol faced challenges in 2008 after the global financial crisis, dropping oil prices, and concerns over using first-generation feedstocks. Even though the RFS mandated blending substantial amounts of cellulosic ethanol (CE) with gasoline in 2010, the first commercial CE production plant started to work in 2014, and the RFS mandates were waived. Saddler et al. [6] conducted a study in 2011 and compared the economic competitiveness of CE and corn ethanol. They showed that if the CE volume mandated by the Energy Independence and

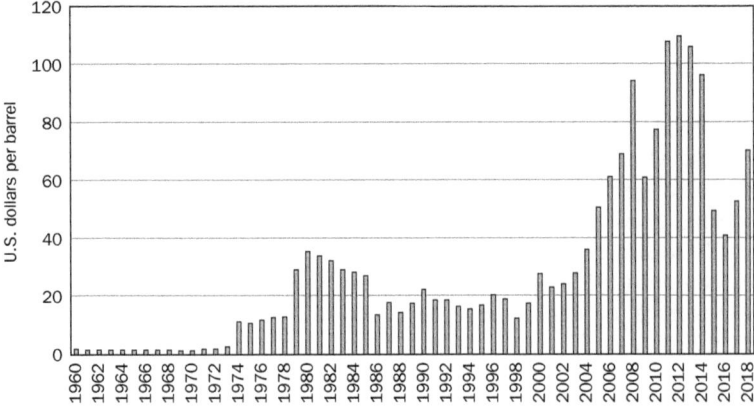

FIGURE 9.1   Average annual OPEC crude oil price from 1960 to 2018 (source: https://www.statista.com)

Security Act (EISA) at 2007 were to be fulfilled, CE could be competitive with corn ethanol by 2020. The oil prices then fell again in 2014 (Fig. 9.1), making a threat for the future of ethanol, especially CE [7, 8].

Figure 9.2 shows global ethanol production by country/region from 2007 to 2015. Global production had a peak in 2015 after a dip in 2012 and 2013. The United States is the world's largest producer of ethanol, having produced more than 15 billion gallons in 2017 alone (Fig. 9.3).

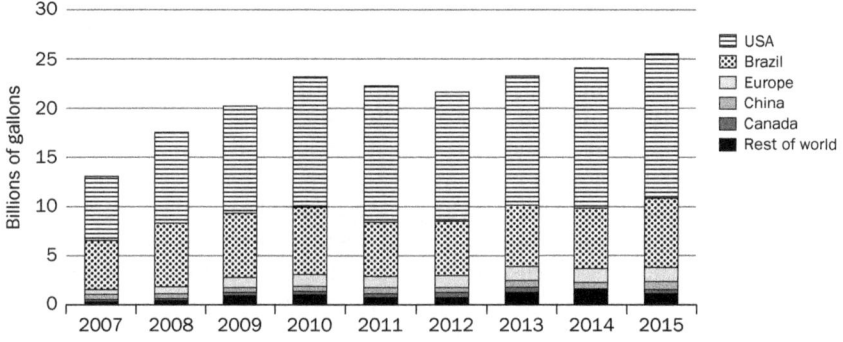

FIGURE 9.2 Global ethanol production by country or region (source: Renewable Fuels Association (http://www.afdc.energy.gov/data)

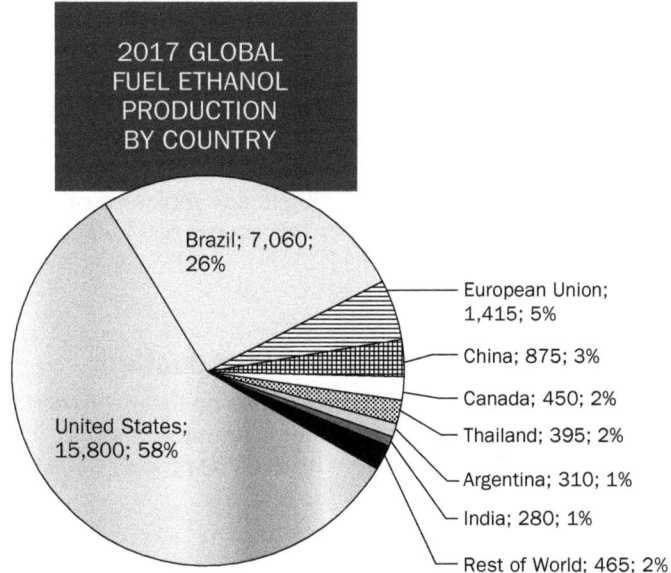

FIGURE 9.3 World ethanol production (country/region; million gallons; share of global production) in 2017 divided by region or country (source: RFA analysis of public and private data sources)

The United States and Brazil produced approximately 85 percent of the world's ethanol in 2017 (Fig. 9.3). Brazil mainly uses sugarcane, while the United States traditionally uses corn kernel as substrate for ethanol production. In Europe, however, the most commonly used substrates for ethanol production are wheat, maize, and sugar beet.

In the United States, Abengoa, Dupont, and POET/DSM are pioneering companies producing ethanol from lignocelluloses with capacity $\geq$20 MGY. They mainly use corn stover as a substrate for ethanol production and generate heat/electricity as co-products. In Brazil, Granbio and Raizer companies produce 10.6 and 21.6 MGY ethanol from bagasse and straw, respectively. In Europe, Beta Renewables company, Italy, produces 13.4 MGY ethanol from grass [9].

For automobiles, ethanol usually blends with gasoline and is used as E85 (85 percent ethanol, 15 percent gasoline), E20, and E10. Blends of gasoline/ethanol with more than 85 percent ethanol can only be used in flexible fuel vehicles (FFVs). Using ethanol as a clean fuel reduces the greenhouse gas emission. Using waste materials as the substrate for ethanol production can help with the disposal of this type of wastes. Moreover, because of some intrinsic properties of ethanol, e.g., low toxicity, high biodegradability, and high solubility in water, its consumption has low environmental impacts.

India imported 77 percent of its fuel in 2015–16 (202.8 MT), and this amount became higher in 2016–17 (213.9 MT). As part of an ambitious target, India planned to reduce the oil import by 10 percent in the next four years, and for this target 1400 million liters of ethanol was supposed to be blended with gasoline for transportation fuel in 2018. However, in 2016/17, only a 2.1 percent ethanol blending target was achieved. The mandated targets by the national policy on biofuels for long terms is not clear, and no penalties have been set to ensure its successful execution. Sugarcane molasses is the major resource for bioethanol production for blending targets in this country. Sugarcane molasses cannot meet the demand for blending, and therefore other crop residues can be explored as second-generation feedstocks [10].

9.3   Overall Process of Bioethanol Production

The process of ethanol production depends on the raw materials used. A general simplified representation of these processes is shown in Figure 9.4, and a brief description of different units of the process is presented in the rest of this chapter.

9.4   Application of Sugar-Based Production

Sugar-based substances (e.g., sugar cane juice and molasses), starchy materials (e.g., wheat, corn, barley, potato, and cassava), and lignocellulosic

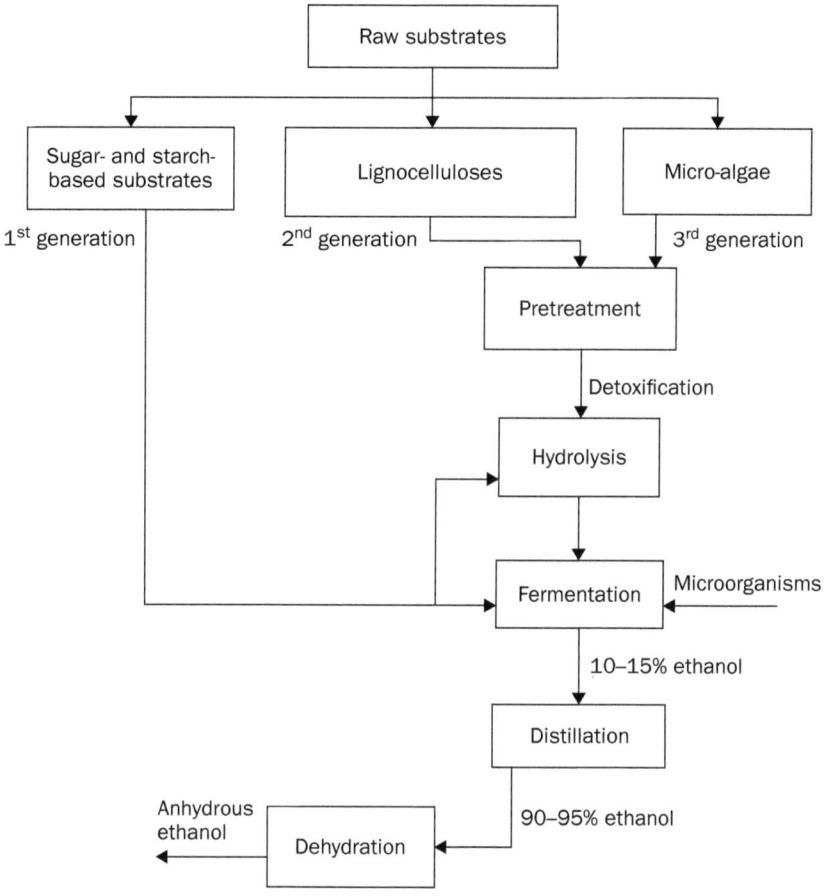

FIGURE 9.4 A general process scheme for ethanol production from different raw materials

materials (e.g., forest residuals, grains, domestic wastes, and municipal solid wastes) are being considered as the raw materials for ethanol production. The dominating sugars available or produced from these popular raw materials are the following:

- Glucose, fructose, and sucrose in sugar-based substances
- Glucose in starchy materials
- Glucose from cellulose as well as mannose, xylose, and arabinose from hemicellulose in lignocellulosic materials

Most of the ethanol-producing microorganisms can use a variety of hexoses such as glucose, fructose, galactose, and mannose, and a limited number of disaccharides such as lactose, cellobiose, and maltose, and rarely their

Substrate	Pretreatment or liquefaction	Hydrolysis or saccharification	Detoxification
Potential sugar substrates	No	No	Typically no
Starchy materials	Yes	Yes	No
Lignocellulose materials	Yes	Yes	Depends on the hydrolysis method

TABLE 9.1   Different types of substrates need the following treatment prior to fermentation

polymers. Therefore, it is necessary to convert the complex polysaccharides such as cellulose and starch to simple sugars. Different types of substrates need treatment, presented in Table 9.1, prior to fermentation.

9.4.1   Sugar production from starchy materials

There are various raw materials that contain starch and are suitable for ethanol production. Corn is the most widely used on an industrial scale for this purpose. The United States is the dominant producer of corn ethanol, and it is expected to produce 36 billion gallons of ethanol by 2022 mostly from corn and corn stover. Certain European countries have recently started to produce corn ethanol. However, there are several other kinds of cereals such as wheat, rye, barley, and sorghum, and crop roots such as potato and cassava, which are used as raw material for ethanol production. The cereals contain about 60–70 percent starch, 8–12 percent proteins, 10–15 percent water, and small amounts of fats and fibers. The compositions of the crop roots are almost identical to the cereals on a dry basis, but the water content of the roots is usually 70–80 percent. The exact composition of each raw material depends on the type and variety of the material used and can be found in the literature (e.g., [3, 11]). The starch of these materials is used as a carbon and energy source, and a part of the proteins as a nitrogen source for the microorganisms.

Starch contains two fractions: amylose and amylopectin. Amylose, which typically constitutes about 20 percent of starch, is a straight-chain polymer of α-glucose subunits with molecular weight that may vary from several thousand to half a million. Amylose is a water-insoluble polymer. The bulk of starch is amylopectin, which is a polymer of glucose too. Amylopectin contains a substantial amount of branches in the molecular chains. Branches occur from the ends of amylose segments, averaging 25 glucose units in length. Amylopectin molecules are typically larger than amylose, with molecular weights ranging up to 1 to 2 million. Amylopectin is soluble in water and can form a gel by absorbing water.

For ethanol production, hydrolysis is necessary to convert the starch into fermentable sugar available to microorganisms. Traditional conversion of starch into the sugar monomers requires a two-stage hydrolysis process: liquefaction of large starch molecules to oligomers and saccharification of the oligomers to the sugar monomers. This hydrolysis may be catalyzed by acid or amylolytic enzymes.

9.4.1.1 Acid hydrolysis of starch

Acid hydrolysis of starchy materials is an old process that is applied in some ethanol industries. Sulfuric acid is the most often applied acid in this process, where starch is converted to low-molecular-weight dextrins and glucose [12]. Main advantages of this process are rapid hydrolysis and less cost for catalyst compared to the enzymatic hydrolysis. However, the acid processes possess drawbacks, including (a) high capital cost for an acid-resistant hydrolysis reactor, (b) destroying sensitive nutrients such as vitamins that are present in the raw materials, and (c) further degradation of sugar to hydroxymethylfurfural (HMF), levulinic acid, and formic acid, which lowers the ethanol yield and inhibits the fermentation process [13].

The acid hydrolysis process can be performed either in batch or in continuous systems. Dilute-acid hydrolysis can also be used as a pretreatment for enzymatic hydrolysis. Prior to enzymatic hydrolysis, it is common to soak the starch or starchy materials in the dilute acid, and then continuously pass it through a steam-jet heater into a cooking tube (called a jet cooker or mash cooker) with a plug flow reactor for a couple of minutes, and then subject it to the enzymatic hydrolysis.

9.4.1.2 Enzymatic hydrolysis of starch

Enzymatic hydrolysis has several advantages compared to acid hydrolysis. First, the specificity of enzymes allows the production of sugar syrups with well-defined physical and chemical properties. Second, the milder enzymatic hydrolysis results in few side reactions and less "browning" [12]. Different types of enzymes involved in the enzymatic hydrolysis of starch are α-amylase, β-amylase, glucoamylase, pullulanases, and isoamylases. The mechanism of action of these enzymes is presented schematically in Figure 9. 5.

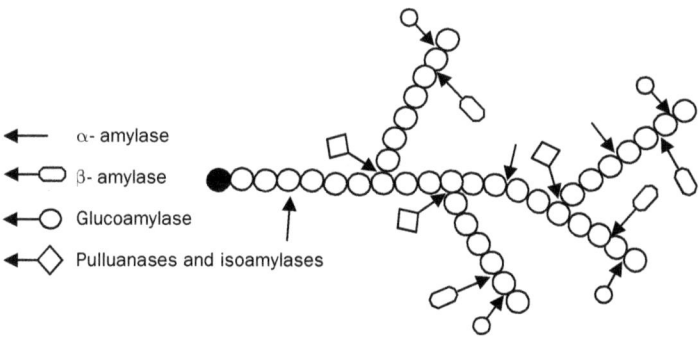

FIGURE 9.5 Mechanism of action of amylase on starch

There are two popular industrial processes from starch materials, i.e., dry-milling and wet-milling. In the dry-milling process, grain is first ground into flour and then processed without separation of the starch from germ and fiber components. In this method, the mixture of starch and other components is processed. Starch is converted to sugars in two stages, liquefaction and saccharification, by adding water, enzymes, and heat (enzymatic hydrolysis). The flour together with water is first heated up to 85–90°C and α-amylases are added. Then, the mixture is heated to 110–120°C and keep for 1 hour to inactivate the α-amylases and reduce the bacterial contamination level. After that, the mixture is cooled down to 70°C with additional glucoamylases supplementation for saccharification [14].

Dry-milling processes produce a co-product, distillers' dried grains with solubles (DDGS), which is used as an animal feed supplement. Without the revenues from that co-product, ethanol from dry-milled corn processing would not be economically feasible [3]. A dry-milling process for alcohol production processes the whole grain, or components derived from the whole grain. Saccharification and fermentation of dry-milled corn result in ethanol and distillers' dried grains (DDG). When DDG are combined with fermentation liquids and dried, they result in distillers' dried grains with solubles (DDGS) as the major feed by-products [15].

In the wet-milling process, grain is steeped and separated into starch, germ, and fiber components. Wet-milling is capital-intensive, but it generates numerous co-products that help to improve the overall production economics [3]. Wet mills produce corn gluten feed, corn gluten meal, corn germ, and other related co-products. In this method, after the grain cleaning, it is steeped and then ground to remove the germ. By further grinding, washing, and filtering steps, fiber and gluten are separated. The starch that remains after these separation steps is then broken down into fermentable sugars by the addition of enzymes in the liquefaction and saccharification stages. The produced fermentable sugars are then subjected to fermentation for ethanol production, like the other fermentable sugars.

Approximately 90 percent of corn ethanol in the United States is produced by dry-milling technology, while wet-milling process was dominant up to the 1990s. This is associated with the higher capital cost for the wet-milling process [16].

9.4.2 Characterization of lignocellulosic materials

Lignocellulosic materials predominantly contain a mixture of carbohydrate polymers (cellulose and hemicellulose) and lignin. Depending on the type of lignocelluloses, minor amounts of acetate, pectin, salt, ash, and minerals are also available in their structure. The contents of cellulose, hemicellulose, and lignin in common lignocellulosic materials are listed in Table 9.2. Different types of carbohydrates (including glucan,

Lignocellulosic materials	Cellulose (%)	Hemicellulose (%)	Lignin (%)
Hardwood stems	40–75	10–40	15–25
Softwood stems	30–50	25–40	25–35
Corn cobs	45	35	15
Wheat straw	30	50	15
Rice straw	32–47	19–27	5–24
Sugarcane bagasse	40	24	25
Leaves	15–20	80–85	0
Paper	85–99	0	0–15
Newspaper	40–55	25–40	18–30
Waste paper from chemical pulps	60–70	10–20	5–10
Grasses	25–40	25–50	10–30

TABLE 9.2 The contents of cellulose, hemicellulose, and lignin in common lignocellulosic materials [17–23]

xylan, galactan, arabinan, mannan), lignin, extractives, and ash content of many lignocellulosic materials have been analyzed and are available in the literature.

9.4.2.1 Cellulose

Cellulose is the main component of most lignocellulosic materials. Cellulose is a linear polymer of several hundred to more than 10,000 glucosyl residues that are linked by ß-1,4 glycosidic bonds. However, each glucose residue is rotated 180° relative to its neighbors, so that the basic repeating unit is in fact cellobiose, a dimer of two glucose units. As glucose units are linked together into polymer chains, a molecule of water is lost, which makes the chemical formula $C_6H_{10}O_5$ for each monomer unit of "glucan." The parallel polyglucan chains form numerous intra- and intermolecular hydrogen bonds, which results in a highly ordered crystalline structure of native cellulose, interspersed with less ordered amorphous regions [24–27].

9.4.2.2 Hemicellulose

Hemicelluloses are heterogeneous polymers of pentoses (xylose, arabinose), hexoses (mannose, glucose, galactose), and sugar acids. Unlike cellulose, hemicelluloses have an amorphous and random structure that make them relatively easily hydrolyzed by acids to their monomer components consisting of glucose, mannose, galactose, xylose, arabinose, and small amounts of rhamnose, glucuronic acid, methyl glucuronic acid, and galacturonic acid. Hardwood hemicelluloses contain mostly xylans, whereas softwood hemicelluloses contain mostly glucomannans. Xylans are the most abundant hemicelluloses. Xylans of many plant materials are heteropolysaccharides with homopolymeric backbone chains of 1,4-linked b-D xylopyranose units. Xylans from different sources, such as grasses, cereals, softwood, and hardwood, differ in composition. Besides xylose, xylans may contain arabinose, glucuronic acid or its 4-O-methyl ether, and acetic, ferulic, and p-coumaric acids. The degree of polymerization of hardwood xylans (150–200) is higher than that of softwoods (70–130) [20, 24, 28].

9.4.2.3 Lignin

Lignin is a complex molecule. Lignin is an aromatic polymer constructed of phenylpropane units linked in a three-dimensional structure. Generally, softwoods and hardwoods are different in lignin composition and softwoods contain more lignin than hardwoods. Lignins are divided into two classes, "guaiacyl lignins" and "guaiacyl-syringyl lignins." Although the principal structural elements in lignin have been largely clarified, many aspects of its chemistry remain unclear [29]. Chemical bonds have been reported between lignin and hemicellulose and even cellulose. Lignins are extremely resistant to chemical and enzymatic degradation. Biological degradation can be achieved mainly by fungi but also by certain actinomycetes [24, 29–31].

9.4.2.4 Pectin

Pectin is a heteropolysaccharide with homogalacturonan backbone that is substituted by side branches composed of galactose and arabinose units. Pectin is located in the cell wall and middle lamellae of non-woody plants, e.g., vegetables and fruits. Pectin is a major plant load bearing after cellulose and holds the cell components together [32]. Depectination was reported to make the structure of lignocelluloses more susceptible to enzymatic hydrolysis [33].

9.4.3   Sugar solution from lignocellulosic materials

There are several possible ways to hydrolyze lignocellulose. The most commonly applied methods can be classified into two groups: chemical hydrolysis and enzymatic hydrolysis. In addition, there are some other hydrolysis methods in which no chemicals or enzymes are applied. For instance, lignocellulose may be hydrolyzed by gamma-ray, electron-beam, or microwave irradiation. However, those processes are commercially unimportant [24].

9.4.3.1   Chemical hydrolysis of lignocellulosic materials

Chemical hydrolysis involves exposure of lignocellulosic materials to a chemical for a period of time at a specific temperature and results in sugar monomers from cellulose and hemicellulose polymers. Acids are predominantly applied in chemical hydrolyzes. Sulfuric acid is the most investigated acid although other acids such as HCl have also been used.

Acid hydrolyzes can be divided into two groups: concentrated-acid hydrolysis and dilute-acid hydrolysis [34].

Concentrated-acid hydrolysis. Hydrolysis of lignocelluloses by concentrated sulfuric or hydrochloric acids is a relatively old process. Concentrated-acid processes are generally reported to give higher sugar and ethanol yield, compared to dilute-acid processes. Furthermore, they do not need high pressure and temperature. Although this is a successful method for cellulose hydrolysis, concentrated acids are toxic, hazardous, and require reactors that are highly resistant to corrosion. The high investment and maintenance costs have greatly reduced the commercial potential for this process. In addition, the concentrated acid must be recovered after hydrolysis to make the process economically feasible. Furthermore, the negative environmental impact strongly limits the application of hydrochloric acid [18, 24].

Dilute-acid hydrolysis. Dilute-sulfuric-acid hydrolysis is a favorable method for either the pretreatment before enzymatic hydrolysis or the conversion of lignocellulose to sugars. This pretreatment method gives high reaction rates and significantly improves enzymatic hydrolysis. Depending on the substrate and the conditions used, up to 95 percent of the hemicellulosic sugars can be recovered by dilute-acid hydrolysis from the lignocellulosic feedstock [3, 19, 22]. Of all dilute-acid processes, sulfuric acid has been the most extensively studied. Sulfuric acid is typically used in 0.5–1.0 percent concentration. However, the time and temperature of the process can be varied. It is common to use one of the following conditions in dilute-acid hydrolysis:

- Mild conditions, i.e., low pressure and long retention time
- Severe conditions, i.e., high pressure and short retention time

In dilute-acid hydrolysis, the hemicellulose fraction is depolymerized at a lower temperature than the cellulose fraction. When higher temperature or longer retention times are applied, the formed monosaccharides are further hydrolyzed to other compounds. It is therefore suggested that the hydrolysis process be carried out in at least two stages. The first stage is carried out at relatively milder conditions during which the hemicellulose fraction is hydrolyzed, and a second stage can be carried out by enzymatic hydrolysis or dilute-acid hydrolysis at higher temperatures during which the cellulose is hydrolyzed [19]. These first and second stages are sometimes called "pretreatment" and "hydrolysis," respectively.

The hydrolyzate of first-stage dilute-acid hydrolysis usually consists of hemicellulosic carbohydrates. The dominant sugar in the first-stage hydrolyzate of hardwoods (such as alder, aspen, and birch) and most agricultural residues such as straws is xylose, whereas first-stage hydrolyzates of softwoods (e.g., pine and spruce) predominantly contain mannose. However, the dominant sugar in the second-stage hydrolyzate of all of the lignocellulosic materials, either by enzymatic or by dilute-acid hydrolysis, is glucose, which is originated from cellulose.

Detoxification of the acid hydrolyzates.   In addition to the sugars, several by-products are formed or released in the acid hydrolysis process. These by-products severely inhibit the downstream bioprocessing, e.g., enzymatic hydrolysis and microbial fermentation steps. The most important by-products are carboxylic acids, furans, and phenolic compounds and other aromatics (Fig. 9.6).

Acetic acid, formic acid, and levulinic acid are the most common carboxylic acids found in the hydrolyzates. Acetic acid is mainly formed from acetylated sugars in the hemicellulose, which are cleaved off already at mild hydrolysis conditions. Since the acid is not further hydrolyzed, formation of acetic acid is dependent on the temperature and pressure of dilute-acid hydrolysis until the acetyl groups are fully hydrolyzed. Therefore, the acetic acid yield in the hydrolysis does not significantly depend on the severity of the hydrolysis process [19, 35].

Furfural and HMF are the only furans usually found in hydrolyzates in significant amounts. They are the hydrolysis products of pentoses and hexoses, respectively [19]. Formation of these by-products is affected by the type and size of lignocellulose as well as the hydrolysis variables such as acid type and concentration, pressure, temperature, and retention time.

A large number of phenolic compounds have been found in hydrolyzates. However, reported concentrations are normally a few milligrams per liter. This could be due to the low water solubility of many of the phenolic compounds, or a limited degradation of lignin during the hydrolysis process. Vanillin, hydroxybenzaldehyde, syringaldehyde, phenol, vanillic acid, and 4-hydroxybenzoic acid are among the phenolic compounds found in dilute-acid hydrolyzate [34].

Enzymatic (e.g., using peroxidase and laccase), physical (e.g., evaporation of volatile fraction and liquid-liquid/liquid-solid extraction of nonvolatile fraction), chemical (e.g., alkali and reducing agent treatment), and microbial treatment (e.g., by *Coniochaeta ligniaria*, *Trichoderma reesei*,

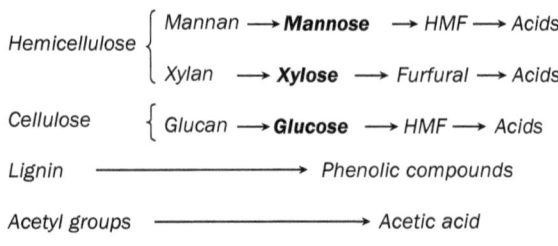

FIGURE 9.6   Formation of inhibitory compounds from lignocellulosic materials during acid hydrolysis

and *Ureibacillus thermosphaericus*) methods have been employed for detoxification of lignocellulosic hydrolyzates. There are other strategies concerning using fermentative microorganisms to reduce the effects of the inhibitors. These strategies include the use of large inocula, selection of inhibitors' resistant species, and genetically modified and culturing of the microorganisms [36–39].

A drawback of many detoxification methods is that a separate step is required. Despite numerous methods for detoxification, only a few of them are subjected to industrial plants for ethanol production because of economic feasibility. Detoxification of lignocellulosic hydrolyzates by overliming is a common method that has been widely used to improve the fermentability [38, 40–43]. In this method, $Ca(OH)_2$ is added to hydrolyzates to increase the pH (up to 9–12) and keep this condition for a period of time (from 15 min up to several days), followed by decreasing the pH to 5 or 5.5. Time, pH, and temperature of overliming are the effective parameters in detoxification [44]. However, the drawback of this treatment is that part of sugars is also degraded during the overliming process. Therefore, it is necessary to optimize the process to achieve a fermentable hydrolyzate without any loss of the sugar [37, 39, 44].

9.4.3.2 Pretreatment prior to enzymatic hydrolysis of lignocellulosic materials

The natural resistance of lignocelluloses to microbial and/or chemical deconstruction was first defined by Lynd et al. [45] as "biomass recalcitrance." Due to the biomass recalcitrance, the yield of commodity products from lignocelluloses is low [46], and therefore a pretreatment step is required to open up the recalcitrant biomass structure and facilitate the biocommodities production mediated by enzymes and/or microbes. An effective pretreatment method can disrupt and/or remove the linkages between cellulose, hemicellulose, and lignin, and reordering hydrogen bonds in cellulose fibers. Degradation of sugars and formation of inhibitory by-products should be minimized in pretreatment. Low heat or power requirements and low-cost facilities are among the desirable traits of an efficient pretreatment. There are a number of pretreatment processes developed in laboratories, including:

- *Physical pretreatment* (mechanical comminution, irradiation, pyrolysis)

- *Physico-chemical pretreatment* (steam explosion or autohydrolysis, ammonia fiber explosion—AFEX, CO_2 explosion, and SO_2 explosion)

- *Chemical pretreatment* (ozonolysis, dilute-acid hydrolysis, alkaline hydrolysis, oxidative delignification, and organosolv process)

- *Biological pretreatment*

However, not all of these methods are technically or economically feasible for large-scale processes. In some cases, a method is used to increase the efficiency of another method. For instance, milling could be applied to create a better steam explosion by reducing the chip size. Furthermore, it should be noticed that the selection of pretreatment method should be compatible with the selection of hydrolysis. For example, when acid hydrolysis is to be applied, a pretreatment with alkali may not be beneficial [34]. Many reviews on the pretreatment of lignocelluloses to improve enzymatic hydrolysis and bioconversion are available in the literature [1, 47–50]. Thermochemical pretreatments have recently dominated the CE field. Dilute acids such as HCl and H_2SO_4 and bases such as sodium hydroxide and sodium carbonate are commonly used for dilute-acid and dilute alkali pretreatments. Hydrothermal pretreatment that only uses water in subcritical or supercritical conditions is favorable among thermochemical pretreatment because it does not need additional chemicals and additional processes for neutralization and removing the inhibitory compounds. Steam-explosion is a hydrothermal pretreatment in which the lignocelluloses are treated with high-pressure steam followed by an explosive decompression, leading to hydrolyzing hemicellulose in a process called "autohydrolysis" [51, 52].

While most of the traditional pretreatment methods suffer from severe reaction conditions, a new category of pretreatment, called cellulose-solvent based pretreatment, has been recently introduced. A variety of solvents, e.g., some ionic liquids, concentrated phosphoric acid, and N-methylmorpholine-N-oxide (NMMO), can dissolve cellulose and the dissolved cellulose is then regenerated by adding an antisolvent without any derivatization. The regenerated cellulose is more reactive to microbial degradation and this technology is used for pretreatment of lignocelluloses. Generally, mild condition is used for this pretreatment and the cellulose solvent can be recycled and reused [1].

9.4.3.3   Enzymatic hydrolysis of lignocellulosic materials

Enzymatic hydrolysis of lignocelluloses can be commercially performed by highly specific cellulase and hemicellulase enzymes (glycosylhydrolases). This group includes at least 80 protein families and some subfamilies [24, 53]. Enzymatic degradation of cellulose to glucose is generally accomplished by synergistic action of three distinct classes of enzymes [3, 27]:

- Endo-1,4-ß-glucanases or 1,4-ß-D-glucan-4-glucanohydrolases that randomly hydrolyze internal ß-1,4-glucosidic bonds in the cellulose microfibril that are commonly measured by detecting the reducing groups released from carboxymethylcellulose (CMC).

- The exo-1,4-ß-D-glucanases, including both the 1,4-ß-D-glucan hydrolases and 1,4-ß-D-glucan cellobiohydrolases. 1,4-ß-D-glucan hydrolases liberate D-glucose from 1,4-ß-D-glucans. 1,4-ß-D-glucan cellobiohydrolases liberate D-cellobiose from 1,4-ß-D-glucans.

- ß-D-glucosidases or ß-D-glucoside glucohydrolyases, which release D-glucose from cellobiose and soluble cellodextrins, as well as an array of glycosides.

Hemicellulases mainly include xylanase, mannanase, ß-glucanases, and xyloglucanases, and debranching enzymes, i.e., α-glucuronidase, α-arabinofuranosidase, α-D-galactosidase, acetyl xylan esterase, and ferulic acid esterase, that act synergically to hydrolyze hemicelluloses [54].

Some bacteria and fungi, e.g., *Acidothermus cellulolyticus*, *Trichoderma reesei*, and *Aspergillus niger*, produce free enzymes, while some others from genera of *Clostridium*, *Acetivibrio*, *Bacteriodes*, and *Ruminicoccus* produce "cellulosomes" that can hydrolyze cellulose, hemicellulose, and pectin [55].

Substrate properties, cellulase activity, and hydrolysis conditions, e.g., temperature and pH, are the factors that affect the enzymatic hydrolysis of cellulose. To improve the yield and rate of the enzymatic hydrolysis, research has focused on optimizing the hydrolysis process and enhancing cellulase activity. Substrate concentration is one of the main factors that affect the yield and initial rate of enzymatic hydrolysis of cellulose. At low substrate levels, an increase of substrate concentration normally results in an increase of the yield and reaction rate of the hydrolysis. However, high substrate concentration can cause substrate inhibition, which substantially lowers the rate of the hydrolysis, and the extent of substrate inhibition depends on the ratio of total substrate to total enzyme [18, 27].

Increasing the dosage of cellulases in the process to a certain extent can enhance the yield and rate of the hydrolysis, but would significantly increase the cost of the process. Cellulase dosage of 10 FPU/g cellulose is often used in laboratory studies because it provides a hydrolysis profile with high levels of glucose yield in a reasonable time (48–72 h) at a reasonable enzyme cost. Cellulase enzyme loadings in hydrolysis vary from 5 to 33 FPU/g substrate, depending on the type and concentration of substrates. The activity of ß-glucosidase is a limiting enzyme in enzymatic hydrolysis of cellulose. Adding supplemental ß-glucosidase can enhance the saccharification yield [56, 57].

Although the exact mechanism of enzymatic hydrolysis by fungal cellulases is still unknown, the kinetics of enzymatic hydrolysis of cellulose can be modeled via three simple steps [18, 58]: (a) adsorption of cellulase enzymes onto the surface of the cellulose, (b) biodegradation of cellulose to fermentable sugars, and (c) desorption of cellulase. Cellulase activity decreases during the hydrolysis. The irreversible adsorption of cellulase on cellulose is partially responsible for this deactivation. Addition of surfactants during hydrolysis is capable of modifying the cellulose surface property and minimizing the irreversible binding of cellulase on cellulose. Tween-20 and Tween-80 are the most efficient surfactants in this regard. Addition of Tween-20, which is a non-ionic surfactant, as an additive in simultaneous saccharification and fermentation at 2.5 g/l has several positive effects in the process. It increases the ethanol yield, reduces the amount of enzyme loading, increases the enzyme activity in the liquid fraction at the end of the process, and reduces the required time to attain maximum ethanol concentration [59].

Novozymes is one of the pioneer companies in producing enzymes for CE production. The company started to develop new generations of cellulases in 2000 and the outcome of their work was introduction of Cellic CTec & Cellic HTec in 2009. The company then produced Cellic CTec2 and Cellic CTec3 and reported consumption of 50 kg Cellic CTec3 for production of 1 ton of ethanol (https://www.novozymes.com/).

9.4.4 Sugar production from microalgae

Third-generation biofuels are produced from microalgae as feedstocks. Microalgae are microscopic algae that are unicellular autotrophic and/or heterotrophic photosynthetic microorganisms and generally found in the ocean and freshwater environment. Microalgae can be grouped into prokaryotic microalgae (cyanobacteria), eukaryotic microalgae, red algae, and diatoms. Microalgae do not require arable land and fresh water and can be grown very fast while they are using atmospheric CO_2. Most microalgae can store substantial amounts of intracellular lipids (up to 70 percent of their body weight). The carbohydrate content of some species is also high and are in the form of starch, cellulose, pectin, hemicellulose, agar, and other carbohydrates that can be converted to monomeric sugars by appropriate enzymatic or acid hydrolysis and fed the ethanol producing microorganisms [60, 61].

Although the use of microalgal biomass for fuel production was mainly focused on biodiesel (and to some extent biogas) production, recent development has initiated the use of microalgae for ethanol production. The technologies for ethanol production include hydrolysis and fermentation of microalgal biomass by bacteria and yeasts, dark fermentation route, and the use of engineered cyanobacteria [62]. In recent years, although the use of engineered cyanobacteria has started in industrial plants, only little information is available in the literature on the efficiency and real advantages and disadvantages of using engineered cyanobacteria.

Even though microalgae seem promising feedstocks for ethanol production, some limitation and challenges have to be overcome for their commercialization. However, there are two key challenges for the widespread applications of microalgae-based carbohydrates.

First, harvesting microalgae accounts for 20–25 percent of the cultivation costs, which is an energy intensive and costly process [63]. Second, a multi-product biorefinery is required to enhance the economic viability of large-scale microalgae processes [64]. However, because of the diverse cellular structure of microalgae, the downstream processing for such biorefinery is more complex than the routine processes, and the obstacles in this context should be addressed. Compared with general lignocelluloses, microalgae contain much lower amounts of lignin. However, a variety of pretreatment may apply to increase the yield of hydrolysis and biochemicals production from microalgae [65].

9.5 Basic Concepts of Fermentation

The general reaction in which ethanol is produced during fermentation is:

$$Sugar(s) \xrightarrow{\ Microorganisms\ } Ethanol + Byproducts$$

In this reaction the microorganisms work as catalysts.

9.5.1 Conversion of simple sugars to ethanol

The catabolic reaction for simple hexose sugars, like glucose and mannose, in fermentation to ethanol can take place anaerobically as:

$$Hexoses(C_6H_{12}O_6) \xrightarrow{\ Microorganisms\ } 2C_2H_5OH(ethanol) + 2CO_2$$

When the entire sugar converts according to the above reaction, the yield of ethanol will be 0.51 g/g of the consumed sugars, meaning that from 1.00 gram of glucose 0.51 gram of ethanol can be produced. This is the theoretical yield of ethanol from hexoses. However, the ethanol yield obtained in fermentation does not usually exceed 90–95 percent of the theoretical, since part of the carbon source in sugars is converted to biomass of the microorganisms and other by-products such as glycerol and acetic acid [13, 66].

A similar reaction for anaerobic conversion of pentoses such as xylose to ethanol might be considered. Xylose is generally converted first to xylulose by a one-step reaction catalyzed by xylose isomerase in many bacteria, or by a two-step reaction via xylitol in yeasts and fungi. It can then be converted to ethanol anaerobically via pentose phosphate pathway and glycolysis. The general reaction can be written as:

$$3C_5H_{10}O_5(Pentoses) \xrightarrow{\ Microorganisms\ } 5C_2H_5OH(ethanol) + 5CO_2$$

In this case, a theoretical ethanol yield from xylose of 0.51 g/g is expected. However, the redox imbalance and slow rate of ATP formation are two major factors that make the anaerobic ethanol production from xylose difficult [67, 68]. There are a few anaerobic ethanol-producing strains from xylose that have been developed in research labs, but no strain is so far available for industrial scale process. Attempts have been made to overcome this problem of xylose assimilation by co-metabolization with, e.g., glucose or working with microaerobic conditions, where oxygen is available at low concentration. The aerobic conversion of xylose to ethanol might be summarized as:

$$Pentoses(C_5H_{10}O_5) + O_2 \xrightarrow{\text{\textit{Microorganisms}}} C_2H_5OH(\textit{ethanol}) + 3CO_2$$

The theoretical yield of ethanol from pentoses in this case is 0.31 g/g. There are a number of microorganisms that are able to produce ethanol aerobically from xylose, where their practical yield of ethanol from xylose and other pentoses is usually lower than this theoretical yield. The challenges in ethanol production from xylose were reviewed by Van Maris et al. [69].

9.5.2 Biochemical basis of ethanol production from hexoses

A simplified central metabolic pathway for ethanol production in yeast and bacteria under anaerobic conditions is presented in Figure 9.7 [24, 70–72].

There are three major interrelated pathways that control the catabolism of carbohydrate in most ethanol-producing organisms:

- The Embden-Meyerhof pathway (EMP) or glycolysis
- The pentose-phosphate pathway (PPP)
- The Krebs or tricarboxylic acid cycle (TCA)

In the glycolysis, glucose is anaerobically converted to pyruvic acid and then to ethanol via acetaldehyde. This pathway provides

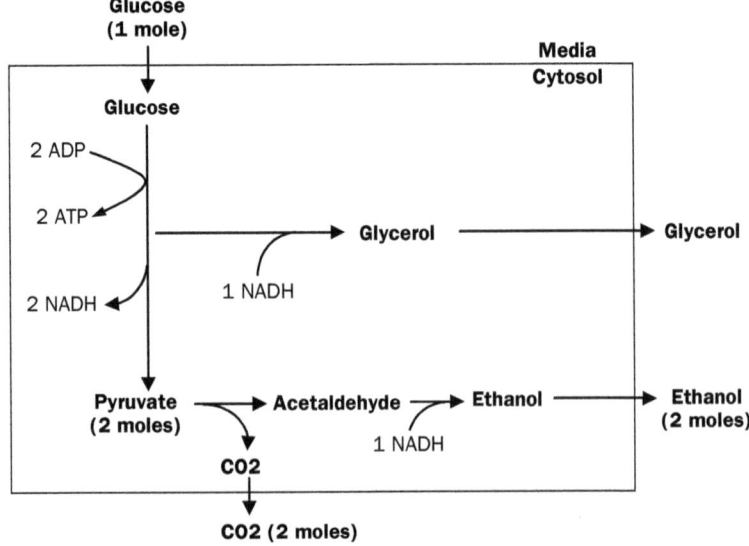

FIGURE 9.7 Central metabolic pathways in yeast under anaerobic conditions

energy in the form of ATP for the cells. The net yield in glycolysis is 2 moles of pyruvate (or ethanol) and 2 moles of ATP from each mole of glucose. This pathway is also the entrance of other hexoses such as fructose, mannose, and galactose into the metabolic pathways. With only 2 ATP formed per glucose catabolized, it can be calculated that large amounts of ethanol (at least 3.7 g ethanol/g biomass) must be formed [24, 73].

The pentose phosphate pathway (PPP) handles pentoses and is important for nucleotide (ribose-5-phosphate) and fatty acid biosynthesis. The PPP is mainly used to reduce $NADP^+$. It is estimated that 6 percent to 8 percent of glucose passes through PPP under anaerobic conditions in *Saccharomyces cerevisiae* [12, 24].

The TCA cycle functions to convert pyruvic and lactic acids and ethanol aerobically to the end products CO_2 and H_2O. It is also a common channel for the ultimate oxidation of fatty acids and the carbon skeletons of many amino acids. In cells containing the additional aerobic pathways, the NADH that forms in the glycolysis results in ATP generation in TCA [12].

Ethanol production from hexoses is redox-neutral, i.e., no net formation of NADH or NADPH occurs. However, biosynthesis of the cells results in net formation of NADH and consumption of NADPH. The pentose phosphate pathway (PPP) is mainly used to reduce $NADP^+$ to NADPH. Oxidation of surplus NADH under anaerobic conditions in *S. cerevisiae* is carried out via the glycerol pathway. Furthermore, there are other by-products, mainly carboxylic acids, which add to the surplus of NADH, such as acetic acid, pyruvic acid, and succinic acid. Consequently, glycerol is also formed to compensate the NADH formation coupled to these carboxylic acids. Thus, formation of glycerol is coupled to the biomass and carboxylic acid formation in anaerobic growth of *S. cerevisiae* [24, 74].

It should be kept in mind that growth of the cells and increasing their biomass is the ultimate goal of the cells. They produce ethanol under anaerobic conditions to provide energy by the catabolic reactions. Glycerol is formed to keep the redox balance of the cells, and carboxylic acids may leak from the cells to the medium. Therefore, the ethanol-producing microorganisms produce ethanol as the major product under anaerobic conditions, while biomass, glycerol, and some carboxylic acids are the by-products.

9.5.3 Biochemical basis of ethanol production from pentoses

In general, yeast and filamentous fungi metabolize xylose through a two-step reaction before it enters the central metabolism (glycolysis) through the pentose phosphate pathway. The first step is a conversion of xylose to xylitol using xylose reductase (XR), and the second step is a conversion of xylitol to xylulose using another enzyme, xylitol dehydrogenase (XDH) [75–77].

Wild-type *S. cerevisiae* possesses the enzymes XR and XDH, but their activities are too low to allow growth on xylose. Although *S. cerevisiae* cannot use xylose, it can use its isomer, xylulose. Thus, if *S. cerevisiae* is to be used for xylose fermentation, it requires a genetic modification to encode XR/XDH [75, 78].

Bacteria have a slightly different metabolic pathway for xylose utilization. Bacteria convert xylose to xylulose in one reaction using xylose isomerase (XI) [15, 79–81].

9.5.4   Microorganisms related to ethanol fermentation

The criteria for an ideal ethanol-producing microorganism are to have (a) high growth and fermentation rate; (b) high ethanol yield; (c) high ethanol and glucose tolerance; (d) osmotolerance; (e) low optimum fermentation pH; (f) high optimum temperature; (g) general hardiness under physiological stress; and (h) tolerance to potential inhibitors presented in the substrate [66, 82]. Ethanol and sugar tolerance allows the conversion of concentrated feeds to concentrated products, reducing energy requirements for distillation and stillage handling. Osmotolerance allows the handling of relatively dirty raw materials with their high salt content. Low pH fermentation combats contamination by competing organisms. High-temperature tolerance simplifies fermentation cooling. General hardiness allows microorganisms to survive stress such as that of handling (e.g., centrifugation) [82]. The microorganisms should also tolerate the inhibitors presented in the medium. A variety of yeasts, bacteria, and filamentous fungi have been identified as ethanol producers.

9.5.4.1   Yeast

Historically, yeasts have been the most commonly used microorganisms for ethanol production. The yeast strains are generally chosen from among *S. cerevisiae, S. ellypsoideuse, S. carlsbergensis, S. fragilis, Schizosaccharomyces pombe, Torula cremoris,* and *candida pseudotropicalis.* Yeast species that can produce ethanol as the main fermentation product are reviewed in the related literature [12, 83]. Most of the bioethanol production plants in Brazil and the United States have exclusively relied on yeasts for ethanolic fermentation.

Among the ethanol-producing yeasts, the "industrial working horse" *S. cerevisiae* is by far the most well-known and most widely used yeast in industry and research for ethanol fermentation. This yeast can grow both on simple hexose sugars, such as glucose, and on the disaccharide sucrose. *S. cerevisiae* is also generally recognized to be safe as a food additive for human consumption and is therefore ideal for producing alcoholic beverages and for leavening bread. The physiology, metabolism, and genetics of this yeast have been thoroughly studied [84]. Up to 93 percent of theoretical 0.51 g-ethanol/g-glucose has been achieved by this yeast, while the rest is mostly converted to cell biomass [85].

For the first-generation ethanol production by *S. cerevisiae*, engineering tools of the yeast have been focusing on minimizing or abolishing the formation of glycerol, a major by-product during ethanolic fermentation. Although some success has been recently achieved in this regard, reducing cell growth and consequently ethanol yield and reduction of the yeast robustness to harsh environment have been sometimes observed for the engineered yeast [84, 86].

For CE production, the major challenge of using *S. cerevisiae* is utilizing D-xylose and L-arabinose. These pentoses are abundant in lignocellulosic hydrolysates and the wild types of the yeast are not able to metabolize them [20, 66]. There have been several research efforts to engineer *S. cerevisiae* to be able to consume xylose [68, 87–89]. Several attempts were made to clone and express various bacterial genes, which is necessary for fermentation of xylose in *S. cerevisiae* [90, 91]. Co-fermentation of a mixture of xylose and cellobiose with an engineered strain was reported to yield an ethanol productivity of 0.65 g/L/h [92]. However, in most studies, pentoses utilization started after glucose exhaustion and its highest conversion yield was well below that of glucose [84].

Alternatively, xylose is converted to ethanol by some other naturally occurring recombinant. Among the wild-type xylose-fermenting yeast strains for ethanol production, *Pichia stipitis* and *Candida shehatae* have reportedly shown promising results for industrial applications in terms of complete sugar utilization, minimal by-product formation, low sensitivity to temperature, and substrate concentration. Furthermore, *P. stipitis* has no absolute vitamin requirement for xylose fermentation and is able to ferment a wide variety of sugars to ethanol [3, 93].

Olsson and Hahn-Hägerdal [36] presented a list of bacteria, yeasts, and filamentous fungi, producing ethanol from xylose. Certain species of the yeasts *Candida, Pichia, Schizosaccharomyces, Kluyveromyces,* and *Pachysolen* are among the naturally occurring organisms. Jeffries and Kurtzman [94] reviewed the strain selection, taxonomy, and genetics of xylose-fermenting yeasts.

Utilization of cellobiose is important in ethanol production from lignocellulosic materials by simultaneous saccharification and fermentation. However, a few ethanol-producing microorganisms are cellobiose-utilizing organisms. The requirement for the addition of β-glucosidase has been eliminated by cellobiose utilization during fermentation since the presentation of cellobiose reduces the activity of cellulase. Cellobiose utilization eliminates the need for one class of cellulase enzymes [3]. *Brettanomyces custersii* is one of the identified yeasts as a promising glucose- and cellobiose-fermenting microorganism for the simultaneous saccharification and fermentation (SSF) of cellulose for ethanol production [95].

High-temperature tolerance could be a good characterization for ethanol production since it simplifies fermentation cooling. On the other hand, one of the problems associated with SSF is the different optimum temperatures for saccharification and fermentation, and there are attempts to find thermotolerant yeasts for SSF. Szczdodrak and Targonski [96] tested 58 yeast strains belonging to 12 different genera and capable of growing and fermenting sugars at temperatures of 40–46°C. They selected several strains belonging to the genera *Saccharomyces, Kluyveromyces,* and *Fabospora* in view of their capacity to ferment glucose, galactose, and mannose at 40°C, 43°C, and 46°C, respectively. *Kluyveromyces marxianus* has been found to be a suitable strain for SSF [97]. This yeast can metabolize cellobiose, xylose, xylitol, arabinose, glycerol, lactose, and inulin [98].

9.5.4.2 Bacteria

A great number of bacteria are able to produce ethanol, although many of them generate multiple end products in addition to ethanol. *Zymomonas mobilis* is a Gram-negative bacterium that has several appealing properties as a biocatalyst for ethanol production. It has a homoethanol fermentation pathway and tolerates up to 120 g/l ethanol. Its ethanol yield is comparable with *S. cerevisiae*, while it has much higher specific ethanol productivity (2.5×) than the yeast. However, the tolerance of ethanol by *Z. mobilis* (ca. 20 g/l) is lower than that of *S. cerevisiae,* since some strain of *S. cerevisiae* can produce ethanol to give concentration as high as 18 percent of the fermentation broth. Continuous removal of ethanol during its production was reported as an effective method to mitigate the inhibition [99]. The tolerance of *Z. mobilis* to inhibitors and low pH is also low. Similarly to *S. cerevisiae, Z. mobilis* cannot utilize pentoses [20, 100, 101], while recombinant strains are able to ferment arabinose and xylose. Several genetic modifications have been performed for the utilization of arabinose and xylose by *Z. mobilis*. However, *S. cerevisiae* has been more welcomed for industrial application, probably because of the industrial problems that may arise in working with bacteria. Separation of *S. cerevisiae* from fermentation media is much easier than *Z. mobilis,* which is an important characteristic for reuse of the microorganisms in ethanol production processes [102].

Using genetically engineered bacteria for ethanol production is also applied in many studies. Ingram et al. [103] have reviewed the metabolic engineering of bacteria for ethanol production. Recombinant *Escherichia coli* is a valuable bacterial resource for ethanol production. The construction of *E. coli* strains to selectively produce ethanol was one of the first successful applications of metabolic engineering. *E. coli* has several advantages as a biocatalyst for ethanol production, including the ability to ferment a wide spectrum of sugars, no requirements for complex growth factors, and prior industrial use (e.g., for production of recombinant protein). The major disadvantages associated with using *E. coli* cultures are a narrow and neutral pH growth range (6.0–8.0), less hardy cultures compared to yeast, and public perceptions regarding the danger of *E. coli* strains. The lack of data on the use of residual *E. coli* cell mass as an ingredient in animal feed is also an obstacle to its application [12].

Interestingly, Dextera and Fu [104] engineered a cyanobacteria that can convert CO_2 to bioethanol. The promising feature of their work is not depending on costly feedstocks for bioethanol production. Besides, they claimed the scaling up potential of this process for both bioethanol production and CO_2 utilization. The Japanese Research Institute of Innovative Technology for the Earth (RITE) developed a microorganism for ethanol production. RITE strain is an engineered strain of *Corynebacterium glutamicum* that converts both pentose and hexose sugars into alcohol. The central metabolic pathway of *C. glutamicum* was engineered to produce ethanol. A recombinant strain that expressed the *Z. mobilis* gene coding for pyruvate decarboxylase and alcohol dehydrogenase was constructed [105]. RITE and Honda jointly developed a technology for production of ethanol from lignocellulosic materials using the stain. It is claimed that application of this strain by application of engineering technology from Honda enables a significant increase in alcohol conversion efficiency, in comparison to conventional cellulosic bioethanol production processes.

9.5.4.3 Filamentous fungi

A great number of molds are also able to produce ethanol. The filamentous fungi *Fusarium, Mucor, Monilia, Ryzypose, Rhizopus, Neurospora,* and *Paecilomyces* are among the fungi that can ferment pentoses to ethanol [68, 106, 107]. Compared with traditional ethanol-producing yeasts and bacteria, fungi are generally more resistant to inhibitors present in fermentation media and can grow in a harsh environment. Zygomycetes are saprophytic filamentous fungi, which are able to produce several metabolites including ethanol. Among three genera of *Rhizopus, Mucor,* and *Rhizomucor, Mucor indicus* (formerly *M. rouxii*) and *Rhizopus oryzae* showed good performances on ethanol productivity from glucose, xylose, and wood hydrolyzate [106, 108, 109]. *M. indicus* has several industrial advantages compared to baker's yeast for ethanol production, such as (a) capability of utilizing xylose, (b) having a valuable biomass for production of chitosan, oil, and mycoprotein, and (c) high optimum temperature of 37°C [110, 111]. Recently, the capability of ethanol-producing zygomycetes for production of hydrolytic enzymes has been reported, which is an interesting feature toward consolidated bioprocessing of these fungi for CE production [112, 113].

Skory et al. [114] examined 19 *Aspergilli* and 10 *Rhizopus* strains for their ability to ferment simple sugars (glucose, xylose, and arabinose) as well as complex substrates. An appreciable level of ethanol has been produced by *Aspergillus oryzae, Rhizopus javanicus,* and *R. oryzae*.

The dimorphic organism *Mucor circinelloides* is also used for the production of ethanol from pentose and hexose sugars. Large amounts of ethanol were produced during aerobic growth on glucose under non-oxygen limiting conditions by this mold. However, ethanol production on galactose or xylose was less significant [115]. A yield as high as 0.48 g/g ethanol from glucose under anaerobic conditions by *M. indicus* has been reported [116]. However, the yield and productivity of ethanol from xylose are lower than that of *Pichia stipitis* [117].

Although filamentous fungi have been industrially used for a long time for several purposes, a number of process engineering problems are associated with these organisms due to their filamentous growth. Problems can appear in mixing, mass transfer, and heat transfer. Furthermore, attachment and growth on bioreactor walls, agitators, probes, and baffles cause heterogeneity within the bioreactor and problems in the measurement of controlling parameters and cleaning of the bioreactor [118, 119]. Besides, ethanol-producing zygomycetes, e.g., *M. indicus*, cannot metabolize xylose in anaerobic condition. Such potential problems might hinder the industrial application of *M. indicus* for ethanol production. However, this fungus is dimorphic and its morphology can be controlled to be yeast-like or filamentous through the fermentation [21, 111, 120].

9.6   Fermentation Process

Fermentation processes, as well as other biological processes, can be classified into batch, fed-batch, and continuous operation. All these methods are applicable in the industrial fermentation of sugar substances and starch materials. These processes are well established, the fed-batch and continuous modes of operation being dominant in the ethanol market. When configuring the fermentation process, several parameters must be considered, including (a) high ethanol yield and productivity, (b) high conversion of sugars, and (c) low equipment cost. The need for detoxification and choosing the microorganism must be evaluated in relation to the fermentation configuration.

Mixing the fermentation broth, which is a non-Newtonian fluid, is also important for having an efficient operation. Mixing is done by mechanical impellers in flat-bottom tanks to avoid solids sedimentation or by the natural mixing caused by CO_2 bubbles formed during the fermentation process [121].

Presentation of a variety of inhibitors and their interaction effects in, e.g., lignocellulosic hydrolyzates, makes the fermentation process more complex than with other substrates for ethanol production [30, 37]. In fermentation of this hydrolyzate, the pentoses should be utilized to increase the overall yield of the process and to avoid problems in wastewater treatment. Therefore, it is still a challenge to use a hexose-fermenting organism such as *S. cerevisiae* for fermentation of the hydrolyzate.

When there is a mixture of hexoses and pentoses in the medium, microorganisms usually take up hexoses first and produce ethanol. As the hexose concentration decreases, they start to take up pentose. Fermentation of hexoses can be successfully performed under anaerobic or microaerobic conditions, with high ethanol yield and productivity.

However, the fermentation of pentoses is generally a slow and aerobic process. If one adds air to ferment pentoses, the microorganisms will start to utilize the produced ethanol as well. It makes the entire process complicated and demands a well-designed and controlled process.

In this section, we discuss different fermentation processes that could be applicable for ethanol production.

9.6.1 Batch processes

In batch processes, all the nutrients required for fermentation are present in the medium prior to cultivation. Batch technology has been preferred in the past due to the ease of operation, low cost of controlling and monitoring system, low requirements for complete sterilization, use of unskilled labor, low risk of financial loss, and easy management of feedstocks. However, the overall productivity of the process is low because of long turnaround times and initial lag phase. After nutrients exhaustion, the cells enter a stationary phase. At the initial fermentation stage, the fermentative cells face high substrate concentration, while a high ethanol concentration inhibits the growth of the cell at the final stage [13, 122].

To improve the performance of batch processes, cell recycle and application of several fermenters are used. Cell recycling is associated with reducing cost and time of inoculum preparation. Recycled cells are more adaptable to the harsh process conditions. In the Brazilian ethanol industry, 70–80 percent of plants use cell recycling. However, in case of hydrolysis of lignocellulosic hydrolysates, because of the residual ligno-celluloses in the fermentation broth, it is impossible to recycle only the microbial cells via conventional methods [123, 124]. Application of several fermenters operated at staggered intervals can provide a continuous feed to the distillation system. One of the successful batch methods applied for industrial production of ethanol is Melle–Boinot fermentation. This process achieves a reduced fermentation time and increased yield by recycling yeast and by application of several fermenters operated at staggered intervals. In this method yeast cells from the previous fermentation are separated from the media by centrifugation and up to 80 percent are recycled [13, 125, 126]. Instead of centrifugation, the cells can be filtered followed by separation of yeast from the filter aid using hydrocyclones [127].

In well-detoxified or completely non-inhibiting acid hydrolyzates of lignocellulosic materials, exponential growth will be obtained after inoculation of the bioreactor. If the hydrolyzate is slightly inhibiting, there will be a relatively long lag phase, during which part of the inhibitors is converted. However, if the hydrolyzate is severely inhibiting, no conversion of the inhibitors will occur, and neither cell growth nor fermentation will occur. A slightly inhibiting hydrolyzate can thus be detoxified during batch fermentation. However, too high concentration of the inhibitors will cause a complete inactivation of the metabolism [34].

Several strategies may be considered for fermentation of hydrolyzate to improve the in-situ detoxification in batch fermentation and obtain higher yield and productivity of ethanol. A high initial cell density, increasing the tolerance of microorganisms against the inhibitors by either adaptation of the cells to the medium or genetic modification of the microorganism, and choosing optimal reactor conditions to minimize the effects of inhibitors are among these strategies.

The volumetric ethanol productivity is low in lignocellulosic hydrolyzates when low cell-mass inocula are used due to poor cell growth. Usually high cell concentration, e.g., 10 g/l dry cells, has been used to find a high yield and productivity of ethanol in different studies. In addition, a high initial cell density helps the process for in-situ detoxification by the microorganisms and, therefore, the demand for a detoxification unit decreases. The in-situ detoxification of the inhibitors may even lead to increased ethanol yield and productivity due to uncoupling by the presence of weak acids, or due to decreased glycerol production in the presence of furfural [37]. Adaptation of the cells to hydrolyzate or genetic modification of the microorganism can significantly improve the yield and productivity of ethanol. Optimization of reactor conditions can be used to minimize the effects of inhibitors. Among the different parameters, it has been found that cell growth is strongly dependent on pH [34, 37].

9.6.2   Fed-batch processes

In fed-batch processes (or semi-batch processes) the substrate and required nutrients are added continuously or intermittently to the initial medium after the start of cultivation or from the halfway point through the batch process. Fed-batch processes have been used to avoid utilizing substrates that inhibit growth rate if present at high concentration, to overcome catabolic repression, to demand less initial biomass, to overcome the problem of contamination, and to avoid mutation and plasmid instability that are found in continuous culture. Furthermore, fed-batch processes do not face the problem of washout, which can occur in continuous fermentation. The major disadvantages of a fed-batch process are the need of additional instruments for control; it requires a substantial amount of operator skill and, in systems without feedback control, where the feed is added on a predetermined fixed schedule, it is difficult to deal with any deviations (i.e., time courses may not always follow the expected profiles) [128]. The fed-batch processes without feedback control can be classified as intermittent fed-batch, constant-rate fed-batch, exponential fed-batch, and optimized fed-batch. The fed-batch processes with feedback control have been classified as indirect- and direct-control fed-batch processes [128, 129].

The fed-batch technique is one of the promising methods for the fermentation of dilute-acid hydrolyzates of lignocellulosic materials.

The basic concept behind the success of this technique is the capability of in-situ detoxification of the hydrolyzates by the cells. Since the yeast has a limited capacity for the conversion of the inhibitors, the achievement of a successful fermentation strongly depends on the feed rate of the hydrolyzate. By adding the substrate at a low rate in fed-batch fermentation, the concentrations of bioconvertible inhibitors such as furfural and HMF in the fermenter remain low, and the inhibiting effect therefore decreases. At a too high feed rate using an inhibiting hydrolyzate, both ethanol production and cell growth can be expected to stop, whereas at a low feed rate, the hydrolyzate may still be converted but at a low productivity, which was experimentally confirmed. Consequently, there should exist an optimum feed rate [24, 34, 37].

Similarly to batch operations, higher optimum dilution rate in fed-batch cultivation can be obtained by (a) high initial cell concentration, (b) increasing the tolerance of microorganisms against the inhibitors, and (c) choosing optimal reactor conditions to minimize the effects of inhibitors. The productivity in fed-batch fermentation is generally limited by the feed rate, which, in turn, is limited by the cell-mass concentration [37].

9.6.3   Continuous processes

Process design studies of molasses fermentation have shown that the investment cost was considerably reduced when continuous rather than batch fermentation was employed, and that the productivity of ethanol could be increased by more than 200 percent. Continuous operations can be classified into continuous fermentation with or without feedback control. In continuous fermentation without feedback control, called a chemostat, the feed medium containing all the nutrients is continuously fed at a constant rate (dilution rate, D) and the cultured broth is simultaneously removed from the fermenter at the same rate. The chemostat is quite useful in the optimization of media formulation and to investigate the physiological state of the microorganism [129]. Continuous fermentations with feedback control are the turbidostat, nutristat, and phuuxostat. A turbidostat with feedback control is a continuous process to maintain the cell concentration at a constant level by controlling the medium feeding rate. A nutristat with feedback control is a cultivation technique to maintain a nutrient concentration at a constant level. A phuuxostut is an extended nutristat that maintains the pH value of the medium in the fermenter at a preset value [129].

When lignocellulosic hydrolyzates are added at a low feed rate in continuous fermentation, low concentration of bioconvertible inhibitors in the fermenter is assured. In spite of a number of potential advantages in terms of productivity, this method has not been much developed in the fermentation of the acid hydrolyzates yet. One should consider the following points in continuous cultivation of acid hydrolyzate of lignocelluloses:

- Cell growth is necessary at a rate equal to the dilution rate to avoid washout of the cells in continuous cultivation.
- Growth rate is low in fermentation of hydrolyzate because of the presence of inhibitors.
- The cells should keep their viability and vitality for a long time.

The major drawback of the continuous fermentation is that, in contrast to the situation in fed-batch fermentation, cell growth is necessary at a rate equal to the dilution rate to avoid washout of the cells in continuous cultivation [37]. The productivity is a function of the dilution rate, and since the growth rate is decreased by the inhibitors, the productivity in continuous fermentation of lignocellulosic hydrolyzates is low. Furthermore, at a low dilution rate the conversion rate of the inhibitors can be expected to decrease due to the decreased specific growth rate of the biomass. Thus, washout may occur even at low dilution rate [34]. On the other hand, one of the major advantages of continuous cultivation is the possibility to run the process for a long time (e.g., several months), whereas the microorganisms usually lose their activity after facing the inhibitory conditions of hydrolyzate. By employing cell-retention systems, the cell-mass concentration in the fermenter, the maximum dilution rate, and thus the maximum ethanol productivity increases. Different cell-retention systems have been investigated by cell immobilization and encapsulation, cell recirculation by filtration, settling, and centrifugation. A relatively old study [130] shows that the investment cost for a continuous process with cell recirculation has been found to be less than for continuous fermentation without cell recirculation.

Biostil® is the trade name of a continuous industrial process for ethanol production with partial recirculation of both yeast and wastewater. The fermenter works continuously and the cells are separated by centrifuge, and a part of the separated cells are returned to the fermenter. Most of the ethanol-depleted beer including residual sugars is then recycled to the fermenter. In this process, besides providing enough cell concentration in the fermenter, less water is consumed and a more concentrated stillage is produced. Therefore, the process has a lower wastewater problem. However, the process needs a special type of centrifuge (which is expensive) to avoid the deactivation of the cells [82, 131].

Application of encapsulated-cell system in continuous cultivation has several advantages compared to either free-cell or traditionally entrapped-cell system in, e.g., alginate matrix. Encapsulation provides higher cell concentrations than free-cell system in the medium, which leads to higher productivity per volume of the bioreactor in continuous cultivation. Furthermore, the biomass can easily be separated from the medium without centrifugation or filtration. The advantages of encapsulation compared to the cell entrapment are having less resistance to the diffusion through the beads/capsules, some degree of freedom in movement of the encapsulated cells, no cell leakage from the capsules, and higher cell concentration [132].

9.6.4   Series-arranged continuous flow fermentation

Ethanol can be produced by using continuous flow fermenters arranged in series with complete sugar utilization or high ethanol concentration. With two fermenters arranged in series, the retention time can be chosen so that the sugar is only partially used in the first, with fermentation completed in the second. Ethanol inhibition is reduced in the first fermenter, allowing a faster throughput. The second, lower-productivity fermenter can now convert less sugar than if operated alone. For high product concentration, the productivity of a two-stage system has been as much as 2.3 times higher than with a single stage [82, 133].

A two-stage continuous ethanol fermentation process with yeast recirculation is used industrially by Danish Distilleries Ltd. of Grena for molasses fermentation (Fig. 9.8). Two fermenters with 170,000-liter volume produce 66 g/l ethanol in 21 h retention time [134].

A seven-fermenter series (70,000 liters each fermenter) system was also used in the Netherlands to produce 86 g/l ethanol in 8 h retention time [136]. A Japanese company used a six-fermenter series (total volume 100,000 liters) with 8.5 h retention time to produce 95 g/l ethanol [137].

9.6.5   Fermentation strategies for fermentation of enzymatic lignocellulosic hydrolyzate

The cellulose fraction of lignocelluloses can be converted to ethanol by separate enzymatic hydrolysis and fermentation (SHF) or via an integrated process, i.e., simultaneous saccharification and fermentation (SSF), non-isothermal simultaneous saccharification and fermentation (NSSF), or simultaneous saccharification and co-fermentation (SSCF). Generally integrated processes are advantageous in term of enhancing ethanol yield by reducing the inhibitory effect of produced sugars on the hydrolytic

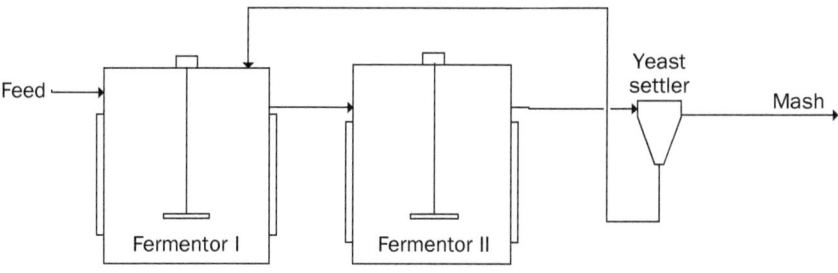

FIGURE 9.8   Two-stage continuous ethanol fermentation process with yeast recirculation [134, 135]

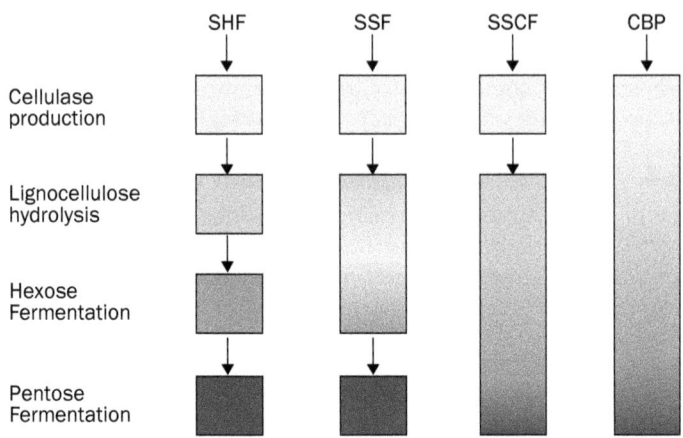

Figure 9.9 Main steps in simultaneous saccharification and fermentation or separate enzymatic hydrolysis and fermentation for ethanol production

enzymes. It is also possible to combine the cellulase production, enzymatic hydrolysis, and fermentation in one step, called consolidated bioprocessing (CBP). There are cost savings because of the reduced number of required vessels in CBP; however, this process is now in the developing stage [138]. A schematic of these processes is shown in Figure 9.9.

9.6.5.1 Separate enzymatic hydrolysis and fermentation (SHF)
In SHF, the first step is the conversion of pretreated lignocelluloses to pentose and hexose sugars. Produced sugars are then converted to ethanol in the second step (fermentation). Enzymatic hydrolysis can be performed in the optimum conditions of the cellulase. The optimum temperature for hydrolysis by cellulase is usually between 45°C and 55°C, depending on the microorganism that produces cellulase. The major disadvantage of SHF is that the released sugars severely inhibit cellulase activity. The activity of cellulase is reduced by 60 percent at a cellobiose concentration as low as 6 g/l. Although glucose also decreases the cellulase activity, the inhibitory effect of glucose is lower than that of cellobiose. Glucose is a strong inhibitor for ß-glucosidase. At a level of 3 g/l of glucose, the activity of ß-glucosidase reduces by 75 percent [53, 139]. Another possible problem in SHF is contamination. The hydrolysis process is rather long (one or possibly several days) and a dilute solution of sugar always has a risk of contamination, even at a rather high temperature such as 45–55°C.

9.6.5.2 Simultaneous saccharification and fermentation (SSF)
SSF combines enzymatic hydrolysis of lignocelluloses and fermentation in one step. As the cellulose converts to glucose, a fermenting microorganism is presented in the medium and immediately consumes the produced glucose. As mentioned, cellobiose and glucose significantly

decrease the activity of cellulase. SSF yields higher ethanol and requires lower amounts of enzyme, because end-product inhibition from cellobiose and glucose formed during enzymatic hydrolysis is relieved by the yeast fermentation [140]. SSF has several advantages compared to SHF, which might be listed as:

- The number of vessels required for SSF is reduced in comparison to SHF, resulting in capital cost saving.
- Less contamination problem during enzymatic hydrolysis, since the presence of ethanol reduces the possibility of contamination.
- Higher yield of ethanol.
- Lower enzyme loading requirement.

On the other hand, SSF has the following drawbacks compared to SHF:

- In SSF, the difference between optimum temperatures of the hydrolyzing enzymes and fermenting microorganisms is usually problematic. For instance, *Trichoderma reesei* cellulases, which constitute the most active preparations, have optimal activity between 45°C and 50°C, whereas *S. cerevisiae* has an optimum temperature between 30°C and 35°C. The optimal temperature for SSF is around 38°C, which is a compromise between the optimal temperatures for hydrolysis and fermentation. Hydrolysis is usually the rate-limiting process in SSF [53]. Several thermotolerant yeasts (e.g., *Candida acidothermophilum* and *Kluyveromyces marxianus*) and bacteria have been used in the SSF to raise the temperature close to the optimal hydrolysis temperature [141].
- Cellulase is inhibited by ethanol. For instance, at 30 g/l ethanol the enzyme activity was reduced by 25 percent [3]. Ethanol inhibition may be a limiting factor in production of high ethanol concentration. However, there has been less attention to ethanol inhibition of cellulase, since practically it is not possible to work with high substrate concentration in SSF because of the problem with mechanical mixing.
- The fermentation process only uses hexoses, and pentose sugars are not used. The unused pentoses may cause problems associated with the wastewater treatment of ethanol plants effluent.

9.6.5.3 *Simultaneous saccharification and co-fermentation (SSCF)*

In addition to the yield, ethanol concentration is also important for the economic feasibility of distillation process. In SSCF, which is a mode of SSF, fermentation of pentose and hexose sugars occur simultaneously.

Unfortunately, the presence of hexoses negatively affects pentose uptake for many microorganisms and the pentose to hexose ratio should increase for starting pentose uptake. By consumption of hexoses in SSCF, the ratio will increase, leading to pentose consumption [138].

Table 9.3 summarizes the results of ethanol production from different lignocelluloses or lignocellulosic-derived sugars using bacteria, fungi, and yeasts, from the recent literature. The advantages of using the microorganisms for ethanol production have been stated as well in this table. As can be seen in the table, using engineered microorganisms for the production of higher ethanol titers has been received great attention in recent years. The research in this area is advancing and the future ethanol production is by using microorganisms that can produce ethanol in a CBP process.

9.6.5.4   *Comparison between enzymatic and acid hydrolysis for lignocellulosic materials*

Between the chemical hydrolyzes methods, two-stage dilute acid hydrolysis, and between the enzymatic processes SSF, are the most promising processes for industrial production of ethanol from cellulosic materials [155, 156]. Advantages and disadvantages of dilute-acid and enzymatic hydrolyzes are summarized in Table 9.4. Enzymatic hydrolysis is carried out under mild conditions, whereas high temperature and low pH result in corrosive conditions for the acid hydrolysis. While it is possible to obtain a cellulose hydrolysis of close to 100 percent by enzymatic hydrolysis after a pretreatment, it is difficult to achieve such high yield with the acid hydrolysis. The yield of conversion of cellulose to sugar with dilute-acid hydrolysis is usually less than 60 percent. Furthermore, the previously mentioned inhibitory compounds are formed during acid hydrolysis, whereas this problem is not so severe for enzymatic hydrolysis. Acid hydrolysis conditions may destroy nutrients sensitive to acid and high temperature such as vitamins, which may introduce the process together with the lignocellulosic materials.

Enzymatic hydrolysis has its own problems in comparison to the dilute-acid hydrolysis. The price of the enzymes is still high [157]. A hydrolysis of several days is necessary for enzymatic hydrolysis, whereas a few minutes are enough for the acid hydrolysis.

9.7   Ethanol Recovery

Fermented broth or "mash" typically contains 7–12 percent (v/v) ethanol. Furthermore, it contains a number of other materials that we can classify into microbial biomass, fuel oil, volatile components, and stillage. The overall composition of fuel oil is found to be a mixture of primary methylbutanols and methylpropanols, formed from α-ketoacids, derived from or leading to amino acids. Depending on the resources used, the important components of fuel oil might be

Bacteria

Yeasts

Microorganisms	Biomasses	Fermentation conditions	Ethanol yield/conc.	Advantages of the microorganism	Refs.
Metabolically engineered *Clostridium thermocellum*	Avicel	Batch cultures at 55°C	20.0 g/l	-4.2 times increase in ethanol yield compared with the wide type strain -High ethanol titers	[142]
Thermoanaerobacterium aotearoense	Xylose and glucose	Batch fermentation	32 mM	-Improving ethanol production via disruption of lactate dehydrogenase by homologous recombination	[143]
Clostridium phytofermentans	Ammonia fiber expansion treated corn stover	CBP	7.0 g/l	-Capable of using in a high solid loading -No need to nutrients supplementation -Tolerance to degradation products of the treatment	[144]
Genetically engineered *Clostridium cellulolyticum*	Acid-pretreated switchgrass and Avicel	CBP	1.3 g/l and 2.7 g/l, respectively	-Development of a high ethanol productivity bacterium from cellobiose, cellulose, and switchgrass	[145]
Recombinant *S. cerevisiae*	Mix arabinose and xylose	Anaerobic batch fermentation	9.3 g/l	-Fermentation of sole arabinose and xylose and mix of them - Lower L-arabitol production	[146]
Kluyveromyces marxianus	Steam exploded barley straw	SSF at 10% solid loading, batch fermentation	67.4% yield or 22 g/l	-Rapidly adoption with media at low inoculum size -Working in a high solid loading	[147]
Engineered *S. cerevisiae*	Mix xylose and cellobiose	Batch fermentation, under oxygen-limited conditions	0.37 g/g	-Co-fermentation of cellubiose and xylose -Improving yield and productivity of ethanol	[148]
Isolated yeasts from industrial environments (*S. cerevisiae* and *Kluyveromyces marxianus*)	hydrolysate of *Eucalyptus globulus* wood	Batch fermentation, anaerobic conditions	> 20 g/l	-Ability to degrade furfural and HMF inhibitors	[149]
Engineered *S. cerevisiae* (SXA-R2P-E)	Dilute-acid treated rice straw and oak	Anaerobic condition, batch mode	20.7 g/l, 0.46 g/g	-Co-fermentation of glucose and xylose -Tolerance to low concentration of acetic acid, furfural, and phenolic	[150]

TABLE 9.3 Recent advances in developments of ethanol-producing microorganisms for high titers ethanol production from lignocelluloses and/or model lignocellulosic materials

	Microorganisms	Biomasses	Fermentation conditions	Ethanol yield/conc.	Advantages of the microorganism	Refs.
Fungi	Peniophora cinerea	glucose, mannose, fructose, sucrose, maltose, and cellobiose	aerobic and semi-aerobic conditions	0.40–0.45 g/g sugar	-Assimilating all the sugars -Exhibition SSF of amorphous cellulose	[151]
	Rhizomucor CCUG 61146 and Rhizomucor CCUG 61147	Glucose, xylose, and mix sugars	Aerobically and micro-aerobically	0.38–0.47, 0.19–0.22, and 0.31–0.38 g/g on glucose, xylose, and mix sugars	-Production of protein and chitosan -Fermentation of mix sugars	[109]
	Phlebia sp. MG-60	Unbleached hardwood kraft pulp and waste newspaper	Batch fermentation, semi-aerobic conditions	71.8% and 51.1%, respectively	-Assimilating glucose, mannose, galactose, fructose, and xylose	[152]
	Flammulina velutipes	Glucose, fructose, mannose, sucrose, maltose, and cellobiose sugarcane bagasse cellulose	Anaerobic condition in batch fermentation	40–60 g/l	-Using high substrate concentration -High conversion rate	[153]
	Engineered Fusarium oxysporum	Glucose	Anaerobic and microaerobic conditions in CBP	69.9%	-Increasing ethanol yield and decreasing acetic acid production	[154]

TABLE 9.3 Recent advances in developments of ethanol-producing microorganisms for high titers ethanol production from lignocelluloses and/or model lignocellulosic materials (Continued)

Parameters	Dilute-acid hydrolysis	Enzymatic hydrolysis
Rate of hydrolysis	Very high	Low
Overall yield of sugars	Low	High and depend upon pretreatment
Catalyst costs	Low	High
Conditions	Harsh reaction conditions, e.g., high pressure and temperature	Mild conditions (e.g. 50°C, atmospheric pressure, pH 4.8)
Inhibitors formation	Highly inhibitory hydrolyzate	Non-inhibitory hydrolyzate
Degradation of sensitive nutrients such as vitamins	High	Low

TABLE 9.4 Comparison between enzymatic and acid hydrolysis for lignocellulosic materials

isoamyl alcohol, *n*-propylalcohol, *sec*-butylalcohol, isobutylalcohol, *n*-butlyalcohol, active amylalcohol, isoamylalcohol, and *n*-amylalcohol. The amount of fusel oil in mash depends on the pH of the fermenter. Fusel oil is used in solvents for paints, polymers, varnishes, and essential oils. Acetaldehyde and trace amounts of other aldehydes and volatile esters are usually produced from grains and molasses. Typically one liter of acetaldehyde and 1–5 liters of fusel oil are produced per 1000 liters of ethanol [13, 82].

Stillage consists of the nonvolatile fraction of materials remaining after alcohol distillation. Its composition depends greatly on the type of feedstock used for fermentation. Stillage generally contains solids, residual sugars, residual ethanol, waxes, fats, fibers, and mineral salts. The solids may be originated from the proteins from the feedstock and spent microbial cells [13].

9.7.1 Distillation

Mash is usually centrifuged or settled to separate the microbial biomass from the liquid and then sent to the ethanol recovery system. Distillation is typically used for separation of ethanol, aldehydes, fusel oil, and stillage [13]. Ethanol is readily concentrated from mash by distillation, since the volatility of ethanol in dilute solution is much higher than the volatility of water. Therefore, ethanol is separated from the rest of the materials and water by distillation. However, ethanol and water form an azeotrope at 95.57 wt% ethanol (89 mol% ethanol) with a minimum boiling point of 78.15°C. This mixture behaves as a single component in a simple distillation, and no further enrichment than 95.57 wt% of ethanol can be achieved by simple distillation [13, 82, 158].

Various industrial distillation systems for ethanol purification are (a) simple two-column systems, (b) three- or four-column barbet system, (c) three-column Othmer system, (d) vacuum rectification, (e) vapor recompression, (f) multi-effect distillation, and (g) six-column reagent alcohol system [13, 82]. These methods are reviewed by Kosaric [13, 82]. The following parameters should be considered for selection of the industrial distillation systems:

- Energy consumption (e.g., steam consumption or cooling water consumption per kg of produced ethanol)
- Quality of ethanol (complete separation of fusel oil and light components)
- How to deal with the problem associated with clogging of the first distillation column and its reboiler because of precipitation or formation of solids. Special design and use of a vacuum may be applied for overcoming the problem in the column. Using open steam instead of application of a reboiler can be used to get rid of the clogging of the reboiler, in spite of the increase in amount of wastewater.
- Simplicity in control of the system
- Simplicity in opening of column parts and cleaning of the columns

Also, the lower capital investment is one of the main parameters in selection of distillation systems.

Recently, the reduction in energy consumption attracted significant attention. These are mainly obtained by using distillation systems working at different pressures. Combing the reboiler of one distillation column with the condenser of another can be obtained when different pressure are applied. Exergy and pinch technology are used to optimize the conditions.

Ethanol is present in the market with different degrees of purity. The majority of ethanol is 190 proof (95 percent or 92.4 percent, minimum) used for solvent, pharmaceutical, cosmetic, and chemical applications. Technical grade ethanol, containing up to 5 percent volatile organic aldehyde, esters, and sometimes methanol, is used for industrial solvents and chemical syntheses. High-purity 200 proof (99.85 percent) anhydrous ethanol is produced for specific applications. For fuel use in mixture with gasoline (gasohol), a nearly anhydrous (99.2 percent) ethanol, but with higher available levels of organic impurities, is used [82].

A simple two-column system is described here while other systems are presented in literature (e.g., [13, 82]). Simple one- or two-column systems with only a stripping and rectification section are usually used to produce lower-quality industrial alcohol and azeotrope alcohol for

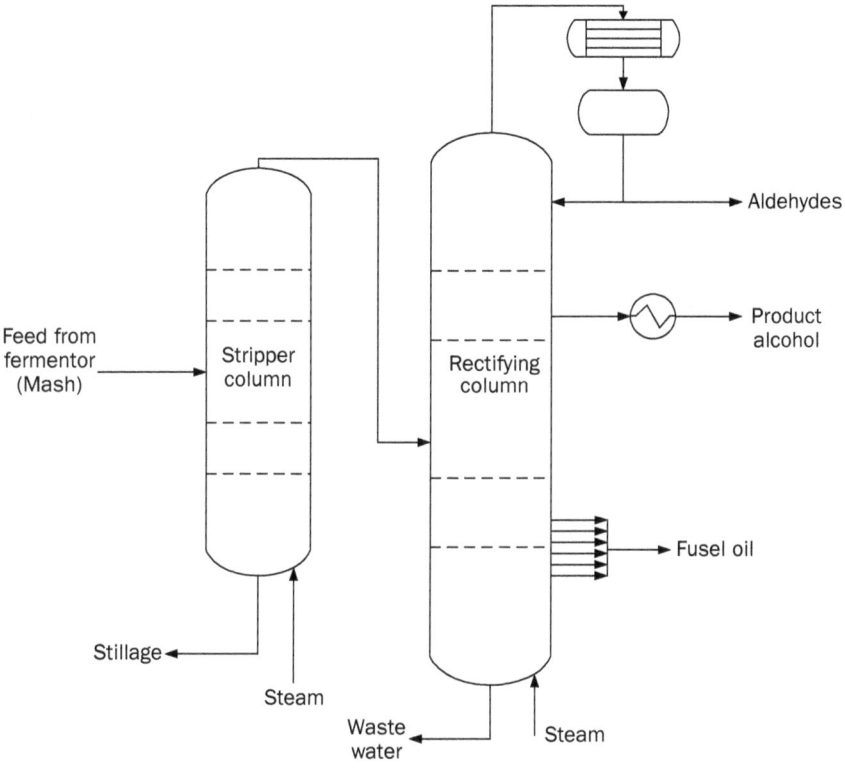

FIGURE 9.10 Two-column system for distillation of ethanol

further dehydration to fuel grade. The simplest continuous ethanol distillation system consists of stripping and rectification sections, either together in one column or separated in two columns (Fig. 9.10).

The produced mash is pumped into a continuous distillation process where steam is used to heat the mash to its boiling point in the stripper column. The ethanol-enriched vapors pass through a rectifying column and are condensed and removed from the top of the rectifier at around 95 percent ethanol. The ethanol-stripped stillage falls to the bottom of the stripper column and is pumped to a stillage tank. Aldehydes are drawn from the head vapor, condensed, and partly used as reflux. Fusel oil is blended from several plates of the rectifying section [13, 82, 159].

With efficient distillation, the stillage should contain less than 0.1 percent ethanol since the presence of ethanol significantly increases the COD of wastewater. For each 1 percent ethanol left in the stillage, the COD of the stillage is incremented by more than 20 g/L. Due to the potential impact of residual ethanol content, therefore, proper control over distillation can greatly affect the COD of stillage [159].

Distillation of ethanol to increase its concentration is an old process and has been used widely for the concentration of ethanol. Ethanol recovery via distillation requires significantly high energy compared to dehydration from azeotrope to >99.5 percent concentration. Some strategies with more energy input efficiency, e.g., double-effect distillation and heat-pump assisted distillation, have recently developed to reduce the energy input and increase the energy efficiency of the distillation process [160].

9.7.2 Alternative processes for ethanol recovery and purification

Since the distillation is a highly energy-consuming process, several processes have been developed for purification of ethanol from the fermentation broth, like solvent extraction, CO_2 extraction process, vapor recompression systems, pervaporation [161], and low-temperature blending with gasoline [13]. Pervaporation is a promising technology for concentration of ethanol produced during the fermentation media. In pervaporation process, molecules with higher affinity for diffusion in a membrane are separated from the feed mixture [162]. Pervaporation needs much less energy than the distillation and a recent study calculated its energy requirement comparable to that of for fermentation on each kg ethanol production basis (2.4 vs. 2.1 MJ per kg of ethanol, respectively) [163].

9.7.3 Ethanol dehydration

To allow the blending of alcohol with gasoline, the water content of ethanol must be reduced to less than 1 percent by volume, which is not possible by distillation. Higher water levels can result in the phase separation of an alcohol/water mixture from the gasoline phase, which may cause engine malfunction. The removal of the water beyond the last 5 percent is called "dehydration" or "drying" of ethanol. Azeotropic distillation was previously employed to produce higher-purity ethanol by adding a third component, such as benzene, cyclohexane, or ether, to break the azeotrope and produce dry ethanol [159]. To avoid the illegal transfer of ethanol from the industrial market into the potable alcohol market, where it is highly regulated and taxed, dry alcohol usually requires the addition of denaturing agents that render it toxic for human consumption, and the azeotropic reagents conveniently meet this requirement [159]. Except in the high-purity reagent-grade ethanol market, azeotropic drying has been nowadays supplanted by molecular sieve drying technology.

9.7.3.1 Molecular sieve adsorption

Nowadays, dehydration of 95 percent ethanol for fuel use is exclusively performed via using molecular sieves in a pressure swing adsorption

(PSA) process. The molecular sieve is a more energy-efficient method than azeotropic distillation. Furthermore, this method avoids the occupational hazards associated with the azeotropic chemical admixtures. In molecular sieve drying, 95 percent ethanol is passed through a bed of synthetic zeolite with uniform pore sizes that preferentially adsorb water molecules that are smaller than ethanol molecules. Approximately ¾ of adsorbed material is water and ¼ is ethanol. The bed becomes saturated after a few minutes and must be regenerated by heating or evacuation to drive out the adsorbed water. During the regeneration phase, a side stream of ethanol/water (often around 50 percent) is produced, which must be redistilled before it can be returned to the drying process [159, 164].

9.8 Integrated Biorefineries for Ethanol Production

The conversion of biomass is more profitable when all the biomass ingredients are converted to biofuels and valuable chemicals via "biorefinery" processes. Notably, even though the biorefinery processes are not limited to green and biochemical technologies, the use of these technologies lead to established sustainable production chains of biofuels and biochemicals [165]. As discussed earlier, second-generation ethanol production is not economically competing with the first-generation production, and uncertainty over the process and investment costs create major obstacles for its commercialization. On the other hand, the economy of the dominant dry mills ethanol processes is dependent on the by-product DDGS (Distiller's Dried Grains and Solubles), which is used as a highly nutritious animal feed. One interesting process is to integrate the dry mills ethanol processes with the second-generation ethanol production. However, care is necessary for the selection of the pretreatment method and fermentative microorganisms when the by-product is going to be used as fodder. The use of edible zygomycetes and ascomycetes fungi are reported for this purpose [166]. Some of these fungi, e.g., *Mucor indicus* and *Neurospora intermedia*, are naturally utilizing pentoses, originated from lignocelluloses, in such processes. The biomass of these fungi are edible and can add nutritionally valuable protein, oil, and carbohydrates to fodder [166].

9.9 Concluding Remarks and Future Prospects

Governmental regulations, economic incentives, and concerns for the environment and depletion of fossil fuels have been the driving force for initiating biomass as the primary source for fuel production. First-generation ethanol production is commercialized for the fuel market; however, future sustainable production of ethanol is feasible only when lignocellulosic biomass is being used as the raw materials for sugars production. Although great advances in the development of the process for ethanol from lignocellulosic materials have been achieved, there are still several challenges left in this process to be faced in the future until the process is fully established. The high price and consumption of cellulase, low concentration of sugar and consequently ethanol, and the necessity of pretreatment and its issues are still the challenges that should be addressed. Consolidated bioprocessing (CBP) is expected to be a promising process for the future sustainable production of ethanol; however, this process is still in the primary stages of development.

References

1. B. Satari, K. Karimi, and R. Kumar. Cellulose solvent-based pretreatment for enhanced second-generation biofuel production: a review. *Sustainable Energy & Fuels.* 2018.
2. J. Baeyens, Q. Kang, L. Appels, R. Dewil, Y. Lv, and T. Tan. Challenges and opportunities in improving the production of bio-ethanol. *Progress in Energy and Combustion Science* **47**, 60–88, 2015.
3. C. E. Wyman. *Handbook on Bioethanol: Production and Utilization.* Washington, DC: Taylor & Francis; 1996.
4. S. Peake and P. Ekins. Exploring the financial and investment implications of the Paris Agreement. *Climate Policy* **2016**, 1–16.
5. H. Rothman, R. Greenshields, and F. R. Calle. *The Alcohol Economy: Fuel Ethanol and the Brazilian Experience.* First ed. London: Francis Printer, 1983.
6. J. D. Stephen, W. E. Mabee, and J. N. Saddler. Will second-generation ethanol be able to compete with first-generation ethanol? Opportunities for cost reduction. *Biofuels, Bioproducts and Biorefining* **6**, 159–176, 2011.
7. F. H. Reboredo, F. Lidon, F. Pessoa, and J. C. Ramalho. The fall of oil prices and the effects on biofuels. *Trends Biotechnol* **34**, 3–6.
8. S. C. Albers, A. M. Berklund, and G. D. Graff. The rise and fall of innovation in biofuels. *Nat Biotechnol.* **34**, 814, 2016.
9. L. R. Lynd, X. Liang, M. J. Biddy, A. Allee, H. Cai, T. Foust T, et al. Cellulosic ethanol: status and innovation. *Curr Opin Biotechnol.* **45**, 202–211, 2017.
10. S. Singh, A. Adak, M. Saritha, S. Sharma, R. Tiwari, S. Rana, et al. Bioethanol Production Scenario in India: Potential and Policy Perspective. In: A. K. Chandel and R. K. Sukumaran, editors. *Sustainable Biofuels Development in India.* Cham: Springer International Publishing, 2017. pp. 21–37.
11. M. Roehr. *The biotechnology of ethanol: classical and future applications.* First ed. Weinheim: Wiley-VCH, 2001.
12. Y. Lin and S. Tanaka. Ethanol fermentation from biomass resources: current state and prospects. *Applied Microbiology and Biotechnology* **69**, 627–642, 2006.

13. N. Kosaric, A. Wieczorirek, G. P. Cosentono, and R. J. Magee. Ethanol fermentation. In: G. Reed, editor. *Biotechnology: A comprehensive treatise.* Verlag-Chemie, 1983. pp. 257–386.
14. M. Vohra, J. Manwar, R. Manmode, S. Padgilwar, and S. Patil S. Bioethanol production: Feedstock and current technologies. *Journal of Environmental Chemical Engineering* 2, 573–584, 2014.
15. M. Gulati, K. Kohlmann, M. R. Ladisch, R. Hespell, and R. J. Bothast. Assessment of ethanol production options for corn products. *Bioresource Technology* **58**, 253–264, 1996.
16. C. Manochio, B. R. Andrade, R. P. Rodriguez, and B. S. Moraes. Ethanol from biomass: A comparative overview. *Renewable and Sustainable Energy Reviews* **80**, 743–755, 2017.
17. J. E. Bailey and D. F. OllisF. *Biochemical engineering fundamentals.* Second ed. Singapore: McGraw-Hill, 1986.
18. Y. Sun and J. Cheng. Hydrolysis of lignocellulosic materials for ethanol production: a review. *Bioresour Technol.* **83**, 1–11, 2002.
19. K. Karimi, S. Kheradmandinia, and M. J. Taherzadeh. Conversion of rice straw to sugars by dilute-acid hydrolysis. *Biomass and Bioenergy* **30**, 247–253, 2006.
20. B. C. Saha. Hemicellulose bioconversion. *Journal of Industrial Microbiology and Biotechnology* **30**, 279–291, 2003.
21. B. Satari, K. Karimi, A. Zamani. Oil, chitosan, and ethanol production by dimorphic fungus Mucor indicus from different lignocelluloses. *J Chem Technol Biotechnol.* **91**, 1835–1843, 2016.
22. B. Satari Baboukani, M. Vossoughi, and I. Alemzadeh. Optimisation of dilute-acid pretreatment conditions for enhancement sugar recovery and enzymatic hydrolysis of wheat straw. *Biosys Eng.* **111**, 166–174, 2012.
23. B. Satari, K. Karimi, and M. Molaverdi. Structural features influential to enzymatic hydrolysis of cellulose-solvent-based pretreated pinewood and elmwood for ethanol production. *Bioprocess Biosystems Eng.* **41**, 249–264, 2018.
24. M. J. Taherzadeh. Ethanol from lignocellulose: physiological effects of inhibitors and fermentation strategies [PhD Thesis]. Göteborg, Sweden: Chalmers University of Technology, 1999.
25. P. Béguin and J.-P. Aubert. The biological degradation of cellulose. *FEMS Microbiology Reviews* **13**, 25–58, 1994.
26. K. Karimi and M. J. Taherzadeh. A critical review of analytical methods in pretreatment of lignocelluloses: Composition, imaging, and crystallinity. *Bioresour Technol.* **200**, 1008–1018, 2016.
27. C. M. Payne, B. C. Knott, H. B. Mayes, H. Hansson, M. E. Himmel, M. Sandgren, et al. Fungal Cellulases. *Chem Rev.* **115**, 1308–1448, 2015.
28. J. S. Van Dyk and B. I. Pletschke. A review of lignocellulose bioconversion using enzymatic hydrolysis and synergistic cooperation between enzymes—Factors affecting enzymes, conversion and synergy. *Biotechnol Adv.* **30**, 1458–1480, 2012.

29. A. J. Ragauskas, G. T. Beckham, M. J. Biddy, R. Chandra, F. Chen, M. F. Davis, et al. Lignin Valorization: Improving Lignin Processing in the Biorefinery. *Science* **344**, 2014.
30. E. Palmqvist and B. Hahn-Hägerdal. Fermentation of lignocellulosic hydrolysates. II: inhibitors and mechanisms of inhibition. *Bioresource Technology* **74**, 25–33, 2000.
31. L. B. Davin, A. M. Patten, M. Jourdes, and N. G. Lewis. Lignins: a twenty-first century challenge. In: M. Himmel, editor. *Biomass Recalcitrance: Deconstructing the Plant Cell Wall for Bioenergy*, 2009. pp. 213–305.
32. B. Satari and K. Karimi. Citrus processing wastes: Environmental impacts, recent advances, and future perspectives in total valorization. *Resources, Conservation and Recycling* **129**, 153–167, 2018.
33. B. Satari, J. Palhed, K. Karimi, M. Lundin, M. J. Taherzadeh, and A. Zamani. Process Optimization for Citrus Waste Biorefinery via Simultaneous Pectin Extraction and Pretreatment, 2017.
34. M. J. Taherzadeh and C. Niklasson. Ethanol from lignocellulosic materials: pretreatment, acid and enzymatic hydrolyses and fermentation. In: B. C. Saha and K. Hayashi, editors. *Lignocellulose Biodegradation*. First ed. Washington DC: American Chemical Society, 2004.
35. M. J. Taherzadeh, C. Niklasson, and G. Lidén. Acetic acid - friend or foe in anaerobic batch conversion of glucose to ethanol by *Saccharomyces cerevisiae*? *Chemical Engineering Science* **52**, 2653–2659, 1997.
36. L. Olsson and B. HahnHagerdal. Fermentation of lignocellulosic hydrolysates for ethanol production. *Enzyme and Microbial Technology* **18**, 312–331, 1996.
37. E. Palmqvist and B. Hahn-Hägerdal. Fermentation of lignocellulosic hydrolysates. I: inhibition and detoxification. *Bioresource Technology* **74**, 17–24, 2000.
38. L. J. Jönsson, B. Alriksson, and N.-O. Nilvebrant. Bioconversion of lignocellulose: inhibitors and detoxification. *Biotechnol Biofuels* **6**, 1–10, 2013.
39. L. J. Jönsson and C. Martín. Pretreatment of lignocellulose: Formation of inhibitory by-products and strategies for minimizing their effects. *Bioresour Technol.* **199**, 103–112, 2016.
40. P. Persson, J. Andersson, L. Gorton, S. Larsson, N. O. Nilvebrant, and L. J. Jonsson. Effect of different forms of alkali treatment on specific fermentation inhibitors and on the fermentability of lignocellulose hydrolysates for production of fuel ethanol. *Journal of Agricultural and Food Chemistry* **50**, 5318–5325, 2002.
41. R. Millati, C. Niklasson, and M. J. Taherzadeh. Effect of pH, time and temperature of overliming on detoxification of dilute-acid hydrolyzates for fermentation by Saccharomyces cerevisiae. *Process Biochemistry* **38**, 515–522, 2002.

42. A. Martinez, M. E. Rodriguez, M. L. Wells, S. W. York, J. F. Preston, and L. O. Ingram. Detoxification of dilute acid hydrolysates of lignocellulose with lime. *Biotechnology Progress* **17**, 287–293, 2001.
43. S. Amartey and T. Jeffries. An improvement in *Pichia stipitis* fermentation of acid-hydrolysed hemicellulose achieved by overliming (calcium hydroxide treatment) and strain adaptation. *World Journal of Microbiology & Biotechnology* **12**, 281–283, 1996.
44. R. Purwadi, C. Niklasson, and M. J. Taherzadeh. Kinetic study of detoxification of dilute-acid hydrolyzates by Ca(OH)2. *Journal of Biotechnology* **114**, 187–198, 2004.
45. L. R. Lynd, C. E. Wyman, and T. U. Gerngross. Biocommodity Engineering. *Biotechnol Prog.* **15**, 777–793, 1999.
46. C. E. Wyman. What is (and is not) vital to advancing cellulosic ethanol. *Trends Biotechnol.* **25**, 2007.
47. R. P. Chandra, R. Bura, W. E. Mabee, A. Berlin, X. Pan, and J. N. Saddler. Substrate pretreatment: the key to effective enzymatic hydrolysis of lignocellulosics. *Adv Biochem Eng Biotechnol.* **5**, 2007.
48. B. Yang and C. E. Wyman. Pretreatment: the key to unlocking low-cost cellulosic ethanol. *Biofuels, Bioprod Biorefin.* **2**, 26–40, 2008.
49. V. B. Agbor, N. Cicek, R. Sparling, A. Berlin, and D. B. Levin. Biomass pretreatment: Fundamentals toward application. *Biotechnol Adv.* **29**, 675–685, 2011.
50. B. Yang, L. Tao, and C. E. Wyman. Strengths, challenges, and opportunities for hydrothermal pretreatment in lignocellulosic biorefineries. *Biofuels, Bioproducts and Biorefining* **12**, 125–138, 2018.
51. C. E. Wyman and B. Yang. Combined Severity Factor for Predicting Sugar Recovery in Acid-Catalyzed Pretreatment Followed by Enzymatic Hydrolysis. In: H. A. Ruiz, M. Hedegaard Thomsen, and H. L. Trajano, editors. *Hydrothermal Processing in Biorefineries: Production of Bioethanol and High Added-Value Compounds of Second and Third Generation Biomass*. Cham: Springer International Publishing, 2017. pp. 161–180.
52. B. Yang, L. Tao, and C. E. Wyman. Strengths, challenges, and opportunities for hydrothermal pretreatment in lignocellulosic biorefineries. *Biofuels, Bioproducts and Biorefining* **12**, 125–138, 2017.
53. G. P. Philippidis and T. K. Smith. Limiting factors in the simultaneous saccharification and fermentation process for conversion of cellulosic biomass to fuel ethanol. *Applied Biochemistry and Biotechnology* **51/52**, 117–124, 1995.
54. D. B. Wilson. Processive and nonprocessive cellulases for biofuel production—lessons from bacterial genomes and structural analysis. *Appl Microbiol Biotechnol.* **93**, 497–502, 2012.
55. E. A. Bayer, J. P. Belaich, Y. Shoham, and R. Lamed. The cellulosomes: multienzyme machines for degradation of plant cell wall polysaccharides. *Annual Review of Microbiology* **58**, 521–554, 2004.

56. K. Stenberg, M. Bollok, K. Reczey, M. Galbe, and G. Zacchi. Effect of substrate and cellulase concentration on simultaneous saccharification and fermentation of steam-pretreated softwood for ethanol production. *Biotechnology and Bioengineering* **68**, 204–210, 2000.

57. M. Linde, M. Galbe, and G. Zacchi. Simultaneous saccharification and fermentation of steam-pretreated barley straw at low enzyme loadings and low yeast concentration. *Enzyme and Microbial Technology* 2006.

58. P. Bansal, M. Hall, M. J. Realff, J. H. Lee, and A. S. Bommarius. Modeling cellulase kinetics on lignocellulosic substrates. *Biotechnology Advances* **27**, 833–848, 2009.

59. M. Alkasrawi, T. Eriksson, J. Borjesson, A. Wingren, M. Galbe, F. Tjerneld, et al. The effect of Tween-20 on simultaneous saccharification and fermentation of softwood to ethanol. *Enzyme and Microbial Technology* **33**, 71–78, 2003.

60. S. A. Jambo, R. Abdulla, S. H. Mohd Azhar, H. Marbawi, J. A. Gansau, and P. Ravindra. A review on third generation bioethanol feedstock. *Renewable and Sustainable Energy Reviews* **65**, 756–769, 2016.

61. M. T. Cesário, M. M. R. da Fonseca, M. M. Marques, and M. C. M. D. de Almeida. Marine algal carbohydrates as carbon sources for the production of biochemicals and biomaterials. *Biotechnol Adv*. **36**, 798–817, 2018.

62. C. E. de Farias Silva and A. Bertucco. Bioethanol from microalgae and cyanobacteria: A review and technological outlook. *Process Biochem*. **51**, 1833–1842, 2016.

63. G. P. Lam, M. H. Vermuë, M. H. M. Eppink, R. H. Wijffels, and C. van den Berg. Multi-Product Microalgae Biorefineries: From Concept Towards Reality. *Trends Biotechnol*. **36**, 216–227, 2018.

64. S. Khanra, M. Mondal, G. Halder, O. N. Tiwari, K. Gayen, and T. K. Bhowmick. Downstream processing of microalgae for pigments, protein and carbohydrate in industrial application: A review. *Food Bioprod Process*. **110**, 60–84, 2018.

65. C. K. Phwan, H. C. Ong, W.-H. Chen, T. C. Ling, E. P. Ng, and P. L. Show. Overview: Comparison of pretreatment technologies and fermentation processes of bioethanol from microalgae. *Energy Conversion and Management* **173**, 1–94, 2018.

66. T. Brandberg T. Fermentation of undetoxified dilute acid lignocellulose hydrolyzate for fuel ethanol production [PhD Thesis]. Göteborg, Sweden: Chalmers University of Technology, 2005.

67. M. Sonderegger, M. Jeppsson, C. Larsson, M. F. Gorwa-Grauslund, E. Boles, L. Olsson, et al. Fermentation performance of engineered and evolved xylose-fermenting *Saccharomyces cerevisiae* strains. *Biotechnology and Bioengineering* **87**, 90–98, 2004.

68. T. W. Jeffries. Engineering yeasts for xylose metabolism. *Current Opinion in Biotechnology Environmental Biotechnology/Energy Biotechnology* **17**, 320–326, 2006.

69. A. J. van Maris, D. A. Abbott, E. Bellissimi, J. van den Brink, M. Kuyper, M. A. Luttik, et al. Alcoholic fermentation of carbon sources in biomass hydrolysates by *Saccharomyces cerevisiae*: current status. *Antonie Van Leeuwenhoek* **90**, 391–418, 2006.
70. M. Jeppsson, B. Johansson, B. Hahn-Hagerdal, and M. F. Gorwa-Grauslund. Reduced oxidative pentose phosphate pathway flux in recombinant xylose-utilizing *Saccharomyces cerevisiae* strains improves the ethanol yield from xylose. *Applied Environmental Microbiology* **68**, 1604–1609, 2002.
71. M. Desvaux, E. Guedon, and H. Petitdemange. Cellulose catabolism by *Clostridium cellulolyticum* growing in batch culture on defined medium. *Applied Environmental Microbiology* **66**, 2461–2470, 2000.
72. I. S. Horváth, M. J. Taherzadeh, C. Niklasson, and G. Lidén. Effects of furfural on anaerobic continuous cultivation of Saccharomyces cerevisiae. *Biotechnology and Bioengineering* **75**, 540–549, 2001.
73. W. H. Kampen. Nutritional requirements in fermentation processes. In: Todaro CL, editor. *Fermentation and Biochemical Engineering Handbook.* Second ed: Noyes publications, 1997.
74. M. J. Taherzadeh, L. Adler, and G. Lidén. Strategies for enhancing fermentative production of glycerol - review. *Enzyme and Microbial Technology* **31**, 53–66, 2002.
75. R. Millati. *Ethanol production from lignocellulosic materials.* Göteborg, Sweden: Chalmers University of Technology, 2005.
76. H. Schneider. conversion of pentoses to ethanol by yeast and fungi. *Critical Reviews in Biotechnology* **9**, 1989.
77. K. Karhumaa, R. Fromanger, B. Hahn-Hagerdal, and M. F. Gorwa-Grauslund. High activity of xylose reductase and xylitol dehydrogenase improves xylose fermentation by recombinant Saccharomyces cerevisiae. *Applied Microbiology and Biotechnology.* 2006.
78. M. Jeppsson, K. Traff, B. Johansson, B. Hahn-Hagerdal, and M. F. Gorwa-Grauslund. Effect of enhanced xylose reductase activity on xylose consumption and product distribution in xylose-fermenting recombinant Saccharomyces cerevisiae. *FEMS Yeast Res.* **3**, 167–175, 2003.
79. Y. C. Bor, C. Moraes, S. P. Lee, W. L. Crosby, A. J. Sinskey, and C. A. Batt. Cloning and sequencing the *Lactobacillus brevis* gene encoding xylose isomerase. *Gene* **114**, 127–132, 1992.
80. X. Zhu, M. Teng, L. Niu, C. Xu, and Y. Wang. Structure of xylose isomerase from *Streptomyces diastaticus* no. 7 strain M1033 at 1.85 A resolution. *Acta Crystallogr D Biol Crystallogr.* **56**, 129–136, 2000.
81. Y. Wang, Z. Huang, X. Dai, J. Liu, T. Cui, L. Niu, et al. The sequence of xylose isomerase gene from *Streptomyces diastaticus* No. 7 M1033. *Chinical Journal of Biotechnology* **10**, 97–103, 1994.
82. B. L. Maiorella. Ethanol. In: Moo-Young M, editor. *Comperehensive biotechnology.* First ed. Oxford: Pergamon Press Ltd, 1985. pp. 861–914.

83. S. H. Mohd Azhar, R. Abdulla, S. A. Jambo, H. Marbawi, J. A. Gansau, A. A. Mohd Faik, et al. Yeasts in sustainable bioethanol production: A review. *Biochemistry and Biophysics Reports* **10**, 52–61, 2017.

84. J. Nielsen, C. Larsson, A. van Maris, and J. Pronk. Metabolic engineering of yeast for production of fuels and chemicals. *Curr Opin Biotechnol.* **24**, 398–404, 2013.

85. O. Kurylenko, M. Semkiv, J. Ruchala, O. Hryniv, B. Kshanovska, C. Abbas, et al. New approaches for improving the production of the 1st and 2nd generation ethanol by yeast. *Acta Biochim Pol.* **63**, 31–38, 2016.

86. I. Papapetridis, M. van Dijk, A. J. A. van Maris, and J. T. Pronk. Metabolic engineering strategies for optimizing acetate reduction, ethanol yield and osmotolerance in Saccharomyces cerevisiae. *Biotechnology for Biofuels* **10**, 107, 2017.

87. S. Govindaswamy and L. M. Vane. Kinetics of growth and ethanol production on different carbon substrates using genetically engineered xylose-fermenting yeast. *Bioresource Technology* **98**, 677–685, 2007.

88. C. Martin, M. Marcet, O. Almazan, and L. J. Jonsson. Adaptation of a recombinant xylose-utilizing *Saccharomyces cerevisiae* strain to a sugarcane bagasse hydrolysate with high content of fermentation inhibitors. *Bioresource Technology.* 2006.

89. L. Ruohonen, A. Aristidou, A. D. Frey, M. Penttila, and P. T. Kallio. Expression of Vitreoscilla hemoglobin improves the metabolism of xylose in recombinant yeast Saccharomyces cerevisiae under low oxygen conditions. *Enzyme and Microbial Technology* **39**, 6–14, 2006.

90. N. Ho, Z. Chen, and A. Brainard. Genetically engineered *Saccharomyces* yeast capable of effective cofermentation of glucose and xylose. *Applied Environmental Microbiology* **64**, 1852–1859, 1998.

91. M. H. Toivari, A. Aristidou, L. Ruohonen, and M. Penttila. Conversion of xylose to ethanol by recombinant *Saccharomyces cerevisiae*: importance of xylulokinase (XKS1) and oxygen availability. *Metabolic Engineering* **3**, 236–249, 2001.

92. S.-J. Ha, J. M. Galazka, S. Rin Kim, J.-H. Choi, X. Yang, J.-H. Seo, et al. Engineered *Saccharomyces cerevisiae* capable of simultaneous cellobiose and xylose fermentation. *Proceedings of the National Academy of Sciences* **108**, 504, 2011.

93. T. M. Long, Y.-K. Su, J. Headman, A. Higbee, L. B. Willis, and T. W. Jeffries. Cofermentation of glucose, xylose, and cellobiose by the beetle-associated yeast *Spathaspora passalidarum*. *Appl Environ Microbiol.* **78**, 5492–5500, 2012.

94. T. W. Jeffries and C. P. Kurtzman. Strain selection, taxonomy, and genetics of xylose-fermenting yeasts. *Enzyme and Microbial Technology* **16**, 922–932, 1994.

95. D. D. Spindler, C. E. Wyman, K. Grohmann, and G. P. Philippidis. Evaluation of the cellobiose-fermenting yeast *Brettanomyces custersii* in the simultaneous saccharification and fermentation of cellulose. *Biotechnology Letters* **14**, 403–407, 1992.

96. E. T. A. Sheikh Idris and D. R. Berry. Selection of thermotolerant yeast strains for biomass production from sudanese molasses. *Biotechnology Letters* **2**, 61–66, 1980.

97. M. Ballesteros, J. M. Oliva, M. J. Negro, P. Manzanares, and I. Ballesteros. Ethanol from lignocellulosic materials by a simultaneous saccharification and fermentation process (SFS) with *Kluyveromyces marxianus* CECT 10875. *Process Biochemistry* **39**, 1843–1848, 2004.

98. P. Majidian, M. Tabatabaei, M. Zeinolabedini, M. P. Naghshbandi, and Y. Chisti. Metabolic engineering of microorganisms for biofuel production. *Renewable and Sustainable Energy Reviews* **82**, 3863–3885, 2018.

99. A. Ajit, A. Z. Sulaiman, and Y. Chisti. Production of bioethanol by *Zymomonas mobilis* in high-gravity extractive fermentations. *Food Bioprod Process* **102**, 123–35, 2017.

100. I. S. Horvath. *Fermentation inhibitors in the production of bio-ethanol.* Göteborg, Sweden: Chalmers University of Technology, 2004.

101. M. X. He, B. Wu, H. Qin, Z. Y. Ruan, F. R. Tan, J. L.Wang, et al. *Zymomonas mobilis*: a novel platform for future biorefineries. *Biotechnology for Biofuels* **7**, 101, 2014.

102. S. Yang, Q. Fei, Y. Zhang, L. M. Contreras, S. M. Utturkar, S. D. Brown, et al. *Zymomonas mobilis* as a model system for production of biofuels and biochemicals. *Microbial Biotechnology* **9**, 699–717, 2016.

103. L. Ingram, P. Gomez, X. Lai, M. Monirruzzamam, B. Wood, L. Yomano, et al. Metabolic engineering of bacteria for ethanol production. *Biotechnol Bioeng.* **58**, 204–214, 1998.

104. J. Dexter and P. Fu. Metabolic engineering of cyanobacteria for ethanol production. *Energy & Environmental Science* **2**, 857–864, 2009.

105. M. Inui, H. Kawaguchi, S. Murakami, A. A. Vertes, and H. Yukawa. Metabolic engineering of *Corynebacterium glutamicum* for fuel ethanol production under oxygen-deprivation conditions. *Journal of Molecular Microbiology and Biotechnology* **8**, 243–254, 2004.

106. B. Satari and K. Karimi. Mucoralean fungi for sustainable production of bioethanol and biologically active molecules. *Appl Microbiol Biotechnol.* **102**, 1097–1117, 2018.

107. J. A. Ferreira, A. Mahboubi, P. R. Lennartsson, and M. J. Taherzadeh. Waste biorefineries using filamentous ascomycetes fungi: Present status and future prospects. *Bioresour Technol.* **215**, 334–345, 2016.

108. R. Millati, L. Edebo, and M. J. Taherzadeh. Performance of *Rhizopus, Rhizomucor,* and *Mucor* in ethanol production from glucose, xylose, and wood hydrolyzates. *Enzyme and Microbial Technology* **36**, 294–300, 2005.

109. R. Wikandari, R. Millati, P. R. Lennartsson, E. Harmayani, and M. J. Taherzadeh. Isolation and characterization of zygomycetes fungi from tempe for ethanol production and biomass applications. *Appl Biochem Biotechnol.* **167**, 1501–1512, 2012.

110. K. Karimi, G. Emtiazi, and M. J. Taherzadeh. Production of ethanol and mycelial biomass from rice straw hemicellulose hydrolyzate by *Mucor indicus. Process Biochemistry* **41**, 653–658, 2006.

111. K. Karimi and A. Zamani. Mucor indicus: Biology and industrial application perspectives: A review. *Biotechnol Adv.* **31**, 466–481, 2013.

112. S. Behnam, K. Karimi, and M. Khanahmadi. Optimization of glucoamylase production by *Mucor indicus, Mucor hiemalis*, and *Rhizopus oryzae* through solid state fermentation. *Turk J Bioch.* **41**, 2016.

113. S. Behnam, K. Karimi, M. Khanahmadi, and Z. Salimian. Optimization of xylanase production by *Mucor indicus, Mucor hiemalis*, and *Rhizopus oryzae* through solid state fermentation. *Biological Journal of Microorganism* **4**, 1–10, 2016.

114. C. D. Skory, S. N. Freer, and R. J. Bothast. Screening for ethanol-producing filamentous fungi. *Biotechnology Letters* **19**, 203–206, 1997.

115. T. L. Lübbehüsen, J. Nielsen, and M. McIntyre. Aerobic and anaerobic ethanol production by *Mucor circinelloides* during submerged growth. *Appl Microbiol Biotechnol.* **63**, 543–548, 2004.

116. A. Sues, R. Millati, L. Edebo, and M. J. Taherzadeh. Ethanol production from hexoses, pentoses, and dilute-acid hydrolyzate by *Mucor indicus. FEMS Yeast Research* **5**, 669–676, 2005.

117. K. Karimi, T. Brandberg, L. Edebo, and M. Taherzadeh. Fed-batch cultivation of *Mucor indicus* in dilute-acid lignocellulosic hydrolyzate for ethanol production. *Biotechnology Letters* **6**, 1395–1400, 2005.

118. P. A. Gibbs, R. J. Seviour, and F. Schmid. Growth of filamentous fungi in submerged culture: Problems and possible solutions. *Critical Reviews in Biotechnology* **20**, 17–48, 2000.

119. M. Papagianni. Fungal morphology and metabolite production in submerged mycelial processes. *Biotechnology Advances* **22**, 189–259, 2004.

120. M. Sharifia, K. Karimi, and M. Taherzadeh. Production of ethanol by filamentous and yeast-like forms of *Mucor indicus* from fructose, glucose, sucrose, and molasses. *J Ind Microbiol Biotechnol.* **35**, 1253–1259, 2008.

121. H. L. Zhang, J. Baeyens, and T. W. Tan. Mixing phenomena in a large-scale fermenter of starch to bio-ethanol. *Energy* **48**, 380–391, 2012.

122. H. Zabed, J. N. Sahu, A. Suely, A. N. Boyce, and G. Faruq. Bioethanol production from renewable sources: Current perspectives and technological progress. *Renewable and Sustainable Energy Reviews* **71**, 475–501, 2017.

123. V. F. Silva, S. C. Nakanishi, S. R. Dionísio, C. E. Rossell, J. L. Ienczak, A. R. Gonçalves, et al. Using cell recycling batch fermentations to validate a setup for cellulosic ethanol production. *Journal of Chemical Technology & Biotechnology* **91**, 1853–1859, 2016.

124. Y. Matano, T. Hasunuma, and A. Kondo. Cell recycle batch fermentation of high-solid lignocellulose using a recombinant cellulase-displaying yeast strain for high yield ethanol production in consolidated bioprocessing. *Bioresour Technol.* **135**, 403–409, 2013.
125. N. Vasconcelos, C. E. Lopes, and F. P. França. Continuous ethanol production using yeast immobilized on sugar-cane stalks. *Brazilian Journal of Chemical Engineering* **21**, 2004.
126. K. Kavanagh and P. A. Whittaker. Application of the Melle-Boinot process to the fermentation of xylose by Pachysolen tannophilus. *Appl Microbiol Biotechnol.* **42**, 28–31, 1994.
127. V. M. da Matta and A. Medronho Rde. A new method for yeast recovery in batch ethanol fermentations: filter aid filtration followed by separation of yeast from filter aid using hydrocyclones. *Bioseparation* **9**, 43–53, 2000.
128. B. McNiel and L. M. Harvey. Fermentation a practical approach. In: Hames BD, editor. *Practical Approach Series.* First ed. Oxford: Oxford University Press, 1990. pp. 113–120.
129. Yujiro Harada, Kuniaki Sakata, Sato S, Takayama S. Fermentation Pilot Plant, 1997.
130. G. R. Cysewski and C. R. Wilke. Process design and economic studies of alternative fermentation methods for the production of ethanol. *Biotechnol Bioeng.* **20**, 1421–1444, 1978.
131. B. Goggin and G. Thorsson. *Operating experience with Biostil in a commercial distillery.* Alfa-Laval, Tumba, Sweden, 1982.
132. F. Talebnia and M. J. Taherzadeh. In situ detoxification and continuous cultivation of dilute-acid hydrolyzate to ethanol by encapsulated *Saccharomyces cerevisiae. Journal of Biotechnology* **125**, 377–384, 2006.
133. T. Ghose and R. Tyagi. Rapid ethanol fermentation of cellulose hydrolyzate. II. product and substrate-inhibition and optimization of fermentor design. *Biotechnology and Bioengineering* **21**, 1401–1420, 1979.
134. K. Rosen. Continuous production of ethanol. *Process Biochemistry* **13**, 25, 1978.
135. R. Purwadi. *Continuous ethanol production from dilute-acid hydrolyzates: Detoxification and fermentation strategy.* Göteborg, Sweden: Chalmers University of Technology, 2006.
136. C. S. Chen. Hawaii ethanol from molasses project- Final report. HNEI-80-03. 1980.
137. I. Karaki, M. Konishi, K. Amakai, and Ishikava. Alcohole production by continuous fermentation of molasses. *Hakko KoyoKaishi.* **30**, 106, 1972.
138. G. Salehi Jouzani and M. J. Taherzadeh. Advances in consolidated bioprocessing systems for bioethanol and butanol production from biomass: a comprehensive review. *Biofuel Res J.* **2**, 152–195, 2015.
139. G. P. Philippidis, T. K. Smith, and C. E. Wyman. Study of the enzymatic hydrolysis of cellulose for production of fuel ethanol by the simultaneous saccharification and fermentation process. *Biotechnology and Bioengineering* **41**, 846–853, 1993.

140. M. J. Taherzadeh and K. Karimi. Enzymatic-based hydrolysis processes for Ethanol. *BioRes.* **2**, 707–738, 2007.
141. D. G. Olson, R. Sparling, and L. R. Lynd. Ethanol production by engineered thermophiles. *Curr Opin Biotechnol.* **33**, 130–141, 2015.
142. D. A. Argyros, S. A. Tripathi, T. F. Barrett, S. R. Rogers, L. F. Feinberg, D. G. Olson, et al. High ethanol titers from cellulose by using metabolically engineered thermophilic, anaerobic microbes. *Appl Environ Microbiol.* **77**, 8288–8294, 2011.
143. Y. Cai, C. Lai, S. Li, Z. Liang, M. Zhu, S. Liang, et al. Disruption of lactate dehydrogenase through homologous recombination to improve bioethanol production in *Thermoanaerobacterium aotearoense*. *Enzyme Microb Technol.* **48**, 155–161, 2011.
144. M. Jin, C. Gunawan, V. Balan, and B. E. Dale. Consolidated bioprocessing (CBP) of AFEX™-pretreated corn stover for ethanol production using *Clostridium phytofermentans* at a high solids loading. *Biotechnol Bioeng.* **109**, 1929–1936, 2012.
145. Y. Li, T. J. Tschaplinski, N. L. Engle, C. Y. Hamilton, M. Rodriguez, J. C. Liao, et al. Combined inactivation of the *Clostridium cellulolyticum* lactate and malate dehydrogenase genes substantially increases ethanol yield from cellulose and switchgrass fermentations. *Biotechnol Biofuels* **5**, 2, 2012.
146. M. Bettiga, O. Bengtsson, B. Hahn-Hagerdal, and M. F. Gorwa-Grauslund. Arabinose and xylose fermentation by recombinant Saccharomyces cerevisiae expressing a fungal pentose utilization pathway. *Microb Cell Fact.* **8**, 40, 2009.
147. M. P. García-Aparicio, J. M. Oliva, P. Manzanares, M. Ballesteros, I. Ballesteros, A. González, et al. Second-generation ethanol production from steam exploded barley straw by *Kluyveromyces marxianus* CECT 10875. *Fuel* **90**, 1624–1630, 2011.
148. S. J. Ha, J. M. Galazka, S. R. Kim, J. H. Choi, X. Yang, J. H. Seo, et al. Engineered *Saccharomyces cerevisiae* capable of simultaneous cellobiose and xylose fermentation. *Proc Natl Acad Sci U S A* **108**, 504–509, 2011.
149. F. B. Pereira, A. Romani, H. A. Ruiz, J. A. Teixeira, and L. Domingues. Industrial robust yeast isolates with great potential for fermentation of lignocellulosic biomass. *Bioresour Technol.* **161**, 192–199, 2014.
150. J. K. Ko, Y. Um, H. M. Woo, K. H. Kim, S.-M. Lee. Ethanol production from lignocellulosic hydrolysates using engineered *Saccharomyces cerevisiae* harboring xylose isomerase-based pathway. *Bioresour Technol.* **209**, 290–296, 2016.
151. K. Okamoto, K. Imashiro, Y. Akizawa, A. Onimura, M. Yoneda, Y. Nitta, et al. Production of ethanol by the white-rot basidiomycetes *Peniophora cinerea* and *Trametes suaveolens*. *Biotechnol Lett.* **32**, 909–913, 2010.
152. I. Kamei, Y. Hirota, T. Mori, H. Hirai, S. Meguro, and R. Kondo. Direct ethanol production from cellulosic materials by the hypersaline-tolerant white-rot fungus *Phlebia* sp. MG-60. *Bioresour Technol.* **112**, 137–142, 2012.

153. T. Maehara, H. Ichinose, T. Furukawa, W. Ogasawara, K. Takabatake, and S. Kaneko. Ethanol production from high cellulose concentration by the basidiomycete fungus *Flammulina velutipes*. *Fungal Biology* **117**, 220–226, 2013.

154. G. E. Anasontzis, E. Kourtoglou, S. G. Villas-Boâs, D. G. Hatzinikolaou, and P. Christakopoulos. Metabolic Engineering of *Fusarium oxysporum* to improve its ethanol-producing capability. *Front Microbiol.* **7**, 2016.

155. M. J. Taherzadeh and K. Karimi. Acid-based hydrolysis processes for ethanol from lignocellulosic materials: A review. . *BioRes.*, 2007. pp. 472–499.

156. M. J. Taherzadeh and K. Karimi. Enzyme-based hydrolysis processes for ethanol from lignocellulosic materials: A review. *BioRes.* **2**, 707–738, 2007.

157. D. Klein-Marcuschamer, P. Oleskowicz-Popiel, B. A. Simmons, and H. W. Blanch. The challenge of enzyme cost in the production of lignocellulosic biofuels. *Biotechnology and Bioengineering* **109**, 1083–1087, 2012.

158. Tadashi Uragami, Kenji Okazaki, Hiroshi Matsugi, Miyata T. Structure and permeation characteristics of an aqueous ethanol solution of organic-inorganic hybrid membranes composed of poly(vinyl alcohol) and tetraethoxysilane. *Macromolecules* **35**, 9156–9163, 2002.

159. A. C. Wilkie, K. J. Riedesel, and J. M. Owens. Stillage characterization and anaerobic treatment of ethanol stillage from conventional and cellulosic feedstocks. *Biomass and Bioenergy* **19**, 63–102, 2000.

160. A. Singh, S. da Cunha, and G. P. Rangaiah. Heat-pump assisted distillation versus double-effect distillation for bioethanol recovery followed by pressure swing adsorption for bioethanol dehydration. *Separation and Purification Technology* **210**, 574–586, 2019.

161. L. M. Vane. A review of pervaporation for product recovery from biomass fermentation processes. *Journal of Chemical Technology & Biotechnology* **80**, 603–629, 2005.

162. P. Shao and R. Y. M. Huang. Polymeric membrane pervaporation. *J Membr Sci.* **287**, 162–179, 2007.

163. L. T. P. Trinh, Y. J. Lee, C. S. Park, and H. J. Bae. Aqueous acidified ionic liquid pretreatment for bioethanol production and concentration of produced ethanol by pervaporation. *Journal of Industrial and Engineering Chemistry* 2018.

164. K. Karimi and Y. Chisti. Bioethanol production and technologies. *Encyclopedia of Sustainable Technologies* 2017. p. 273-84.

165. F. Cherubini. The biorefinery concept: Using biomass instead of oil for producing energy and chemicals. *Energy Conversion and Management* **51**, 1412–1421, 2010.

166. P. R. Lennartsson, P. Erlandsson, and M. J. Taherzadeh. Integration of the first and second generation bioethanol processes and the importance of by-products. *Bioresour Technol.* **165**, 3–8, 2014.

Questions

1. What are the main steps in bioconversion of first-, second-, and third-generation feedstocks to ethanol?

2. What are the main incentives for transforming to the next-generation feedstock for fuel ethanol production?

3. What is the difference between cellulase and amylase enzymes? What is the differences between exoenzyme and endoenzyme?

4. What are the major fermentation inhibitors in cellulosic ethanol and how they are formed?

5. If 12 g/L ethanol is produced from a medium containing 25 g/L starch, what is the yield (percent of maximum theoretical) of ethanol?

6. What are the challenges in bioconversion of second- and third-generation feedstocks that limit their use in industrial processes? What are the recommendations for improving the process economy?

7. When 400 m^3 CO_2 is produced in a fermenter, how much ethanol should be produced? How much molasses with 55 percent fermentable should be consumed?

8. If you are interested in the production of 10,000-liter ethanol, how much corn with 65 percent starch should be hydrolyzed? Assume that the hydrolysis yield of starch is 95 percent and ethanol yield is 95 percent of theoretical yield.

CHAPTER **10**

Hydrogen Energy

L. M. Das

10.1 Introduction

Thousands of years ago our ancestors identified wood to produce fire, which was used for several purposes such as cooking and heating. Wood was identified as an energy liberating material. Almost 3000 years ago humankind used coal as a source of heat and light and had adopted commercial mining. Coal was successfully adopted for running steam engines. With the growth of civilization, we used several other sources to derive energy for improved living standards. During the 18th century research by eminent scientists had provided significant information on combustion and products of combustion. During the 19th century research was focused mainly on synthesized hydrocarbons and valuable information was obtained as to how energy was available from these hydrocarbons. The focus during the 20th century was directed to identify naturally available sources of hydrocarbons such as oil and natural gas. Abundance of availability and enormous potential of hydrocarbon-based natural sources ensured an aggressive entry of oil and petroleum-based fuels into our regular lifestyle. By this time, these naturally occurring sources had almost become indispensable to our daily life. Rapid utilization of these sources led to gradual depletion and search for viable alternatives became urgent, and it was impossible to think of maintaining the existing living pattern that had evolved over the centuries. Use of petroleum-based fuels, in the meantime, had demonstrated the damaging effects of environmental degradation. Several significant features include greenhouse gas emission, global warming, smog, and El Niño. Acid rain had become widespread and frequent in several parts of the earth. Such a situation made it more important to search for viable alternatives to the depleting natural sources that also threatened the safe survival of human beings, plants, and animals on earth. Faster depletion of natural sources of energy (fossil

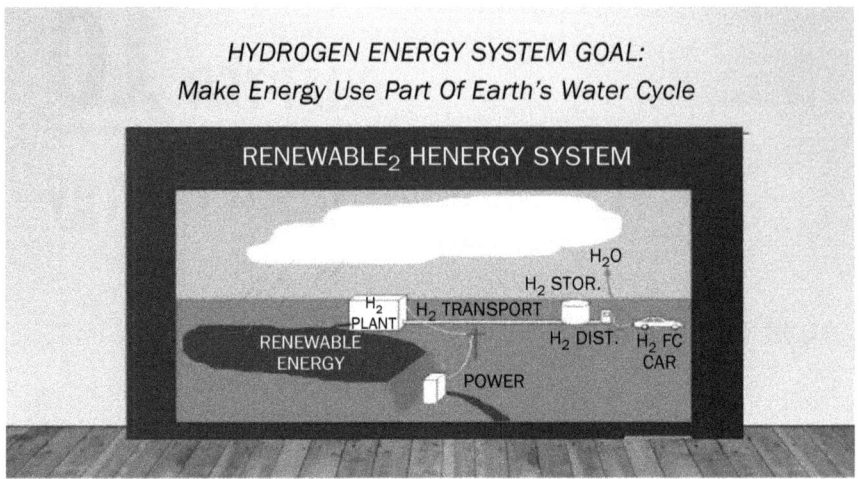

FIGURE 10.1   Renewable characteristics of hydrogen energy

fuels), as well as rapid environmental degradation arising due to combustion of fossil fuels, are the two major challenges of the 21st century. A solution to such challenges lies in identifying sustainable energy sources that produce practically no pollutants. A wide and diverse array of energy sources such as solar, hydro, wind, geothermal, tidal energy, and several biomass routes are being tried in several parts of the world for applications that were hitherto monopolized by fossil fuels such as oil, coal, natural gas, and nuclear energy. Since these are being deleted, our energy sources need to be "renewable" in character. It is perhaps relevant to mention here that hydrogen and electricity are two such forms of energy systems that are perpetual in character and independent of energy sources [1, 2]. Hydrogen energy is gifted with the unique features that make it an ideal energy carrier [3]. It is one such promising option that provides an eventual freedom from difficulties caused due to depletion of natural fossil sources of energy and the environmental pollution.

Hydrogen is a completely renewable fuel as shown in Figure 10.1. The most distinctive features of hydrogen energy are that it can be produced from water and a host of fossil and nonfossil sources. It can be produced from water and upon combustion it again gives out water. Similarly, it can be produced from electricity and can be converted into electricity at a much higher efficiency.

10.2   Occurrence of Hydrogen

It is important to note that hydrogen is the most abundant element in the universe and makes up to about 75 percent of its mass. It is believed

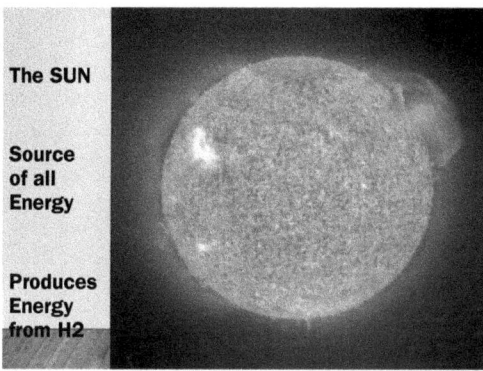

The SUN

Source
of all
Energy

Produces
Energy
from H2

FIGURE 10.2   Sun, source of all energy contains hydrogen

that sun has, at present, about 70 percent hydrogen and 28 percent helium by mass (Fig. 10.2).

The sun, at present, is approximately about 4.5 billion years old and it is believed that hydrogen gets converted to helium in the sun's core.

Hydrogen is also a major component in Jupiter where it is believed to be existing under high pressure. Stars and galaxies in the universe (including the earth's sun) have hydrogen, in plasma state, as a major constituent.

As far as occurrence of hydrogen in the earth is concerned, only about 1 percent of the earth's total mass contains hydrogen. Moreover, hydrogen on earth is not found in a free state. This is so because hydrogen, being so light, escapes the influence of earth's gravity. It is commonly observed in some volcanic gases.

However, hydrogen occurs in quite a large quantity in the earth as a component of several compounds. It is estimated that hydrogen is the 10th most abundant element existing as a component in numerous chemical compounds. Very reactive, it forms a vast number of chemical compounds that exist on earth. Water, which happens to be a compound of hydrogen and oxygen, covers almost 2/3rds of earth. Other common compounds in which hydrogen occurs are a vast number of minerals apart from hydrocarbons such as natural gas and many petroleum products. Last, but not the least, almost 70 percent of the human body is water. An overview of discussions on hydrogen energy broadly comprises of its production methods, modes of handling, transport and distribution, storage, and utilization. But other issues related to hydrogen combustion and safety need attention, as these have often been observed to be deterrent to wide-scale implementation of hydrogen energy. Figure 10.3 explains the large number of sources from which hydrogen can be produced and can be transferred and subsequently used in the transport sector with higher efficiency and greater degree

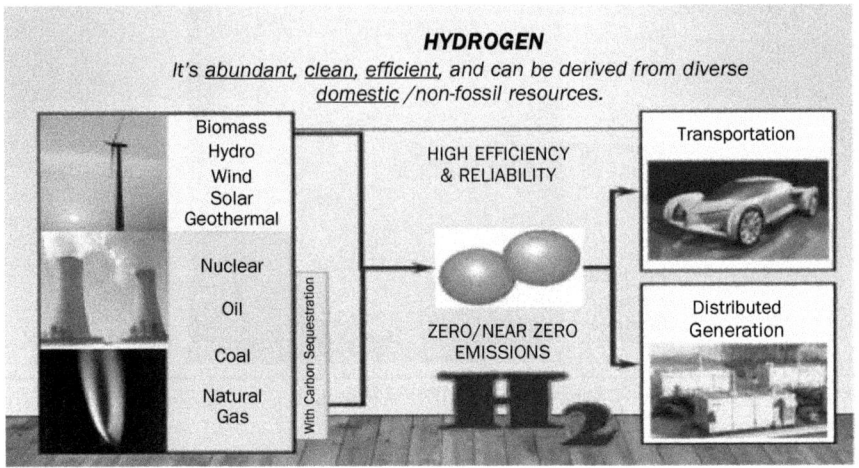

FIGURE 10.3 Production sources of hydrogen and its end use in transportation and power sector

of reliability. One of the most attractive features of such applications is the environmental compatibility with almost zero emission. Technical descriptions on uses of hydrogen in such applications related to vehicles and decentralized energy systems are discussed later in this chapter. An appropriately designed hydrogen-specific energy system is capable of avoiding the problem of environmental pollution at each stage starting with its production, storage, transport, and end use.

10.3 History of Hydrogen Energy

Hydrogen was discovered almost four centuries ago. The valuable experimental contributions of Robert Boyle, Cavendish, Antoine Lavoisier, William Nicholson, and Sir Anthony Carlisle in discovery of hydrogen and evaluating its properties are well-known. Cavendish had named it the inflammable air and subsequently Lavoisier called it hydrogen. However, long before Robert Boyle was successful in collecting hydrogen as a unique gas by way of dissolving iron in diluted hydrochloric acid. *Hydrogenium* in Greek is a combination of *hydro* meaning "water" and *genium* meaning "to give birth to."

This nomenclature of hydrogen gives a general idea that water hydrogen can be produced from water. This is precisely the definition of electrolysis that has been elaborated later in the chapter. At this point it would be appropriate to have a closer look at the physical and thermodynamic properties of hydrogen. Any system designed for specific hydrogen utilization must consider the physico-chemical characteristics of hydrogen.

10.4 Physical Properties

Hydrogen has one proton and one electron; the most common isotope, protium (^{1}H), has no neutrons. It is the lightest element with a density of 0.08988 g/A. Its density is less than air. Its melting point is $-259.14°C$ and boiling point of $-252.87°C$. Hydrogen is the most abundant chemical substance in the universe. Even then monoatomic hydrogen is low on earth primarily due to hydrogen's tendency to form covalent bonds with many elements. Hydrogen is abundantly found on earth in the form of chemical compounds such as hydrocarbons and water.

Hydrogen is the simplest of all elements consisting of only one proton in its nucleus and only one electron. It has the symbol H and atomic number 1 with average atomic weight 1.0079 amu. It is the only element that has different names for its isotopes. It has three isotopes denoted by ^{1}H (Protium), ^{2}H(Deuterium), and ^{3}H(Tritium). Protium occurs in 99.8 percent abundance. It has no neutrons. Deuterium contains one electron, one proton, and one neutron (mass number = 2). Deuterium and its compounds are often used as non-radioactive labels in chemical experiments and in solvents for 1H-NMR spectroscopy. Tritium contains one proton, two neutrons, and one electron (mass number = 3). It is radioactive in character and decays into helium-3 through beta decay with a half-life of 12.32 years. A distinctive property of hydrogen is that it has two distinct oxidation states ($+1$, -1). Therefore hydrogen can act as both an oxidizing and a reducing agent.

Under normal pressure and temperature (room temperature) hydrogen is colorless, odorless, tasteless, nontoxic, and nonpoisonous. It is also the lightest element (approximately 1/15th air). Because of its high diffusivity it rises and gets dispersed quickly. It is also highly soluble in alcohol, water, and ether. Hydrogen becomes a transparent odorless liquid at a temperature of 252.7°C. Liquid hydrogen is noncorrosive and is about 1/14th as heavy as water. However, when converted from liquid to gas, hydrogen expands about 840 times.

According to Sieverts' law, solubility of a diatomic gas in metal is proportional to the square root of the partial pressure of the gas in thermodynamic equilibrium. Solubility of hydrogen decreases with increasing temperature. Impurities in metals and hydrogen strongly affect the solubility, thereby decreasing the solubility or even cancelling it sometimes.

Absorption of hydrogen in metals and metal alloys plays a crucial role in the storage of hydrogen in hydrides. Palladium can absorb up to about 2800 times its own volume of hydrogen. Palladium and palladium-silver alloys demonstrate more solubility potential for hydrogen. Usually silver content in such alloys is above 40 percent by weight.

Hydrogen is nonpoisonous, so it does not inflict any kind of physiological effects. However, under some circumstances, hydrogen gets accumulated at a specific location and displaces the oxygen, thereby reducing the oxygen concentration to below 18 percent by volume. Threat of asphyxiation occurs only under such situation. Sometimes cold gaseous or liquid hydrogen causes numbness in the skin and also is responsible for frostbite.

As far as quality of hydrogen is concerned, hydrogen produced electrolytic ally is typically more than 99.5 percent pure volume. However, impurities in this case are mainly nitrogen, oxygen, and water. Removal of impurities is usually carried out by way of (i) catalytic combustion (oxygen), (ii) drying system (water), (iii) adsorption, and (iv) diffusion. Adequate emphasis needs to be stressed on the purity of hydrogen to be liquefied because at liquid hydrogen temperatures all impurities except helium become solid and often lead to blockage of refrigeration unit. To achieve liquefaction of gases, energy needs to be removed until the condensation, and this is quite often carried out either by cooling down to low temperatures or through external means such as liquid nitrogen or magneto-caloric effect or internally by way of irreversible or isentropic decompression.

10.5   Chemical Properties of Hydrogen

Hydrogen possesses the unique ability of forming compounds with almost all the elements except the noble gases. Hydrogen with carbon also forms many hydrocarbon fuels such as methane (CH_4), ethane (CH_6), propane (C_3H_8), and butane (C_4H_{10}) and alcohols such as methanol (C_3HOH), and ethanol (C_2H_5OH). Many of these hydrocarbon fuels and alcohols are being considered as a substitute to petroleum fuels (such as gasoline and diesel) to combat the energy crisis and environmental degradation. Some specific properties of hydrogen are given in Table 10.1 and Table 10.2 as compared to gasoline and methane. Such information is relevant because both gasoline and methane (the major combustible component of natural gas is methane) are fuels being used in vehicles.

10.6   Hydrogen Production

It is an interesting fact that despite its abundant availability in the universe, only traces of free hydrogen are found on earth. Therefore, it is not a primary energy source and is an energy carried like electricity, so it must be produced from its compounds. Production of a small amount of hydrogen in laboratory scale is relatively simple. However, production of vast amounts of hydrogen in a large scale depends on several factors such as the amount of hydrogen required as well as the raw material/source available. A brief description of several routes of hydrogen production is described (see Table 10.3). Some of these are already commercial

Properties	Hydrogen	Methane	Gasoline
Molecular weight	2.016	16.043	107
Density of gas at NTP (g m^{-3})	83.764	651.19	4400
Heat of combustion (low) (kJ g^{-1})	119.93	50.02	44.5
Heat of combustion (high) (kJ g^{-1})	141.86	55.53	48
Specific heat (cp) of NTP gas (J g^{-1}K^{-1})	14.86	2.22	1.62
Viscosity of NTP gas (g cm^{-1}s^{-1})	0.0000875	0.00011	0.00005
Specific heat ratio (γ) of NTP gas	1.383	1.308	1.05
Gas constant (R) (cm^2 atm g^{-1} K^{-1})	40.703	5.11477	0.77
Diffusion coefficient in NTP air (cm^2 s^{-1})	0.61	0.16	0.005

TABLE **10.1**   Thermodynamic properties of hydrogen, methane, and gasoline (generally accepted values from literature)

Properties	Hydrogen	Methane	Gasoline
Limits of flammability in air (volume %)	4.0–75.0	5.3–15.0	1.0–7.6
Stoichiometric composition in air (volume %)	29.53	9.48	1.76
Minimum energy for ignition in air (mJ)	0.02	0.29	0.24
Auto-ignition temperature (K)	858	813	501 to 744
Flame temperature in air (K)	2318	2148	2470
Burning velocity in NTP air (cm s^{-1})	265–325	37–45	37–43
Quenching gap in NTP air (cm)	0.064	0.203	0.2
Percentage of thermal energy radiated from flame to surrounding (%)	17–25	23–32	30–42
Diffusivity in air (cm^2 s^{-1})	0.63	0.2	0.08
Normalized flame emissivity (2000K, 1 atm)	1.00	1.7	1.7
Limits of flammability (equivalence ratio)	0.1–7.1	0.53–1.7	0.7–3.8

TABLE **10.2**   Combustion properties of hydrogen, methane, and gasoline (generally accepted values from the literature)

and some are still in the process of development. Hydrogen can be produced from a variety of sources, such as fossil fuels (natural gas and petroleum), coal by several methods of the gasification process, renewable resources (solar and wind energy), nuclear energy, biomass, and other relevant emerging technologies for hydrogen production.

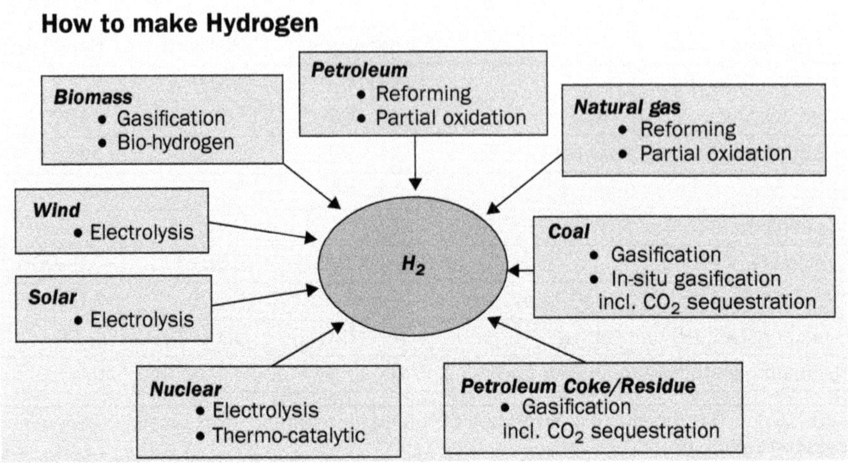

How to make Hydrogen

FIGURE 10.4   Methods of hydrogen production from various sources

Figure 10. 4 below gives a general idea about the several sources of producing hydrogen through various routes/processes.

10.6.1   Hydrogen from fossil fuels

Hydrogen can currently be produced from natural gas by means of three different chemical processes: (i) steam reforming (steam methane reforming [SMR]), (ii) partial oxidation (POx), and (iii) autothermal reforming (ATR).

10.6.2   Steam reforming (SR)

At present there are several methods of producing hydrogen ranging from experimental lab scale to well-established industrial processes. However, reformation of fossil fuels happens to be the most common method. Reformation is a chemical process by which hydrogen is produced by breaking hydrocarbon molecules. It involves the endothermic conversion of methane and water vapor into hydrogen and carbon monoxide. The schematic diagram of hydrogen production from natural gas by steam reformation is shown below in Figure 10.5 and the chemical reaction in Equation 10.1. In such cases heat is often supplied from the combustion of some of the methane feed-gas. The process typically occurs at temperatures of 700°C to 850°C and pressures of 3 to 25 bars. The product gas contains approximately 12 percent CO, which can be further converted to CO_2 and H_2 through the water-gas shift reaction as shown in Equation 10.2.

$$CH_4 + H_2O + heat = CO + 3H_2 \tag{10.1}$$

$$CO + H_2O = CO_2 + H_2 + heat \tag{10.2}$$

	Processes	Raw material	Source of Energy
Electrochemical	Electrolysis	• Water	• Electricity from renewable energy source (e.g., wind, geothermal, solar, hydro) • Electricity from nonrenewables (e.g., fossil fules, nuclear)
Thermochemical	Reforming	• Natural gas • Hydrocarbons • + Water	• Combustion of natural gas/ syngas • Concentrating solar thermal
	Gasification	• Coal • Carbonaceous Materials • Biomass • + Water	• Combustion of coal/biomass/ carbonaceous materials/syngas • Concentrating solar thermal
	Decomposition	• Natural gas • Fossil fuel hydrocarbons • Biomethane • Biohydrocarbons	• Natural gas combustion • Concentrating solar thermal
	Thermolysis	• Water	• Concentrating solar thermal
	Thermochemical cycles	• Water	• Concentrating solar thermal • Nuclear heat
Photochemical	Photosynthesis	• Water	• Solar radiation, artificial light
	Photobiological	• Microbial (e.g., algae) • + Water	• Solar radiation

TABLE 10.3 Process and raw material for hydrogen production in a specific process

In practice, several hydrocarbons have been used as feed stock for steam reformation (SR). The available literature gives out a description on the kinetics of methanol on a Cu-based catalyst [4, 5]. Introduction of a catalyst obviously gets the reaction accelerated. As reported in literature,

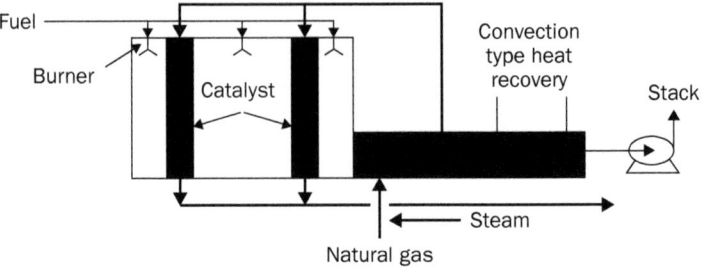

Figure 10.5   Steam reforming of natural gas to produce hydrogen

the catalysts help in three different modes of action [6] such as activity, selectivity, and stability. Activity indicates how the reaction proceeds in the presence of a catalyst. Selectivity of the catalyst shows measure of desirable product with respect to the feedstock. Quite often the selectivity gets affected by undesirable secondary reactions, which tell upon the efficiency. Sometimes formation of coke results in the degradation of catalyst, which is the third mode of action.

10.6.3   Partial oxidation (POx)

POx is usually generally adopted with higher hydrocarbons or under the circumstances where pure oxygen is available [7]. However this process is susceptible to coke formation and therefore is always carried out at high temperatures. POx of natural gas is the process whereby hydrogen is produced through the partial combustion of methane with oxygen gas to yield carbon monoxide and hydrogen. During this process, heat is produced in an exothermic reaction. In the absence of any need to externally heat the reactor, partial oxidation helps in having a more compact design. In such cases CO produced is further converted to H_2. Equations 10.3 and 10.4 explain the process and Figure 10.6 shows the process of partial oxidation for producing hydrogen from natural gas.

$$CH_4 + 1/2O_2 = 2H_2 + \text{heat} \tag{10.3}$$

$$CO + H_2O = CO_2 + H_2 + \text{heat} \tag{10.4}$$

Quite often the response time of the POx reformer is greatly affected and it produces low levels of hydrogen concentration and relatively much higher levels of CO.

10.6.4   Autothermal reforming (ATR)

Autothermal reforming can be considered as a combination of both steam reforming and partial oxidation. This is usually achieved by bringing the two reforming reactions into close thermal contact.

Sometimes they are also placed into a single catalytic reactor. The second configuration of placing into a single catalytic reactor has been

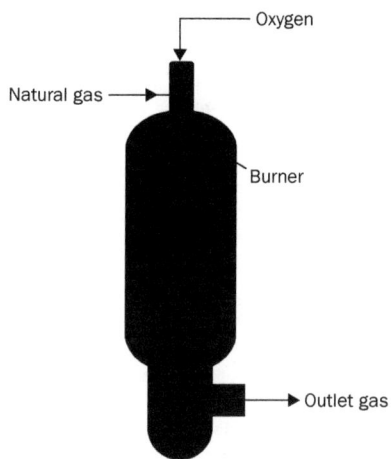

FIGURE 10.6   Partial oxidation of natural gas to produce hydrogen

the most efficient heat transfer system. In case of ATR total reaction is exothermic, so it releases heat. The outlet temperature from the reactor is in the range of 950°C to 1100°C and the gas pressure could be as high as 100 bar. Here the CO produced is converted to H_2 through the water-gas shift reaction. A shortcoming of this process is that it becomes necessary to purify the output gases, and this adds significantly to plant costs and reduces the total efficiency.

$$CH_4 + H_2O + heat = CO + 3H_2 \tag{10.5}$$

$$CH_4 + 1/2O_2 = CO + 2H_2 + heat \tag{10.6}$$

$$CO + H_2O = CO_2 + H_2 + heat \tag{10.7}$$

A comparative evaluation among the three processes shows that SMR has a higher efficiency and exhibits better emission characteristics as compared to ATR and POx. However, the system is quite complex and sensitive to the quality of natural gas. ATR and POx systems have lower efficiency but are relatively simpler in design. A schematic diagram of ATR is shown in Figure 10.7.

10.6.5   Hydrogen by gasification

The gasification of carbonaceous, hydrogen-containing fuels has been an effective method of hydrogen production [8]. A broad schematic diagram is shown in Figure 10.8.

It has been reported in literature that gasification has the highest conversion efficiency as compared to other modes of solid fuel conversion technologies [9]. By way of gasification it is possible to obtain syngas comprising of CO, CO_2, H_2, CH_4, H_2O, and other constituents in low concentrations [10].

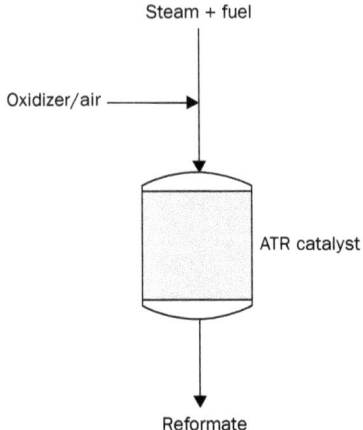

FIGURE 10.7 Autothermal reforming

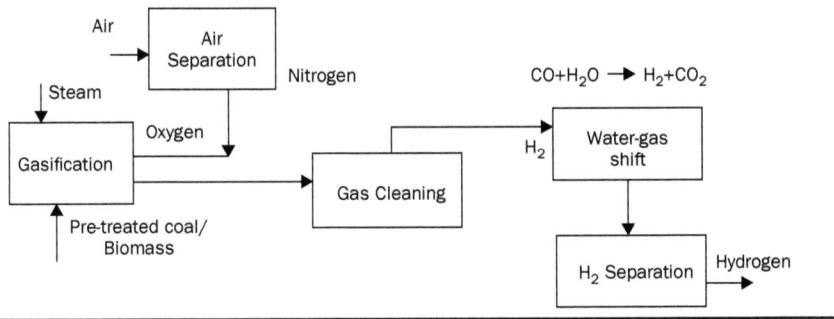

FIGURE 10.8 Hydrogen production by gasification

Hydrogen can be produced from coal through several routes of gasification processes (e.g., fixed bed, fluidized bed, or entrained flow). Figure 10.9 gives a broad schematic diagram of gasification by coal/biomass to produce hydrogen. Usually high-temperature entrained flow processes are preferred to process where carbon conversion to gas is maximized. This helps in avoiding the formation of significant amounts of char, tars, and phenols. Equation 10.8 gives a typical reaction for the process showing carbon getting converted to carbon monoxide and hydrogen.

$$C(s) + H_2O + \text{heat} \quad CO + H_2 \qquad (10.8)$$

As this reaction is also endothermic, it needs additional heat as is required with methane reforming. Water-gas shift reaction as shown in Equation 10.9 is responsible for converting CO to CO_2 and H_2

$$CO + H_2O = CO_2 + H_2 + \text{heat} \qquad (10.9)$$

Hydrogen production from coal is a commercially mature technology [11] but it is quite complex as compared to production of hydrogen from

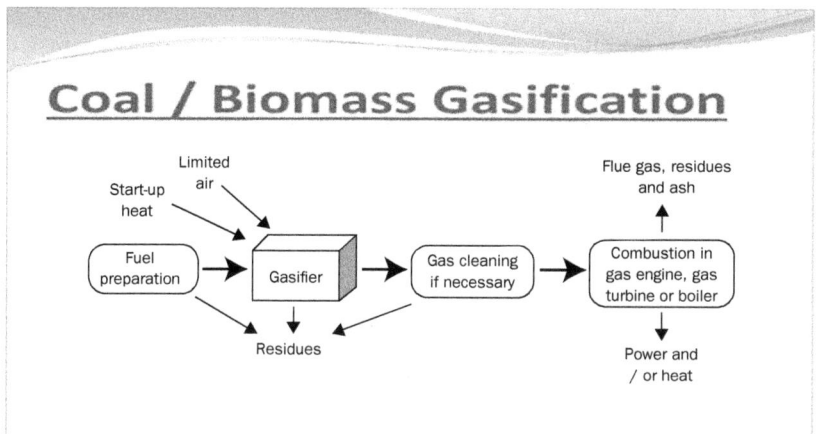

FIGURE 10.9 Schematic diagram of coal/biomass gasification

natural gas. Even though the cost of hydrogen produced is higher, it is still widely adopted primarily because of adequate availability of coal in several parts of the world. Fossil sources such as natural gas, coal, and oil have hydrocarbon in them. Therefore, all methods of hydrogen production from fossil sources will have carbon dioxide. However, the amount of CO_2 depends on the amount of hydrogen content in the feed stock. In this context it is essential that CO_2 be captured and stored so that a sustainable production of hydrogen is achieved. This process is known as de-carbonization. In the entire chain of events during hydrogen production, there could be three stages to capture CO_2. These are pre-combustion, post-combustion, and oxy-fuel combustion. In a post-combustion process CO_2 can be removed from the exhaust gas of the combustion process. For example, in a conventional steam turbine or CCGT (combined cycle gas turbine) power plant, such methods can be adopted via the "amine" process. In a pre-combustion process, CO_2 is captured when producing hydrogen by any process. Oxyfuel-combustion is accomplished when fossil fuel is converted to heat in a combustion process in a conventional steam turbine or CCGT power plant by using pure oxygen as an oxidizer. During this process, mainly CO_2 and water vapor are produced in the flue gases. CO_2 can be easily separated by condensing the water vapor. As far as post-combustion and oxyfuel-combustion systems are concerned, it is possible to produce electricity in steam and CCGT power plants and subsequently this electricity could be used for water electrolysis. Capture and storage of CO_2 and its utilization characteristics depend on several factors. One possibility is to use the captured CO_2 to an energy conversion process and generate electricity for carrying out electrolysis of water. Under such circumstances, the energy conversion process itself may have a relatively lower efficiency and might affect the overall efficiency. The other feasible option is to store the captured CO_2 in geological formations such as oil and gas fields and in aquifers. But the effectiveness of this method of storing CO_2 for a long period is yet to be established. Besides, questions are often raised on the effectiveness of de-carbonization by permanent storage of CO_2 in this method. CO_2 must be critically investigated from the environment point of view. Transportation systems for CO_2 depend on the specific sites selected for production plant and storage. Production of hydrogen by gasification has been a well-established technology. Gasification, by definition, is a process in which either coal or biomass is converted into gaseous components by applying heat under pressure and in the presence of steam. Subsequently, a series of chemical reactions are carried out to produce synthetic gas that reacts with steam to enhance the quantity of hydrogen produced. Figures 10.10 and 10.11 show block diagrams for coal gasification and biomass gasification separately.

Several types of biomass have been successfully used to produce hydrogen by way of gasification. But biomass is quite often bulky. The other important characteristics are that biomass is difficult for storage and deteriorates over time. Last but not the least it has a lower energy density (GJ per unit of weight or volume). Sometimes combustion becomes difficult. Biomass used for hydrogen production includes agricultural residues. The materials made of glass, plastic, and several types are not biomass as they are essentially nonrenewable in nature.

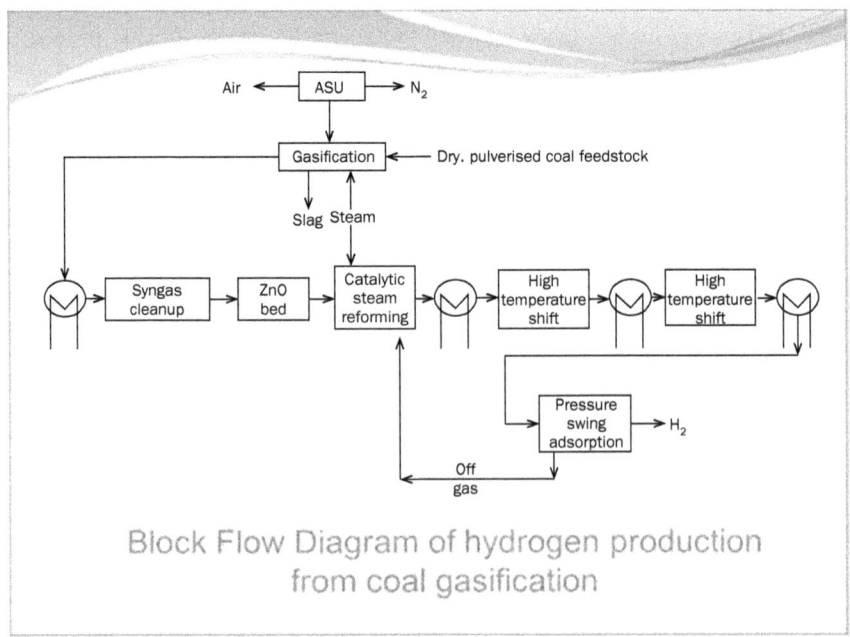

Block Flow Diagram of hydrogen production from coal gasification

Figure 10.10   Hydrogen production by coal gasification

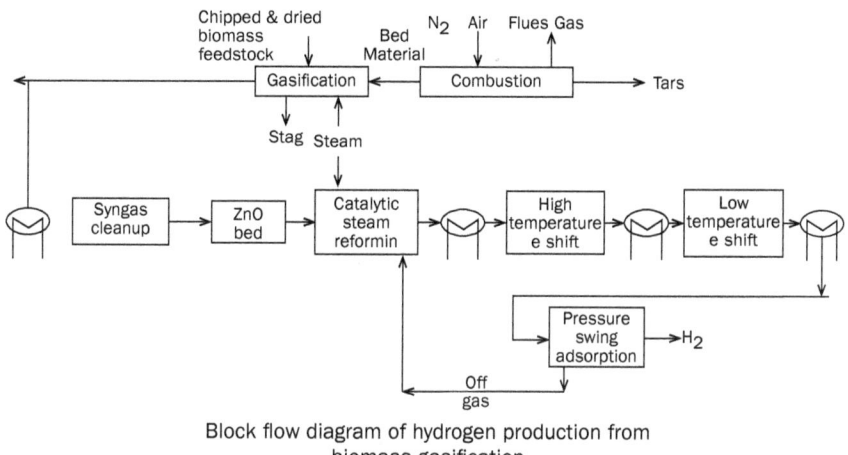

Block flow diagram of hydrogen production from
biomass gasification

FIGURE 10.11   Hydrogen production by biomass gasification

Municipal solid waste (MSW) broadly falls into the category of bio-mass as it consists essentially of the biodegradable fraction of residues of agriculture and food as well as forest residues, energy crop residues, organic solid wastes, and firewood (charcoal). Usually waste heat from gasification is used to dry bio-mass for bringing down the humidity level of 15 percent. Often air rather than pure oxygen has been used and the process of gasification is carried out with steam resulting in product gas consisting of about 40 percent hydrogen.

MSW and biomass are known to be the largest contributor of anthro-pogenic methane emission into the atmosphere in the entire world. Land fill gas (LFG) generated from MSW has been a proven technology in many parts of the world. One of the prospective routes is to generate hydrogen from LFG reformation. Biomass feedstock could be mixed wood chips and several industrial wastes. The process of gasification varies depending on the character of biomass.

Coal obviously is not the ideal fuel for the future due to its CO_2 emissions characteristics. However, when used for hydrogen production, this environmental pollution is drastically reduced because downstream process can be so designed to capture CO_2. In combination with biomass, overall CO_2 emission level gradually reduced. Perhaps biomass along with high-energy intensity fuels can be a prospective option for obtain-ing low-emission fuels.

Coal gasification has been a well-known process to produce hydro-gen. In view of the present thrust on combating environment pollution, underground coal gasification is adopted in several situations (Fig. 10.12). In this process, conversion of coal to a syngas (consisting of hydrogen, carbon monoxide, and methane) is achieved by igniting the coal. This ignition is achieved by way of introducing required amount

UNDER GROUND COAL GASIFICATION

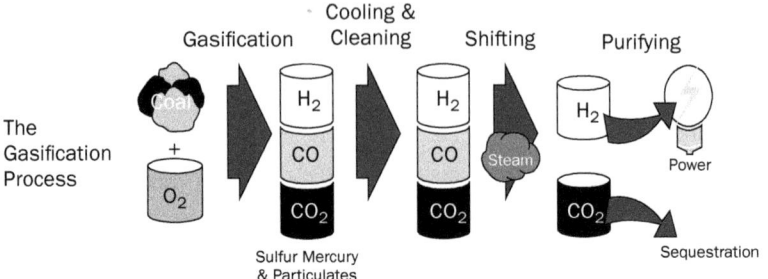

FIGURE **10.12** Underground coal gasification process

of steam and oxygen into the seam. Conditions of partial combustion are ensured by mixing coal with oxygen, air, or steam at high temperatures. This way of gasification has several routes of practical utilization such as hydrogen production surface for heating, power generation, etc. There are two ways of carrying out the underground coal gasification (UCG). The first method uses vertical wells and a "reverse" combustion so as to open up the internal pathways in the coal. The other method uses dedicated in-seam boreholes, adapted from oil and gas production. The establishment of UCG is an economically viable option in some locations. The composition of the syngas is particularly suited to CO_2 capture and the high pressure from deep UCG will require a smaller and less costly plant. Thus, storing CO_2 in nearby coal seams is an attractive feature. Coal gasification plants have a much lower emission of CO_2 (up to about 20 percent as compared to conventional coal power plants). UCG in combination with CCS (CO_2 capture and storage) is a promising and economically feasible technology toward carbon abatement.

10.6.6 Ammonia dissociation

In the present context of emphasis on hydrogen energy, potential of ammonia as a promising hydrogen carrier is being studied as the next generation carbon free technology. High energy density and ease of liquefaction are strong points in its favor. However, use of ammonia as a hydrogen carrier has not matured into a commercial scale primarily in absence of an efficient process for decomposing ammonia to hydrogen and nitrogen. The ammonia decomposition process is highly endothermic, and it requires a high temperature of about 400°C. Choice of an appropriate catalyst is also of significance in these reactions. In the process of hydrogen production from ammonia, ammonia is dissociated with the aid of a catalyst at a temperature of about 1000°C into its components: hydrogen and nitrogen. The reaction results in about 75 percent hydrogen and 25 percent nitrogen. Nitrogen is produced in a mononuclear form (N) and not N_2. Active research is in progress to carry out the chemical reactions at lower, even room, temperature [12].

10.6.7   Hydrogen from methane hydrate

It has been observed that methane hydrate exists in vast quantities in certain locations such as in sea sediments and the geological structures below them. It is also found in and below artic regions. This could be reformed in an endothermic process to get hydrogen and carbon dioxide [13]. Hydrogen produced from such sources could be clean, if the carbon dioxide is separated by any of the existing technological routes and subsequently sequestered.

10.6.8   Hydrogen from biomass

There is a lot of interest presently in the research and development of producing hydrogen from biomass mainly because it has probably the largest potential to meet the energy requirement. Biomass and the fuels derived from biomass can supply a sustainable source of hydrogen [14].

It has already been discussed earlier that a hydrogen-containing gas is produced from biomass by way of gasification of coal. As has been discussed earlier, municipal solid waste or garbage is another important source of biomass: biodegradable fraction of waste, agricultural residues, residues of food industry as well as biodegradable fraction of several industrial waste. However, there are hardly any major commercial plants that produce hydrogen from biomass in a large scale. Available literature shows that steam gasification (direct or indirect), entrained flow gasification, and technologically advanced concepts such as gasification in supercritical water, application of thermo-chemical cycles, or the conversion of intermediates (e.g., ethanol, bio-oil) are the possible routes to produce hydrogen from biomass. But there is still a long way to go in practical demonstration of plants using these technologies for large-scale hydrogen production. At present the biomass feed stock are heterogeneous in character. The quality and quantity of fuel (hydrogen produced) depend on the consistency of the biomass feed stock. Research efforts in the past have clearly shown that heterogeneous and low-quality fuels need more sophisticated conversion systems. The process and mechanism of production of hydrogen from biomass also depend to a large extent on the geographical location and climatic variations. There is a need to rationalize the production and preparation of biomass to ensure more consistent, higher-quality fuels with better homogeneity. It is also important to improve the economics of production processes and have well-defined logistics of biomass feedstock handling and transport. A thorough understanding of these factors would help in identifying the characteristics of feedstocks that can adopt the technology. As far as gasification of biomass is concerned, raw gas handling and subsequent system-specific clean-up is crucial. Methods of producing hydrogen from biomass could be by thermochemical process such as gasification, high pressure conversion in presence of water, and pyrolysis. Figure 10.13 gives a schematic diagram of producing hydrogen from biomass by way

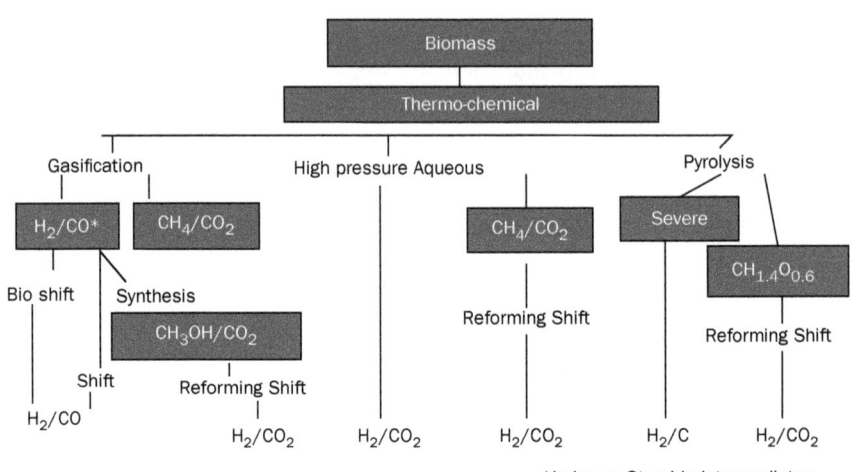

FIGURE 10.13   Thermochemical conversion of biomass to produce hydrogen

of thermochemical conversion. The conversion technologies could be either by direct production route or conversion of storable intermediates. Gasification coupled with water gas shift reaction is widely used. Such processes have several types of feed stocks, dedicated crops, agricultural and forest product residues, etc. The storable intermediates are H_2/CO, CH_4/CO_2, CH_3OH/CO_2. Subsequently, shift reactions help convert these intermediates into H_2/CO_2. High pressure aqueous processes yield H_2/CO directly or via storable CH_4/CO_2. Similarly, pyrolysis of biomass and a subsequent set of chemical reactions yield H_2 and CO_2.

10.6.9   Biological methods of hydrogen production

Biological routes of hydrogen production have been known for a long time. Lots of research and development activities have been carried highlighting critical limiting factors to achieve the process efficiently [15]. One of the major obstacles has been due to low energy content of solar irradiation, which governs the photosynthetic processes. Besides, the cost of the photobioreactor has been high and the conversion efficiencies for direct biophotolysis have been as low as 1 percent. Figure 10.14 gives a general overview of the several biological routes of hydrogen production from biomass. Broadly speaking, these processes involve organic compounds and microorganisms. The principal biological processes are (i) anaerobic digestion and (ii) fermentation.

Anaerobic digestion yields directly H_2 and CO_2 mixture. Fermentation yields ethanol and CO_2 and via reforming shift H_2 and O_2 are produced. One of the major shortcomings of this process is that hydrogen yield is relatively much lower compared to production of hydrogen by SMR.

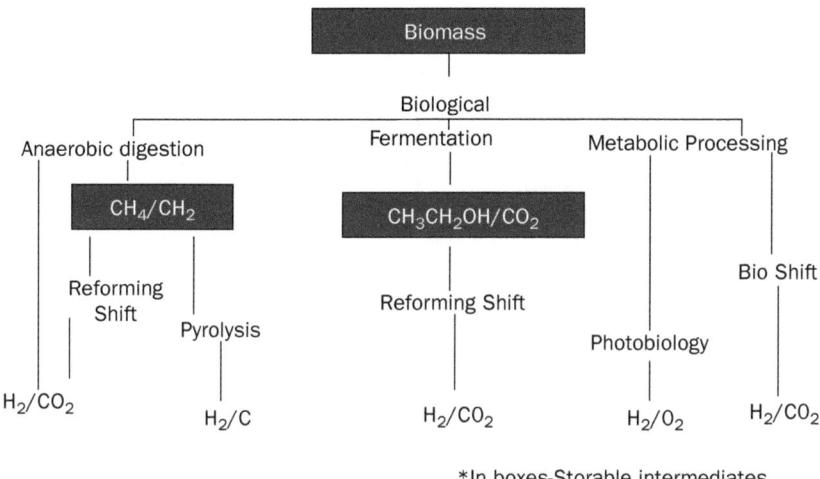

FIGURE 10.14 Biological routes of hydrogen production

Biophotolysis

Biophotolysis is a water-splitting process that can be carried out in biological systems. In this process, light is used as the energy source and produces molecular O_2 and H_2. Research activities are being carried out on hydrogen production by biophotolysis based on microalgal photosynthesis. Studies on hydrogen metabolism of marine microalgae and photosynthetic bacteria have reported useful information [16]. Photobiological production or Biophotolysis of hydrogen is based on two steps:

$$\text{Photosynthesis: } 2H_2 O\ 4H^+ + 4e^- + O_2$$

Hydrogen Production: $4H+ + 4e- 2H_2$ catalyzed by hydrogenase such as green algae and cyanobacteria. This is a relatively novel area of research and a lot of research and development (R&D) activities need to be pursued in this field. However, this has the potential of evolving a long-term solution of hydrogen production. Figure 10.15 illustrates the principle of photobiological hydrogen production.

10.6.10 Hydrogen by electrolysis

Electrolysis is a well-developed technology that has been commercialized for more than almost a century and a half. Hydrogen, by electrolysis (Fig. 10.16) is produced by splitting water molecules into hydrogen and oxygen by passing direct current through them. Hydrogen is produced characteristically pure in nature:

$$H_2O + \text{electricity} = H_2 + 1/2O_2$$

Principle of photo-biological hydrogen production

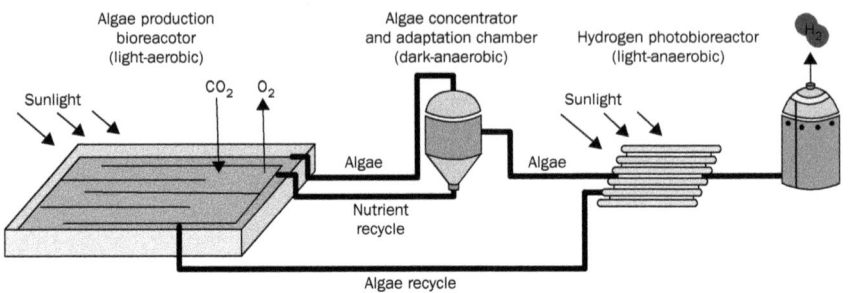

FIGURE 10.15   Photo biological hydrogen production

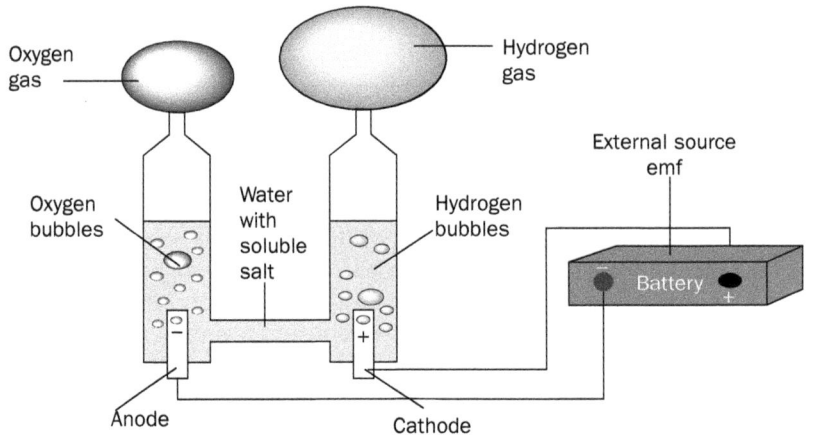

FIGURE 10.16   Electrolysis of water to produce hydrogen

The total energy required to carry out water electrolysis is increasing slightly with temperature while the required electrical energy decreases. It is therefore better to adopt high temperature when high temperature heat is available from several other industrial processes. This method has a significant global relevance since most of the electricity produced in the world uses fossil energy and has relatively lower efficiency. High-temperature electrolysis is usually based on technology from high-temperature fuel cells. Significant improvement has been reported in efficiency by steam electrolyzing at high temperatures. The attractive features of high-temperature electrolysis are that it not only needs a lower level of electrical energy but also produces a higher efficiency over conventional technologies.

A detailed life cycle analysis (LCA) by researchers [17] has given significant information that global warming potential (GWP) and the acidification potential (AP) are about one-sixth and one-third, respectively, of the corresponding value given by hydrogen production by steam reforming of natural gas. It is well-known that electrical energy required to split water at 1000°C would be much lesser than electrolysis carried out at 100°C. This indicates that a high-temperature electrolyzer can operate at a much higher efficiency than a conventional low-temperature electrolyzer. An application of this is evident in a solid oxide electrolyzer cell (SOEC). An electrolyzer based on the solid oxide fuel cell (SOFC) operates at 700°C to 1000°C. At this temperature level, the electrode reactions are more reversible, and fuel cell reaction can be easily reversed to an electrolysis reaction. Presently no commercial method of hydrogen production has been developed by using geothermal energy, and it is possible to develop systems in which some of the electricity consumed by the electrolyzer can be replaced with the heat available from geothermal sources. Use of solar or natural gas sources can also help in substantially reducing electricity consumption.

10.6.11   Alkaline electrolysis

It is well-known that the major challenges for widespread use of water electrolysis have been high energy consumption. It is also reported that the cost related to durability and reliability for prolonged maintenance and safe operation is quite high. Studies have been carried out [18] on alkaline water electrolysis in which an electrical circuit analogy of three types of resistance (such as electrical resistance, the reaction resistance, and the transport resistance) have been introduced in the study of electrolysis system. A detailed thermodynamic analysis has been carried out to optimize the resistance level for higher efficiency on alkaline water electrolysis. Alkaline electrolysis is a mature technology of hydrogen production that has been successfully used for industrial application, which permits remote operation. This method is usually suitable for stationary applications. Alkaline electrolyzers use an aqueous KOH solution (caustic) as electrolyte that circulates through the electrolytic cells. The sequence of chemical reactions that take place inside the alkaline electrolytic cell are given below.

$$\text{Electrolyte: } 4H_2O = 4H+ + 4OH$$
$$\text{Cathode: } 4\,H+ + 4e- \, 2H_2$$
$$\text{Anode: } 4OH- \, O_2 + 2H_2O + 4e-$$
$$\text{Sum: } 2H_2O = O_2 + 2H_2-$$

10.6.12   Polymer electrolyte membrane (PEM) electrolysis

Polymer exchange membrane (PEM) electrolysis is the electrolysis of water in a cell equipped with a solid polymer electrolyte. The function of solid polymer electrolyte mainly consists of conduction of protons, separation of product gases, and electrical insulation of the electrodes.

PEM electrolysis is an acidic polymer membrane and does not need a liquid electrolyte electrolyzer. It can be designed to operate at a pressure level of several hundred bars. This is a relatively sound technology that can be adopted in both stationary and mobile applications. The principle of PEM electrolysis is given below in following equations:

$$\text{Anode: } H_2O = 1/2O_2 + 2H{+} + 2e{-}$$
$$\text{Cathode: } 2H{+} + 2e{-} \ H_2$$

One of the major constraints of this technology lies in the limited life period of the membrane. However, it is relatively safer due to the absence of KOH electrolytes. Higher densities and higher operating pressures are critical to its design and add to the cost. This technology has not been matured to a wide commercial level primarily because of higher cost, shorter life time, and relatively poor efficiency as compared to alkaline electrolyzers. There is scope to enhance the performance of the PEM fuel cell significantly by further improvement on materials development and cell stack design.

PEM electrolysis has also been considered as a feasible alternative for producing hydrogen from renewable energy sources. Several aspects of this technology have been studied and reported [19] considering grid independent and grid assisted hydrogen generation. The paper also discusses several other configurations of the system such as using an electrolyzer for peak shaving, as well as integrated systems both grid connected and grid independent.

10.6.13 Onsite electrolysis

In this method, KOH is usually added to water at 30 percent weight to improve its conductivity as electrolyte. This electrolyzer has a relatively much lower efficiency compared to others, but it is still widely used since it's been widely developed and involves much less cost. Considering the application potential for fueling a vehicle within a short period, it is required that a number of electrolyzers should be connected in parallel. Figure 10.17 shows the schematic diagram of an onsite electrolysis.

10.6.14 Hydrogen production using solar energy

There are several ways by which the solar energy can be useful for hydrogen production. Figure 10.18 shows various thermal routes of hydrogen production using solar energy.

10.6.15 Solar thermal methane splitting

In this process, the high temperature of the concentrated solar energy can be used to split methane into hydrogen and carbon. An effective thermochemical cycle can be achieved from the high temperature level in concentrated solar energy. This method of thermochemical

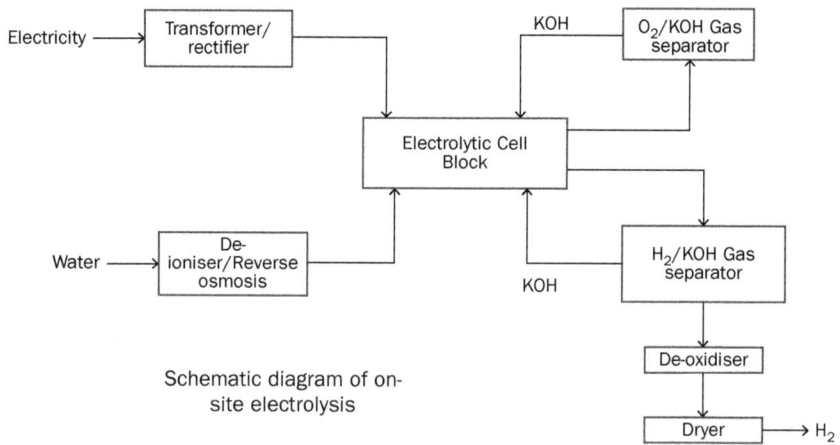

Figure 10.17   Schematic diagram of onsite electrolysis

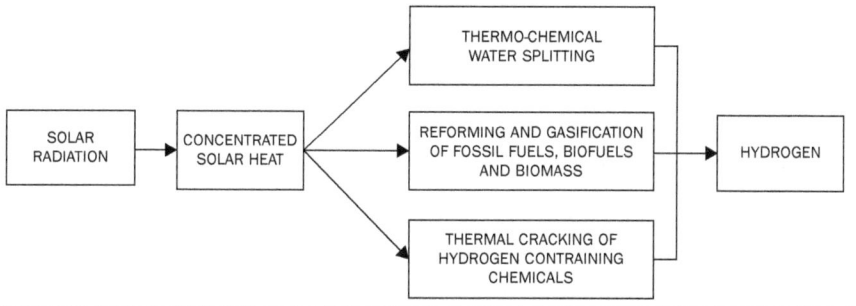

Figure 10.18   Major thermal routes of hydrogen production using solar energy

hydrogen production process causes a fast rate of reaction and the production rate of hydrogen also increases. Besides, this process is highly environment-friendly.

Studies have been carried out [20] on single-step thermal decomposition (pyrolysis) of methane without catalysts. It is possible to obtain hydrogen-rich gas and high-grade carbon black (CB) from concentrated solar energy. This study also reports the design features of a solar chemical reactor to achieve thermal splitting of methane at high temperature so as to produce hydrogen.

Hydrogen production is also possible by way of the solar thermal dissociation of ZnO/Zn. In this process, concentrated solar radiation is used as source of high thermal energy supplying the required amount of process heat Zn acts as the energy carrier thereby producing hydrogen via hydrolysis of ZnO. The only point of concern is that the process needs a large amount of argon for quenching Zn and O_2. However, hydrolysis need not be conducted at the solar plant as Zn can be transported to the location of electrolysis. There is another method of producing hydrogen from the solar carbo-thermal reduction of Zn/ZnO where no argon is required to quench resulting gases. In such processes the reaction temperature is relatively less at about 1500 K. In this process broadly carbonaceous material (coal/charcoal) is added to reaction chamber thereby facilitating solar carbo-thermic reduction of ZnO (SCR). In this process CO is put through the shift reactions thereby causing higher amounts of hydrogen production per kg of Zn.

Production of hydrogen from water using solar energy by thermal dissociation of ZnO(s) into Zn(g) and O_2 at 2300 K using concentrated solar energy as the source of process heat has been reported in literature [21]. This has been essentially an endothermic step with using solar energy. The researchers have also reported the results of a non-solar, exothermic step by way of hydrolysis of Zn(l) at 700 K to get hydrogen.

10.6.16   Solid oxide electrolyzer

High-temperature solid oxide electrolyzer cell (SOEC) has been reported to possess great potential for efficient production of hydrogen. YSZ and LSGM have been found to be suitable electrolyte materials for SOEC working at high and intermediate temperatures, respectively [22]. In general, solid oxide system has the unique feature that the cell uses some ceramic materials that can act both as an electrolyzer as well as a separate membrane so that hydrogen and oxygen are separated.

10.6.17   Sea water electrolyzer

Abundant water available in the oceans is probably the everlasting and apparently infinite source of hydrogen production through the method of electrolysis. However, the high level of sodium chloride (NaCl) in seawater creates environment-related problems. A new process for chlorine-free seawater electrolysis has been proposed in a recent study [23] where, initially, separation of Mg^{2+} and Ca^{2+} ions takes place from seawater by nanofiltration. In the next step the NF permeate is dosed into the electrochemical system so as to obtain hydrogen and oxygen gases along with NaCl precipitate.

10.6.18   Photo-electrolysis (photolysis)

Photo-electrolysis is the process whereby light is used to split water directly into hydrogen and oxygen. Photovoltaic (PV) systems coupled to electrolyzers are being used and they offer the flexibility of getting either electricity from photovoltaic cells or hydrogen from the electrolyzer. Figure 10.19 shows the process of photolysis. It has already been described in techniques for earlier that feeds for bio-hydrogen are water for photolysis where hydrogen is produced by some bacteria or algae. Bio-photolysis is predominantly a biological process adopted for production of hydrogen gas [24].

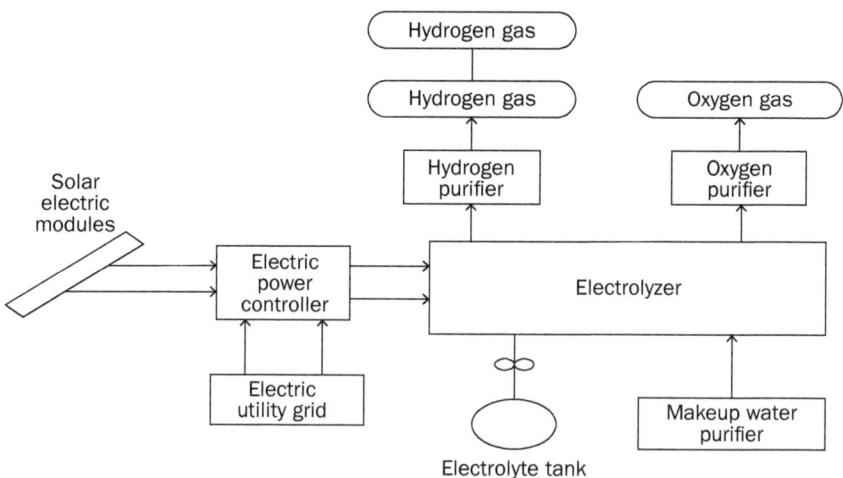

Figure 10.19 Principle of photolysis

10.6.19   Other methods of hydrogen production

Hydrogen from H₂S in Black Sea waters

The Black Sea is the world's largest water body that contains a large amount of hydrogen sulfide. Interestingly it has inflow of salty Mediterranean Sea water in the deep layers and an inflow of river water on the top layer. The sea is characteristically stratified and cannot even be mixed by wind. The hydrogen sulfide layer in the Black Sea occurs at about 200 meters below the surface. It is reported that total sulfide pool in the Black Sea is around 4.6×10^9 tons. H_2S is an environmental pollutant and therefore appropriate steps need to be taken during production of hydrogen from this source. One-step thermal decomposition methods have been found to be more suitable for this purpose of generating hydrogen from the H_2S. It is also believed that superadibatic partial combustion is a promising future [25]. It has been approximately estimated that 90 percent of Black Sea water is anaerobic in character. H_2S concentration is almost constant at 9.5 mg/l at a depth of 1500 m. It is therefore possible to extract H_2S at relatively higher temperature and at the same time by bringing down the pH of the water to 4. This process can be accomplished in two stages. In the first stage extraction of H_2S could be carried out from seawater and then produce hydrogen and polysulfides by way of electrochemical production.

Hydrogen production using enterobacter cloacae

Enterobacter cloacae are an anaerobic, non-spore-forming bacteria that grow on a range of organic substrates. They produce hydrogen only under anaerobic condition. Research activities have been successful developing a process for continuous hydrogen production using immobilized Enterobacter cloacae IIT-BT 08 on lignocellulosic solid matrices [26]. Tests have been carried out and it has been observed that this process is relatively more efficient than conventional continuous hydrogen processes using packed bed reactor.

Production by reforming natural gas and bio-derived liquids using dense ceramic membrane

Even though water dissociation is a well-established technology for hydrogen production, it is well-known that dissociation of water gives low concentration of hydrogen and oxygen. To improve the concentration level, a mixed-conducting (i.e., electron- and ion-conducting) membrane such as oxygen transport membrane (OTM) is used to separate oxygen and hydrogen.

Joseph Schwartz, Hankwon Lim, and Raymond Drnevich of Praxair, Inc. carried out a study on the integrated ceramic membrane system for hydrogen production. The results of this project [27] showed excellent membrane performance and it could be useful for hydrogen production.

Plasma Reforming

Studies have been carried out on potential application of plasma catalysis, the combined effect of plasma catalysis along with thermal catalysis, on hydrocarbon reforming for Hydrogen generation (28). It has been found that performance of plasma catalysis is better than the combined effect of plasma-alone and catalysis-alone. Thus, it will be useful particularly for hydrogen generation.

Thermal plasma technology can be used to produce hydrogen and a variety of hydrogen-rich gases from a broad range of fuels including heavy hydrocarbons fuels. Plasma reformers have been used to produce hydrogen-rich gases with gasoline, diesel, oil, biomass, natural gas, jet fuel, etc. The conversion efficiency to hydrogen-rich gases have been observed to be close to 100 percent. In this process, the plasma conditions (high temperatures and a high degree of ionization) are used to ensure that thermodynamically desirable chemical reactions occur at a much faster rate without a catalyst. So, the process does not have any problems related to catalyst sensitivity and deterioration. The reaction has a much faster response time: almost a fraction of a second.

Hydrogen from nuclear energy

Several possible techniques such as thermochemical processes, water electrolysis, or high-temperature steam electrolysis can be used as an energy source in centralized hydrogen production. Usually high-temperature reactors (such as the gas-cooled, molten-salt-cooled, and liquid-metal-cooled reactors) are required for hydrogen production because higher operating temperatures are necessary for efficient thermochemical and electrochemical hydrogen production using nuclear energy [29].

It is often observed that during off-peak hours, nuclear power plants generate more electricity than normally required for the grid consumers, and this is perhaps the appropriate time to use the excess electricity to generate hydrogen. The amount of heat generated in the nuclear reactors can be made use of to produce hydrogen from conventional sources such as water, coal, biomass, and natural gas. The present methods of producing hydrogen from nuclear energy consist of three processes: (i) electrolysis, (ii) high temperature steam electrolysis, and (iii) thermochemical water splitting cycles. Hydrogen produced by light water reactors (LWR) does not need any cleaning for chemical impurity. It would be relevant to recall here that hydrogen produced by way of coal or steam reforming needs to be cleaned for chemical impurity. Process of thermochemical water splitting is divided into different phases of partial reactions and each such reaction is at a relatively lower temperature of 550–850°C. In this reaction, water and heat happen to be the process inputs and water is split by using only heat. Hydrogen and oxygen are process output. The efficiency of this process is generally reported to be in the range of 40 percent to nearly 60 percent. Iodine sulfur (I-S), calcium-bromine (Ca-Br), and copper-chlorine (Cu-Cl) happen to be the more common among thermochemical processes.

With the background technical information on existing and prospective routes of hydrogen production, it is important to discuss that a successful centralized hydrogen production requires large market demand and design and construction of hydrogen transmission and distribution infrastructure and pipeline. Care must be taken at this point for storage of CO_2. Keeping the long-term sustainability in vision, it is perhaps important to design a centralized hydrogen production from high-temperature processes that need to be based on renewable energy. In such systems, effective use of waste heat can also be considered to ensure sustainability, thereby reducing the task for capture and storage of CO_2.

10.7   Hydrogen Transportation, Delivery, and Distribution

Most of the hydrogen produced at present all over the world is in the proximity of its consumption; it does not need to be transported or shipped over long distances. The meager amount of such hydrogen that is produced onsite is used in chemical industries.

Most hydrogen production, at present, is from natural gas and some amount is generated from other fossil sources such as coal and oil. Electrolysis currently contributes little to hydrogen production. The present mode of hydrogen transport from production point to the point of use is by pipeline and on the road in cryogenic liquid tanker trucks or gaseous tube trailers. It is also being transported by rail. In absence of safe and long pipeline infrastructure, hydrogen in bulk quantity is usually transported as a liquid in super-insulated trucks commonly referred to liquid tankers. It is relevant to mention here that to liquefy hydrogen, it must be cooled to cryogenic temperature by a liquefaction process. Usually, the liquid hydrogen trailers are referred to by their volume capacity. The trailers that are commonly used to transport liquid hydrogen now have a gross volume capacity in the range of 28,400 to 49,200 liters. A liquid tanker truck can transport a much larger mass of hydrogen compared to a gaseous tube trailer. Therefore, transport of hydrogen in liquid form is more economical than gaseous mode of transport. However, one of the major challenges of liquid transport is mainly due to boil-off during delivery. It is essential that both pressurization and refrigeration must be effectively accomplished to avoid boiling off at the low cryogenic temperature at which hydrogen is transported.

A hydrogen transport and delivery infrastructure system should be designed to deliver safely the required amount of hydrogen from the point of production to the dispenser or refueling station. The infrastructure should have a series of elements such as pipelines, trucks, storage facilities, compressors, and dispensers that are installed in sequence for fuel delivery.

Hydrogen in gaseous form is mainly transported by cylinders in trucks (referred to as tube trailers) or through appropriately designed pipelines. Usually, gaseous hydrogen is produced in a relatively much lower pressure (about 20 to 30 bars); it is compressed to a much higher pressure (about 180 bars or more) into the long gas cylinders and then transported. These tube trailers have lengths up to about 12 meters and they can carry many hydrogen gas cylinders. Pipeline transport of hydrogen is adopted from the point of hydrogen production to the point of utilization or demand, which is within approximately about 200 km at best. Hydrogen is dispensed to hydrogen stations or other places depending on the demand and existing infrastructure. Hydrogen used for stationary applications is usually supplied by truck to the storage facility. It is also a common feature to carry hydrogen in cylinders (just as LPG/propane cylinders) to the storage facility or to point of use. Currently, the hydrogen transport and delivery depend on several challenges before its widespread application. Figure 10.20 shows a broad outline of hydrogen transport from production to its end use application point.

A national hydrogen transport infrastructure needs to be built and delivery cost of hydrogen must be reduced. The other critical factors include the purity of hydrogen and transporting it safely without any leakage. An integrated overall system must be optimized based on the techno-economic criteria of hydrogen production option and hydrogen transport/delivery option. Development of onsite hydrogen production from renewable sources might change the picture substantially.

10.7.1 Hydrogen storage and distribution

The benefits of hydrogen energy and its uses have already been discussed. It is important to note the storage method of hydrogen energy determines its application mode. With a lot of development on hydrogen technologies, hydrogen storage continues to be one of the major stumbling blocks in the widespread use of hydrogen energy in specific

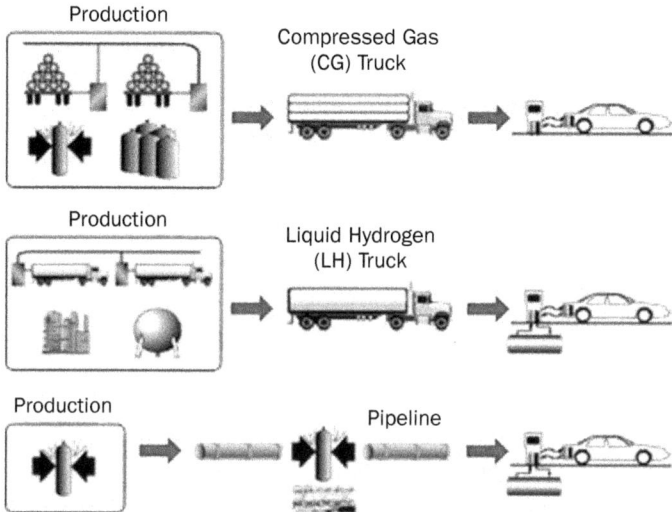

Figure 10.20   From hydrogen production until end use application

Compressed Fuel Storage	• Cylindrical tanks • Quasi-conformable tanks
Liquid Hydrogen Storage	• Cylindrical tanks • Elliptical tanks • Cryotanks • HP liquid tanks
Solid State Conformable Storage	• Hydride materials • Carbon adsorption
Chemical Hydrides	• Off-board recycling

Table 10.4   Different methods of hydrogen storage

applications such as vehicles. Hydrogen can be stored as compressed gas, cryogenic liquid, and in solid form such as chemical or physical combination with materials, such as metal hydrides and carbon materials. Presently compressed gas storage is used for large-scale land transportation, whereas liquid hydrogen storage has been used for a long period for air and space transportation. Metal hydride storage at present is restricted to small-scale transportation. Metal hydride has also a lot of potential for vehicular application. Some metals readily absorb gaseous hydrogen under conditions of high pressure and moderate temperature to form metal hydrides. Metal hydrides, when heated at low pressure and relatively high temperature, release hydride. The major concern of hydrides is their low mass and volumetric energy density, which makes hydride vehicles much heavier and less efficient than comparable gasoline.

Compressed hydrogen storage

It has been reported in literature that hydrogen in the form of compressed gas was transported in 1880s in wrought-iron metal cylinders to inflate war balloons [30]. Compressed hydrogen storage offers the simplest mode of storing hydrogen. The highest energy density (120 MJ/kg) and the energy content of hydrogen per volume (0.0107 kJ/m^3) are two relevant properties in this context. It has been estimated that at ambient conditions of temperature and pressure (300 K, 1 bar), 1 kg of hydrogen gas occupies 11 m^3. This is a major constraint because it implies an enormous reduction of volume of hydrogen gas. It becomes essential to compress the gas inside pressure vessels. There are four types of cylinders prescribed for hydrogen storage depending on the specific applications. Table 10.5 gives the characteristics of the four types of cylinders referred to as Type 1, Type 2, Type 3, Type 4, and Type 5.

There are well-defined recommendations of minimum wall thickness for aluminum and steel cylinders [31, 32] respectively with the corresponding diameters (around 300 mm). Three main types of hydrogen storage tanks are used. These are made of steel, aluminum core encased in fiber glass (composite), and plastic core encased in fiber glass (composite). Compressed hydrogen gas stored in metallic pressure vessels increases the weight to a large extent. This feature of storage lowers the gravimetric efficiency. Composite polymer pressure vessels are used at a pressure level of about 68.94 MPa, thereby facilitating the system to attain a high level of volumetric and gravimetric efficiency. High pressure gaseous hydrogen (HPGH) is a mature technology and has got some definite advantages. It has been estimated that liquefaction of hydrogen consumes 30 percent to 40 percent of the lower heating value (LHV) of hydrogen whereas the energy required for hydrogen compression up to 35–70 MPa is approximately in the range of 5 percent to 20 percent of LHV. Almost 80 percent of the hydrogen refueling stations worldwide have been adopting HPGH storage technique. Considering several factors such as compression energy consumption, driving range, infrastructure investment and several other factors, the pressure for on-board hydrogen systems is mostly in the range of 35 MPa to 70 MPa.

Type 1: all metal cylinder
Type 2: load-bearing metal liner hoop wrapped with resin-impregnated continuous filament
Type 3: non-load-bearing metal liner axial and hoop wrapped with resin-impregnated continuous filament
Type 4: non-load-bearing non-metal liner axial and hoop wrapped with resin-impregnated continuous filament
Type 5 (Other): type of construction not covered by Types 1 to 4 above.

TABLE 10.5 Hydrogen storage tank classification

As far as the design of the high-pressure composite cylinders for hydrogen storage are concerned, these are made a high molecular weight polymer or aluminum liner that serves as a hydrogen gas permeation barrier. Aluminum-lined, composite-wrapped (Type 3) vessels are common because of the moderate weight and it does not allow hydrogen permeation.

Liquid hydrogen storage is another way of storage of hydrogen. Liquid hydrogen has long since been used in rocket engines and for space applications such as Apollo missions and Voyager missions to the moon and Viking missions to Mars. Hydrogen gas can be changed to the liquid state by pressurizing hydrogen gas at its boiling point of −235°C. Design features of the containment tanks for storage of liquid hydrogen are different from a gasoline tank. It has been observed that many materials, coming in contact with liquid hydrogen, become brittle and their physical dimensions are decreased. Moreover, sometimes it so happens that the oxygen in air gets condensed in the liquid hydrogen, thereby creating chances of explosion. Liquid hydrogen is stored in cryogenic storage systems. The material should be appropriately designed to withstand the cold temperature. It has been reported that quite often the seals of the container are not able to maintain effective sealing at low temperature. It is important that the containers must effectively prevent heat leakage because heat leakage from the surrounding objects into the tank has often been reported to cause evaporation. LH_2 technologies of course have advanced substantially now, thereby allowing rapid fueling and no evaporative losses.

Liquid hydrogen storage for on-board vehicular application has a major drawback due to boil-off losses. It has been briefly discussed earlier that boil-off is one of the most common problems encountered with hydrogen storage in liquid form. Quite often, liquid hydrogen vaporizes at extreme low temperature and escapes from the tank. As a result, the tank pressure increases due to the vapor in the tank. The continuous process of pressure building increases the weight of the tank. Unless the continuous rise in pressure is checked, this might even result in exploding the tank.

As a precautionary measure, pressure relief valves are installed to reduce the pressure inside the tank and prevent any possible explosion. Use of such pressure relief valves has been observed to result in some hydrogen leakage, which is referred to as boil-off loss.

Boil-off occurs typically at a pressure of 1 MPa or less. Cryogenic liquid hydrogen tanks have been designed to prevent both convective and radiative heat transfer losses. Linde has designed a system called Cool H_2 whereby the boil-off can be delayed [33]. Boil-off loss rate depends on the quantity of liquid hydrogen stored in the storage tank. Surface area of the storage vessel is a crucial factor because the amount of heat leakage depends on the surface area. The design of the liquid hydrogen storage tanks is made with a double wall construction so as to

Metal Hydride Formation Mechanism

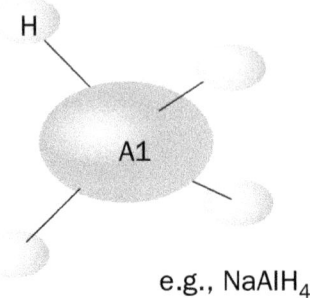

e.g., NaAlH$_4$

FIGURE 10.21   Typical structure of a metal hydride

maintain low temperature with thermal insulation. Tanks are built up with a multi-layer insulation (MLI). MLI consists of a thin metal layer that is also effective in controlling both radiation and thermal irradiation between the layers.

Liquid hydrogen tanks have been installed in the trunk or rear of the vehicle and have been successfully tested. A broad description of the design features of such tanks is given here. These tanks are of weight about 50 to 60 kg and are usually made of aluminum alloy and fitted with vapor-cooled shields. Glass fiber or carbon fiber composites are used in the inner vessel to reduce the evaporation rate loss to about 1.5 percent per day. They are optimally instrumented with the relevant equipment such as safety valves, tubes, heat exchangers, and other subsystems located at appropriate locations.

Slush hydrogen is a liquid-solid hydrogen mixture at the triple point with a lower temperature and a higher density than liquid hydrogen. It is formed by further refrigeration of the liquid by bringing liquid hydrogen down to nearly melting point (14.01 K or –259.14°C), which results in increase in density about 16–20 percent. In the slush hydrogen state, solid hydrogen particles with particle diameter of several mm are contained in liquid hydrogen. It has higher density and refrigerant heat capacity than liquid hydrogen. Slush hydrogen is a cryogenic solid–liquid two-phase fluid with better features than that of solid or liquid hydrogen independently.

Studies have indicated that slush hydrogen should be pressurized and be kept free from impurities. Quite often, for certain applications, it needs to be continuously upgraded [34]. It has also been observed by the researchers that at lower flow rates slush has a higher viscosity than the liquid.

Hydrogen storage by metal hydrides is the other option of storing hydrogen (Fig. 10.21). Metal hydrides are one of the safest methods for

FIGURE 10.22   Volume comparison of different hydrides for vehicular application

storing hydrogen as it avoids the leakage that can be a problem of concern with compressed gas and liquid hydrogen.

Still, there is big gap to hydride. Hundreds of intermetallic alloys have been developed to store hydrogen [37, 38] for several applications, and the hydrogenation potentials of several of these alloys have also been studied [35, 36].

Hydride as is well known is an intermetallic alloy phase that is capable of absorbing and retaining a large amount of hydrogen by way of chemical bonding. The metal hydrides are formed by the reaction between a metallic phase and hydrogen.

Corresponding metal hydrides get formed when metals are exposed to certain level of temperature and pressure such that they absorb a vast amount of hydrogen to form hydrides. Figure 10.21 shows the metal hydride formation mechanism during hydrogen absorption.

Chemical hydrogen storage is adopted in which hydrogen is generated through a chemical reaction. The common chemical reactions (hydrolysis) are carried out on chemical hydrides with water or alcohols to produce hydrogen. Chemical storage technique is described by the following example. The reaction of sodium borohydride with water produces hydrogen.

$$NaBH_4 + 2H_2O = NaBO_2 + 4H^+$$

Figure 10.22 gives a general idea of volume comparison of 4 kg hydrogen storage shown in different ways, with size relative to the size of a car.

10.7.2   Hydrogen in glass microspheres

This mode of storage technique considers the fact that glass is sufficiently impervious to hydrogen at ambient temperature. However, its porosity can allow hydrogen to be released at conditions of higher temperature. Hydrogen storage in microspheres has some distinct advantages. Microsphere storage allows a much higher level of internal pressure as compared to conventional storage cylinders. Some studies have shown promising results for this technique. Apart from overall better safety conditions, it has been found to have a better weight density (hydrogen weight/total storage weight) than hydrides. Energy requirements are almost one-eighth of that required for liquefaction. However, there are some design criteria that need to be taken into consideration because the rate of hydrogen transfer depends on several factors such as dimension of the microcapsule and its material of composition. Hydrogen transfer rate also depends on the temperature and the pressure gradient across the microcapsule wall. The size diameter of the micro capsules and the choice of microsphere material are to be considered on the basis weight content of the hydrogen to be stored in it. The storage time and the filling conditions depend on the gas permeability of the wall. Tensile strength of the material and the physical dimension of the microcapsule need to be considered to determine the maximum storage pressure.

10.7.3   Hydrogen storage in zeolites

Storage of hydrogen in zeolites has been another option. It is known that the gas molecules cannot penetrate the zeolite at normal conditions of temperature and pressure. However, the molecules are forced into the channel at higher levels of pressure and temperature and get entrapped after the system has been brought back to the ambient temperature. Hydrogen is usually entrapped in small cages (e.g., sodalite cages). Interestingly, the trapped molecules are not able to escape at this low pressure and it is possible for the gas can be kept at a relatively high internal pressure because there is no equilibrium between the absorbed molecules and the surrounding atmosphere. Hydrogen from the zeolite can be released to a temperature higher than the temperature of encapsulation. It is perhaps relevant to mention here that despite hydrogen inside the zeolite cages, a little amount of hydrogen often gets leaked. The storage capacity is dependent on the composition and framework structure of the zeolite such as specific surface area, storage temperature, pore volume, etc. These conditions are critical to the ability for high adsorption. It is important to know under what conditions the entrapped hydrogen can be released at a suitable rate. It has been reported that the maximum storage capacity of the zeolite system is much less than metal hydride.

Use of zeolites for hydrogen storage has been investigated. Zeolites of different characteristics such as pore architecture and composition have been reported [39]. This study indicates that storage capacity is dependent on the framework structure as well as composition of the zeolite. In their study the researcher observed that the highest storage capacity was obtained with zeolite having highest concentration of sodalite.

10.8   Safety aspects of hydrogen

Hydrogen has unfortunately been considered by many as unsafe because of the devastating potential of hydrogen bombs and the sad Hindenburg disaster. The cause of the Hindenburg accident is still not known. However, it is suspected that hydrogen, when vented, got ignited by the electrostatic charges present in the atmosphere because of the thunderstorm. Human error and subsequent inappropriate handling of hydrogen have been the cause of several accidents [40]. Some of these have been due to corrosion, embrittlement at storage tanks at low pressure conditions, material failure, boiling liquid, and expanding vapor explosion. It would be appropriate to mention here that hydrogen has got some properties that are radically different from conventional fuels. Table 10.6 gives a list of typical combustion features of hydrogen that need to be closely monitored from a safety point of view. A judicious handling of these properties makes hydrogen safer than gasoline depending on the situation. So, it needs to be handled in a different way. Wider flammability limit, lower minimum ignition energy, and high flame speed are three crucial properties that need to be properly handled to avoid any untoward safety-related problems. Hydrogen is odorless, and in some applications, an odorant is added to enable its detection. Selection of odorant is important because sometimes some odorants can contaminate some hydrides. Table 10.6 gives a comparative picture of hydrogen with respect to gasoline.

Hydrogen flame is nearly invisible in daylight, but it is visible in the dark and also in subdued light. The nearly invisible characteristics of hydrogen flame has been the major cause of several accidents during tests. It is desirable to use some colorant so that the flame is visible (Fig. 10.23) and appropriate safety steps could be conveniently adopted.

Care must be taken that the colorant does not contaminate the hydride if hydride is used for that specific application. It has been reported that several incidents have affected persons due to hydrogen leakage, fire, or explosion of hydrogen air mixture [42, 43]. However, a close analysis of the properties of hydrogen often indicates that a person can stay closer to hydrogen fire as compared to a gasoline fire without getting burned. Gaseous Hydrogen (GH_2) diffuses rapidly with air turbulence, thereby increasing the rate of dispersion.

Hydrogen is the smallest and lightest element in nature. So it diffuses rapidly to the surroundings to noncombustible proportions. In this context it is relevant to recall that low molecular weight of hydrogen results in high molecular diffusivity. Hydrogen diffuses three to eight times faster than air. Roof vents and forced ventilation have proven to be successful in minimizing the accumulation of gaseous fuels within enclosures.

Hydrogen Safety–Undeserved Reputation as a Dangerous Fuel		
Temperamental Properties that need to be closely monitored		
	Hydrogen	**Gasoline**
Diffusivity in air (cm^2/s)	0.63	0.08
Minimum Ignition energy (MJ)	0.02	0.24
Stoichiometric ratio %	29.6	1.76
Flame speed (m/s)	2.7	0.35
Quench. dist. (cm)	0.07	0.02
Auto-ignition temp. (°C)	580	400

TABLE 10.6 Some significant safety related properties of hydrogen

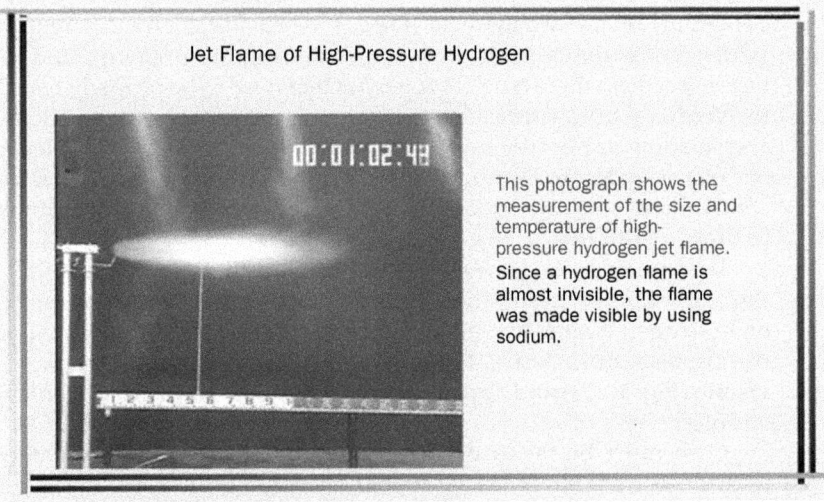

Jet Flame of High-Pressure Hydrogen

This photograph shows the measurement of the size and temperature of high-pressure hydrogen jet flame. Since a hydrogen flame is almost invisible, the flame was made visible by using sodium.

FIGURE 10.23 Colorant added to have visible hydrogen flame [41]

On a volume-of-gas basis, hydrogen has a lower explosion potential as compared to methane and propane. But on a mass-of-fuel basis, hydrogen has the maximum explosion potential. However, volume basis criteria are more pertinent to storage safety. In any hydrogen operating system it is desirable that hydrogen storage and supply system should have the least number of potential leakage points in pipe connections, seals on components, evacuation valves, and pressure relief vent ports. etc.

Hydrogen fire burns rapidly and radiates little heat. There is practically no radiation threat from hydrogen fire. It would perhaps be relevant to mention here that there was no report of death in the Hindenburg accident due to hydrogen burns. It is also little known that Hindenburg had already made 10 successful round-trip transatlantic flights before this infamous accident.

A comparative picture showing the safer aspect of hydrogen as compared to gasoline was studied by researchers at the University of Miami [44]. Figure 10.24 compares an intentional hydrogen tank release and a small gasoline fuel line leak [44]. As evident in the figure, the hydrogen flame has begun to subside after 60 seconds whereas the gasoline fire is intensifying. After 100 seconds, it was observed that all of the hydrogen had gone and the car's interior was undamaged (the maximum temperature inside the back window was 67°F). On the other hand, the gasoline car continued to burn for several minutes and was destroyed totally by fire.

Researchers at the University of Miami had studied computer models of different geometries of kitchen stoves to evaluate a comparative safety features of hydrogen with respect to methane and propane. Intentional leaks of the gases were studied. Stoves were supposed to have no fans and therefore there was no convection. Computer models generated to show cloud formation in the kitchen a few minutes after the leak started. The results were in favor of hydrogen where the thin cloud got diffused straight through the vent and no combustible cloud was formed. A small combustible cloud was seen with a methane-fuelled stove whereas the large cloud was seen with propane. It showed that hydrogen was the least danger whereas propane posed the greatest threat.

In the event of leakage of liquid hydrogen, it evaporates and gets dispersed quickly. A gasoline leakage remains a threat for a relatively much longer period. Observation alone cannot be considered as a reliable technique for detecting hydrogen-air fires. Severity of fire cannot be assessed from such observations. Sometimes a fire can result from an accident in which a small leak can also develop. As far as liquid hydrogen is concerned, it must be considered that sometimes rapid evaporation occurring in liquid hydrogen spill results in a flammable mixture over a relatively longer stretch of distance. In such cases, even though ignition sources may not be present at the location of spill or leak, fire could still occur if the flammable mixture moves forward to an ignition source. The significant fire damage parameters as thermal radiation, flame engulfment, and smoke inhalation. As has been discussed earlier, the problem of safety emerges from some of the unique properties of the fuel such as flammability limit, ignition energy, and flame speed. Ignition of explosive mixtures of hydrogen occurs with low energy input of the order of 1/10 that of gasoline air mixture. Moreover, it has been observed that hydrogen easily dissolves/diffuses through several materials, thereby causing a degradation of the mechanical properties of the materials. The hazards associated with the use of hydrogen can be broadly characterized as physiological (frostbite, respiratory ailment, and

A picture from a video which compared fire from a leak in a gasoline engine car and the same kind of leak from a hydrogen car. The pictures are taken at one minute after ignition

The hydrogen flame has begun to subside, the gasoline fire is intensifying. After 100 seconds, all the hydrogen was gone and the interior of the car was undamaged. The gasoline car continued to burn for a long time and was totally damaged.
[Dr. Michael R. Swain 2001]

Figure 10.24   Effect of hydrogen fire and gasoline fire [44]

asphyxiation), physical (phase changes, component failures, and embrittlement), and chemical (ignition and burning). It has been observed that hydrogen released in a confined area alters the composition of air by displacing oxygen. This is a situation that leads to problems of asphyxiation.

Let us now look at the combustion behavior of hydrogen. In a specific system, the primary hazard with any form of hydrogen is due to inadvertently producing a flammable mixture. Hydrogen's transition from deflagration-to-detonation is a crucial phenomenon as this is the phase of simple combustion to explosion. Hydrogen's higher burning velocity is often responsible for this causes this situation of deflagration-to-detonation transition (DDt).

After hydrogen is ignited, the reaction can proceed either by deflagration (subsonic propagation) or detonation (supersonic propagation). It has been observed that deflagration in a closed volume increases the pressure level substantially to almost eight times the initial pressure. In the other situation of combustion, detonation can occur from a low energy ignition source when hydrogen-air mixtures of 18 percent to 60 percent volume that are well-mixed and confined. The phenomenon of hydrogen detonation is rare. However, when it occurs, the situation gives rise to rapid rate of pressure rise. In some cases it has been observed that pressure-relief devices become ineffective. Hydrogen gets accelerated to its detonation velocity much quicker and in a much shorter distance as compared to gasoline, propane, and methane.

Safety of hydrogen also depends on its phase. It has been estimated that a single volume of liquid hydrogen expands to about 850 volumes of gas at STP when vaporized and about 2000-meter elevation, this expansion is approximately 1000 volumes of gas at ST. Selection of the appropriate material is important for hydrogen use. A very common piping material such as carbon steel has been observed to be affected by hydrogen use. The different components used in a hydrogen system need to be carefully designed: pressure relief devices, valves, regulators, electric equipment. Storage of hydrogen in hydrogen gas cylinders and adequate ventilations and appropriate alarms in hydrogen utilization are some of the critical design factors that need to be critically analyzed. It is strongly recommended that only the regulator intended for hydrogen gas must be used in the cylinder. While carrying tests with hydrogen systems in the lab during the stage of development it is recommended that all the components and transfer lines be purged with an inert gas such as nitrogen. Diatomic hydrogen, even though is the most stable form of hydrogen, gets dissociated with high energy that often permeate a surface coating structural damage. These hydrogen atoms are small and often permeate a container surface causing structural damage. This phenomenon, which is referred to as embrittlement, is more prevalent when metal containers are used.

10.9   Hydrogen Detection Technologies

In view of its typical combustion properties such as low density $(0.0899$ kg/m^3), boiling point (20.39 K), high diffusion coefficient $(0.61$ cm^2/s in air), and buoyancy it is essential that presence of hydrogen and its concentration must be identified in the neighborhood of its application at area. Besides, it is a colorless, odorless, and tasteless flammable gas and cannot be detected by human senses. From the point of view of safety and to rule out any condition of explosion, it is essential to detect and identify the location of hydrogen leakage. Hydrogen sensors, which are basically transducers, have been developed for this purpose. Broadly the sensing elements are built on the principle that interaction of hydrogen with the sensing element cause changes in several parameters such as temperature, refractive index, as well as some electrical and mechanical properties. A transducer is used to transform all these into an electrical signal. Sensors can be installed in enclosed areas to detect concentrations of hydrogen approaching 4 percent (the lowest flammability range) and are useful to identify concentration level. BMW's prototype hydrogen cars have been designed in a way such that windows and sunroofs open if hydrogen leakage takes place. Design of sensor technology is based on several different principles such as catalytic and thermal conductivity, electrochemical resistance, work function, and mechanical, optical, and acoustic sensing.

10.10   Hydrogen Embrittlement

The exact mechanism of hydrogen embrittlement is not well known. Broadly it is the penetration of atomic hydrogen into the metal structure. The mechanisms proposed for hydrogen embrittlement are based on slip interference by dissolved hydrogen due to accumulation of hydrogen near dislocation sites. There are several techniques that have proven to be fruitful for solving the problems related to embrittlement.

Embrittlement can be broadly divided into two types: internal hydrogen embrittlement (IHE) and external hydrogen embrittlement (EHE). IHE occurs mainly due to the residual hydrogen from various manufacturing processes such as electroplating and pickling [45]. EHE occurs mainly due to hydrogen from external sources in a hydrogen-rich environment such as stress corrosion cracking [46].

Hydrogen embrittlement occurs frequently during pickling operations when the rate of corrosion of the base metal causes a high rate of hydrogen evolution. Addition of suitable inhibitors has proven to be effective in reducing the base metal corrosion during pickling. The other well-known technique is based on using clean steel. Rimmed steel has many voids that cause embrittlement, and this can be reduced using clean steel, which has fewer voids. Hydrogen embrittlement in steels has been observed to be almost a reversible process. Therefore the mechanical properties of the treated material do not substantially vary if hydrogen is removed. This is an advantageous feature and usually steel is baked to relatively lower temperature of about 100–150°C to remove hydrogen. High strength steels are found to be susceptible to high rate of embrittlement. Ni and Mo alloys do not have the problem of embrittlement. It is often recommended to use nickel-based alloys as they have low hydrogen diffusion rate and so embrittlement is not a problem.

Proper welding is a relevant factor in connation; in the problem of embrittlement it is always desirable to use low-hydrogen welding rods to reduce the embrittlement. Water and water vapor, which happen to be the major source of hydrogen, need to be avoided and care must be taken to maintain dry conditions during welding.

Certain metals (such as martensitic and ferritic) are more susceptible to embitterment than highly alloyed austenitic steels. Metal degradation by hydrogen needs to be investigated by way of using suitable material for hydrogen structures/sealing. Choice of appropriate material for use in specific locations in hydrogen operating systems is critical. Ferric steels and nickel-based steels have been found to be highly susceptible, whereas austenitic steels are moderately affected, and low aluminum alloys and copper alloys are almost unaffected. Carbonaceous steel should be avoided as hydrogen can react with carbon in the steel to form CH_4 bubbles in the metal at high temperature.

10.11   Utilization of Hydrogen

Hydrogen in form of liquid and gas is used for a wide range of industrial applications such as producing chemicals, foods, and making plastics. The other applications of hydrogen include gas welding, petroleum refinery, ultraviolet lamps, and gas chromatography. It is also used for making ammonia for fertilizers and methanol. Hydrogen is used as a fuel for rockets where liquid hydrogen, in combination with liquid oxygen, produces a powerful explosion. Use of hydrogen as a fuel has a long history. Town gas, which was used as a fuel in early part of 20th century, had about 50 percent hydrogen in its composition. NASA uses hydrogen for its space shuttle. Besides, it is used in hot air balloons as well as in disastrous applications such as bombs used for mass destruction. Hydrogen in some form is useful in some applications and not replaceable by other elements or gases.

In atomic absorption spectroscopy hydrogen is used as fuel to generate heat in the process of experimentation. The ability of hydrogen to ignite is used in producing the neutral atoms in the process. As far as use of hydrogen in gas welding is concerned, in this process the temperature level goes up to about 4000°C, thereby causing the metals to melt and join the broken surfaces. Atomic hydrogen welding is the other important distinctive feature of this welding. In this welding process, hydrogen, besides generating heat, also helps prevent the highly reactive metals from reacting with other elements such as nitrogen and carbon during the process. Hydrogen gas is widely used in the petroleum industry to remove sulfur content. It is also useful for hydrocracking in which long-chain hydrocarbons are broken into shorter ones. Other uses in petroleum refinery include hydroisomerization (to convert normal paraffin to isoparaffin) and also, for dearomatisation (to convert aromatic to cycloalkanes). The isotope of hydrogen is used to make heavy water (D_2O), which is used in nuclear reactors as a coolant.

Hydrogen has several uses in chemical analysis. It is used in hydrogen electrode for titrations by potentiometry in which hydrogen gas, at constant speed and under a pressure of one atmosphere pressure, is passed into the reference electrode. However, it is important to note that purity of the hydrogen has been found to affect the analysis of these potentiometric titrations. These methods especially are used in atomic absorption spectroscopy where the hydrogen is used as fuel to generate heat in the process of experimentation.

In any reaction, it is known that addition of hydrogen is termed as reduction while removal is called oxidation. Hydrogen is also used as a reducing agent for both the electro-negative and electro-positive elements. This electro-positive nature is used in redox reactions. In gas chromatography, hydrogen is used as a mobile phase to separate volatile substances from a mixture.

Presence of hydrogen has been found to be useful in structural identification. Sometimes it is not easy to study the nature of bonds between atoms and molecules in many compounds. It becomes difficult to identify the structure of such compounds. Nuclear (NMR) has been one of the useful techniques adopted to find out the molecular structure from character of protons and getting the desired information about the molecular structure. Sometimes it is referred to as Proton NMR. Often the presence of hydrogen in many types of bio-molecules has been observed to be useful in carrying out this technique, thereby helping in structural identification.

Hydrogen, in combination with oxygen, forms H_2O_2 (hydrogen peroxide). Hydrogen is a common sterilizing agent used in clinics and hospitals. There are several other pharmaceutical and medical applications of hydrogen peroxide such as for cleaning of wounds, cuts, and other damaged tissue portions. Its application is also being studied in research activities for testing the antioxidant potential of enzymes like catalase.

As is well-known one of the first uses of hydrogen gas was in flying of hot balloons. This was attractive because of hydrogen's lightweight. However, subsequently hydrogen was replaced by helium mainly because of hydrogen's explosive characteristics under certain conditions. Helium is highly nonreactive. Figure 10.25 shows the hydrogen-operated flying balloons.

As has been indicated earlier, hydrogen has some distinctive properties such as lowest molecular weight, high flame speed, and minimum ignition energy. Hydrogen combustion with oxygen provides the highest specific impulse in relation to the amount of fuel consumed. Hydrogen has been the signature fuel for the American space program.

Hydrogen can develop one of the most powerful weapons such as hydrogen bombs (Fig. 10.26) built on the principle of nuclear fusion of hydrogen atoms isotopes. However, this should never be used for the destruction of human beings and societies.

One of the most significant and potential applications of hydrogen is in the internal combustion engines widely used in the developing countries for transport and decentralized energy sector. Use of hydrogen as an automobile fuel is described afterward in greater detail. A specially designed hydrogen engine is capable of improved performance and practically zero-emission characteristic.

There are two types of internal combustion engines: spark ignition (SI) and compression ignition (CI) engines. Both of these types of internal combustion engines, presently operated by liquid petroleum fuels (mainly gasoline, diesel), have widely penetrated our present-day lifestyle engine and presently form the backbone of global transport infrastructure. Safe survival of these engines has been put to question because of the rapid rate of fossil fuel depletion. Alternative substitutes to petroleum fuel need to be expeditiously sought. The other dimension of the

FIGURE 10.25   Hydrogen-operated balloons

FIGURE 10.26   Schematic explosion from a hydrogen bomb

problem emerges from environmental degradation due to combustion of conventional fuels. This problem further strengthens the effort to identify nonpolluting renewable fuels.

Any alternative fuel cannot be arbitrarily chosen for use in both the SI and CI engines without any modification. This is because the combustion mechanisms in gasoline and diesel engines are different from each other. A more elaborate technical discussion in this aspect is not within the scope of the present level of conversation. However, hydrogen happens to be the appropriate substitute fuel to be used in SI engines. Hydrogen possesses the properties that if appropriately used can provide a lasting solution to both energy crisis (due to starvation of fossil fuels) and environmental degradation (caused by the pollutants coming out of the exhaust of vehicles). It is important to have a closer look at the physico-chemical characteristics of hydrogen vis-a-vis gasoline as given earlier.

The effects of some of the significant properties that affect the engine design and operating condition are discussed here. Gasoline and diesel petroleum fuels remain in liquid form at room temperature whereas hydrogen in form of a gas even at a much lower temperature of $-253°C$ heat of combustion of hydrogen fuel per unit mass is relatively much higher than gasoline. But it has a much lower mass density. This property results in relatively lower output of a hydrogen engine.

Minimum Ignition Energy as a Function of Equivalence Ratio for Hydrogen and Methane

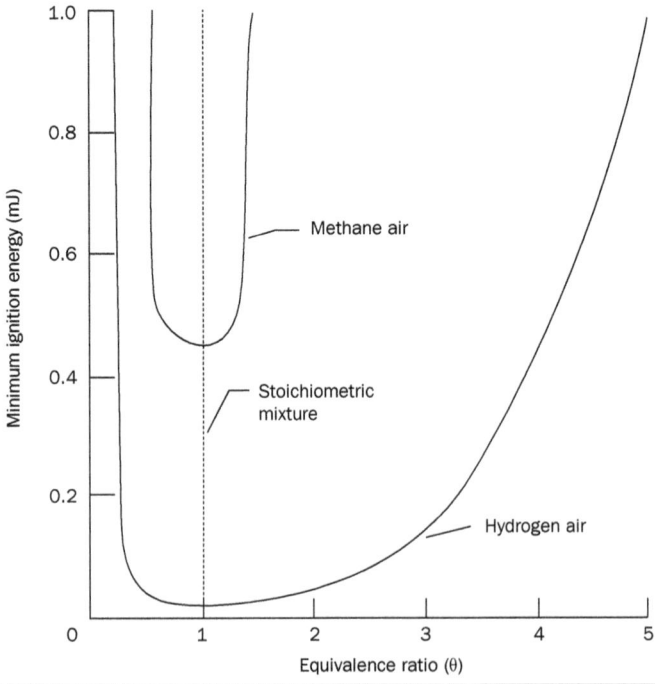

FIGURE 10.27　Minimum ignition energy of hydrogen and methane

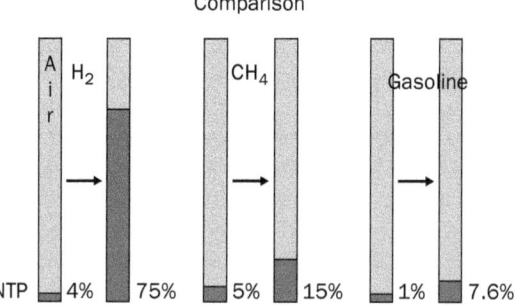

FIGURE **10.28** Flammability range of hydrogen as compared to other fuels

The critical combustion-related properties such as wider flammability range, flame speed, auto ignition temperature stoichiometric fuel/air ratio, quenching distance, and minimum ignition energy need to be closely looked into while developing a hydrogen engine. Figure 10.28 represents a comparative picture of minimum ignition energy of hydrogen with methane. It may be relevant to mention here that compressed natural gas (CNG) with methane as its major combustible component is being used as a regular transportation fuel in several countries. The minimum ignition energy has some significance when hydrogen is considered as a transportation fuel.

Figure 10.28 shows the flammability range of hydrogen with respect to methane and gasoline. As can be observed this figure, flammable range of hydrogen in air is about 10 times that of methane and gasoline. This feature makes it possible that a relatively much lesser amount of hydrogen is required as compared to gasoline or CNG to operate an engine or vehicle. Equivalence ratio is a crucial governing parameter for the hydrogen engine because hydrogen in an engine can sustain combustion from a low equivalence ratio (ultra-lean mixture) to a mixture strength above the stoichiometric (Fig. 10.29). This versatile feature of hydrogen is beneficial for engine application. By way of selecting the appropriate window of equivalence ratio for engine/vehicle operation performance, exhaust emission and combustion characteristics can be optimized.

A closer look at the hydrogen engine evolution technology indicates it is not a radically new concept. In 1829 Rev. W. Cecil presented his paper titled "On Hydrogen Gas as a Moving Power in Machinery" before the Cambridge Philosophical Society. Subsequently. Rudolf Erren, Weil, Oehmichen, Ricardo, King, and his coworkers had carried out extended experimental activities from which important information about improves performance as well as the problem of backfire were reported. As it appears, hydrogen engine research suffered a setback and R&S activities because of plentiful supply of petroleum fuel along with the disastrous consequence of backfire. It has been reported in literature that the severity of backfire resulted ranged from simple misfire to total destruction of test rig. However these earlier tests had clearly brought out the improved performance and lower emission features of the hydrogen engine. After a gap of several decades, hydrogen again received attention in the 1970s when major threats from fossil fuel depletion and environmental degradation were realized. Many researchers from academia, industries, and research organizations all over the world started hydrogen engine work with a renewed spate of interest. Detailed discussion on all these aspects is beyond the scope of this chapter. However, relevant information is available in several research papers [47–57].

Backfire has been the major stumbling block to the development and growth of the hydrogen engine. Backfire or flashback is a preignition

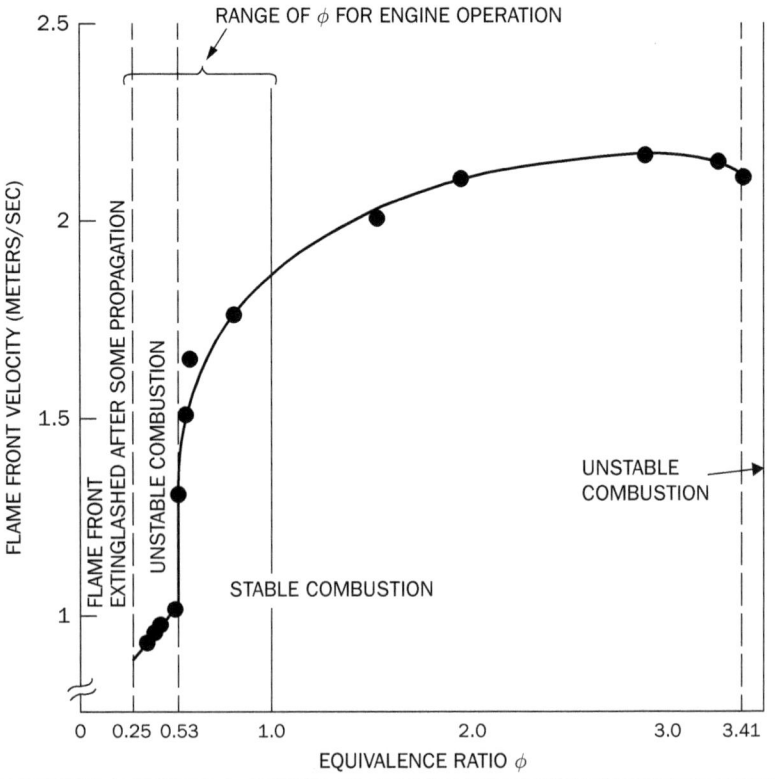

Figure 10.29   Range of equivalence ratio for stable combustion in engines

Mixture formation	Classification	Hydrogen flow timing	Supply pressure
Continuous carburetion (CC)	Pre-IVC	Continuous flow	A little above atmospheric
Continuous manifold injection (CMI)	Pre-IVC	Continuous flow	Slightly greater than atmospheric
Timed manifold injection (TMI)	Pre-IVC	Hydrogen flow commences after opening of the intake valve but completed prior to intake valve closure	$1.4 \pm 5.5 \ \text{kgf/cm}^2$
Low pressure direct cylinder injection (LPDI)	Pre-IVC	Hydrogen flow commences after the intake valve closure and is completed before significant compression pressure rise	$2.0 \pm 8.0 \ \text{kgf/cm}^2$

TABLE 10.7   Fuel induction techniques

problem that occurs during the intake stroke, when the inlet fuel-air mixture comes in contact with a hot spot (source of high thermal energy of sufficient magnitude or intensity) to initiate combustion when the valve is open. The causes of backfire are due to presence of hot particles/deposits in the cylinder, hot spots on the cylinder walls and spark plugs, cross talk between spark plug wires, and several other situations. Lean operation, exhaust gas recirculation is useful in controlling the frequency and severity of backfire. However, an appropriate fuel induction technique based on the typical combustion characteristics of hydrogen can eliminate the problem of backfire and other undesirable combustion phenomena. Various modes of fuel induction technique [58] were experimented in the Indian Institute of Technology to adopt the most reliable and conveniently implementable system to get rid of backfire. The flow timing and the supply pressures are critical parameters on the basis of which an injection system can be built for the hydrogen engine. Flow timing can be with respect to inlet valve closure, either before IVC (pre-IVC) or after IVC (post-IVC). Table 10.7 shows the flow timing and supply pressure ranges investigated in an exhaustive series of tests [58].

The benefits of different configurations of several modes of fuel injection have been investigated by several researchers [59–63]. High pressure direct injection (HPDI) at the end of compression stroke puts the injector to severe thermal environment in terms of pressure and temperature. Moreover late injection offers much lesser time for mixing and it is difficult to get a homogeneous mixture. Timed manifold injection that offers the flexibility of initiating the fuel delivery sometimes after the opening of intake valve and gets it completed prior to closure of the intake valve [64].

The overall performance characteristics of a hydrogen engine show that it can have much higher thermal efficiency because the engine can run on a relatively much higher compression ratio than a conventional gasoline engine. However, the intrinsic feature of lower output can be compensated to a large extent by adopting direct hydrogen injection and with super charging. Figures 10.30 and 10.31 show the performance characteristics (brake specific fuel consumption and thermal efficiency) of a hydrogen engine that was operated at different ranges of compression ratio under a wider range of equivalence ratio.

As far as the exhaust emission characteristics of hydrogen engine are concerned, it is perhaps relevant to mention here that there is practically

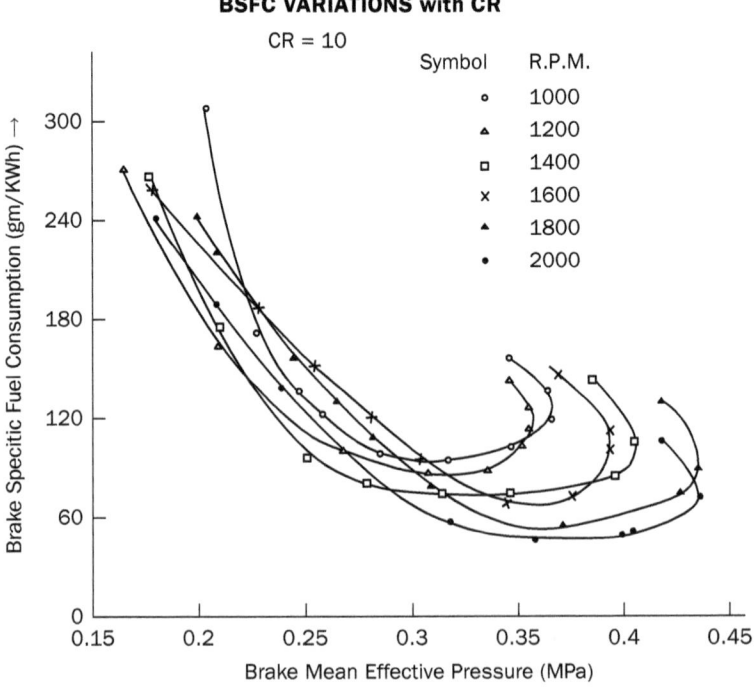

Figure 10.30 Effect of compression ratio on brake specific fuel consumption

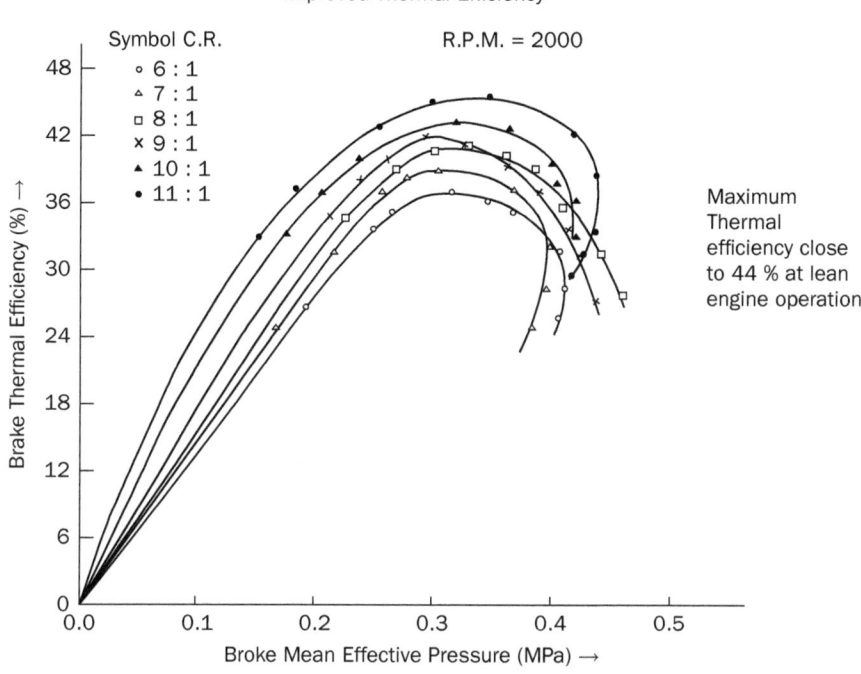

Improved Thermal Efficiency

Symbol	C.R.
○	6 : 1
▵	7 : 1
□	8 : 1
×	9 : 1
▲	10 : 1
●	11 : 1

R.P.M. = 2000

Maximum Thermal efficiency close to 44 % at lean engine operation

FIGURE 10.31 Brake thermal efficiency at various compression ratio

no CO, CO_2, HC, smoke, SOx, lead or other toxic metals, ozone, benzene, and formaldehyde. Sometimes traces of CO and HC are found due to lube oil oxides of nitrogen, which happen to be the major pollutant from hydrogen engine exhaust. Tests have shown that water injection, exhaust gas recirculation, and engine operation in lean mode have been effective in reducing NOx level at an equivalence ratio below 0.6. The lean operation creates an environment of lower temperature accompanied by the slower chemical reactions that weaken the kinetics of NOx formation [65]. Lean operation can have an engine configuration close to zero emission features as shown in Figure 10.32.

Hydrogen–oxygen reaction mechanism has pronounced influence in hydrogen combustion [66]. Lots of information is available with respect to rocket and some other space applications. Hydrogen combustion in engine plays a critical role. In engine combustion detonation and knocking occur depending on the engine configuration. However, it has been reported in literature that backfire happens to be the most dominant phenomenon. As has been discussed earlier, backfire in engine plays a critical role. As discussed earlier, backfire has been the Achilles' heel to growth of engine technology and its subsequent penetration into

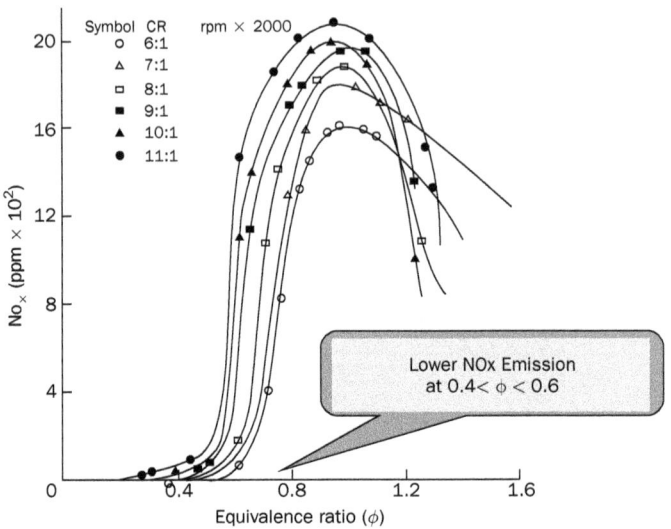

FIGURE **10.32** Variation of oxides of nitrogen with equivalence ratio

the field causes of backfire and effective methods to control them have been discussed [55, 58, 60, 67, 68].

In view of a set of properties of hydrogen being different from conventional fuels, it is important to have closer look at the combustion features. In the initial stages of hydrogen engine research, it was conclusively realized that the effect of the fuel's combustion features needs to be closely monitored so as to avoid any undesirable combustion phenomena; several studies were carried out to study the significance of hydrogen combustion [69, 70, 71], which have provided ample valuable information for subsequent research. The discussion in this chapter is limited to engine combustion noise, maximum pressure, and rate of pressure rise, important parameters for smooth combustion in a hydrogen engine. It has been reported that it is possible to achieve smooth combustion without high rate of pressure rise in a TMI-operated hydrogen engine [63, 65] as has been shown in the typical pressure crank angle diagram (Fig. 10.33).

10.12   Hydrogen Use in CI Engines

Hydrogen, in view of its combustion-related properties, is ideally suited for application in the spark ignition engines. Diesel is being used in many developing countries in transport sector (cars, buses), mechanized agricultural system (tractors, pumping irrigation), and decentralized energy application. Hydrogen use in diesel engines has been explored by several investigators at different phases of engine research [72–77].

PRESSURE CRANK ANGLE DIAGRAM-H_2

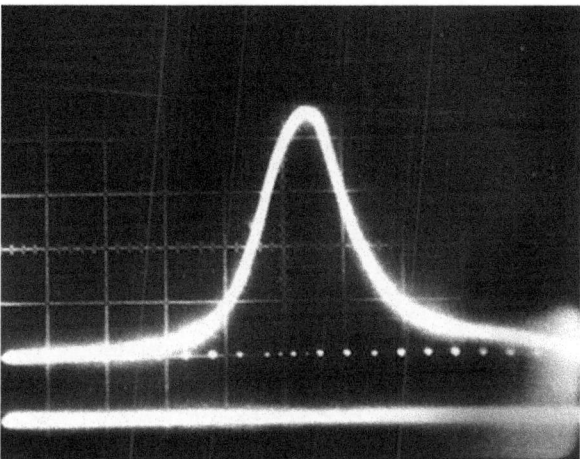

FIGURE 10.33 Pressure crank angle diagram of a hydrogen engine

Auto ignition of hydrogen is about 576°C and therefore, compression ignition of hydrogen is difficult even at a much higher compression ratio. This experience has been encountered by several instigators. However, some researchers have been successful in using triggering devices such as spark plug and glow plug to run diesel engine on hydrogen. Ikegami and co-workers [72, 73] had carried out exhaustive tests on hydrogen engines using pilot injection. They had also observed that even leakages of hydrogen from the injector could cause ignition in a diesel engine.

Therefore, the route of hydrogen utilization in diesel engines must be by way of dual-fuelling. In such systems, one of the major constraints is that higher percentage of hydrogen substitution under full load conditions that give rise to rapid rate of pressure and severe knocking. After a series of exhaustive tests have been experimentally established that use of diluents (inert gas such as helium, argon, and nitrogen) as well as water can be effective in preventing conditions leading to high pressure rate.

Abnormal conditions of knocking have been successfully reduced in a hydrogen-diesel dual fuel engine by this technique [77–79]. Figure 10.34 shows the performance and emission characteristics of a diesel engine in dual fuel mode with different amounts of hydrogen substitution. These data were generated keeping in mind the knocking condition of the engine.

Figure 10.35 shows optimum amount of each diluent that has proven effective in controlling knocking conditions in the dual fuel engines. Water vapor only up to 2640 ppm has been able to achieve full-load energy substitution of about 66 percent [78].

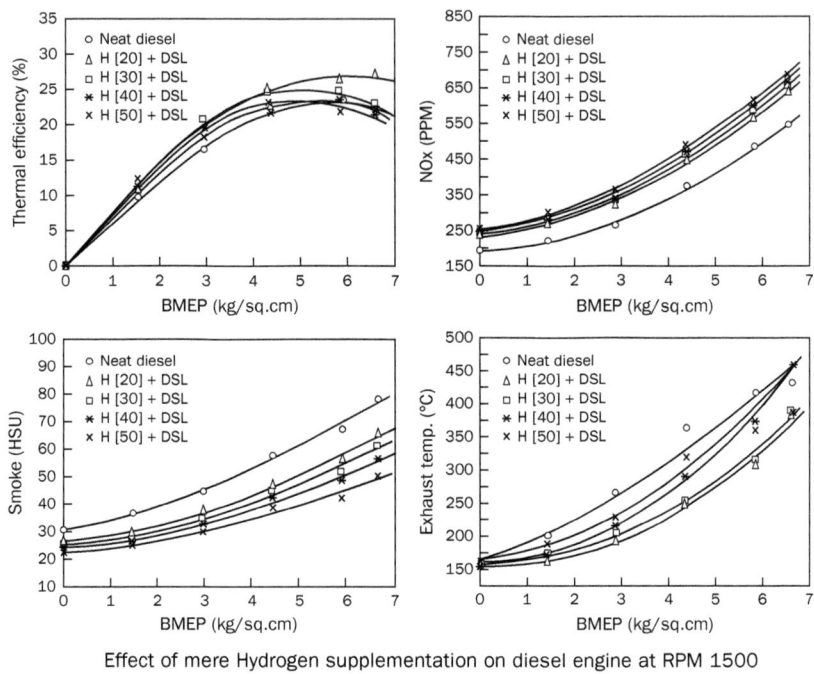

Effect of mere Hydrogen supplementation on diesel engine at RPM 1500

FIGURE 10.34   Performance and emission features of a hydrogen diesel dual fuel engine

Effect of Diluents

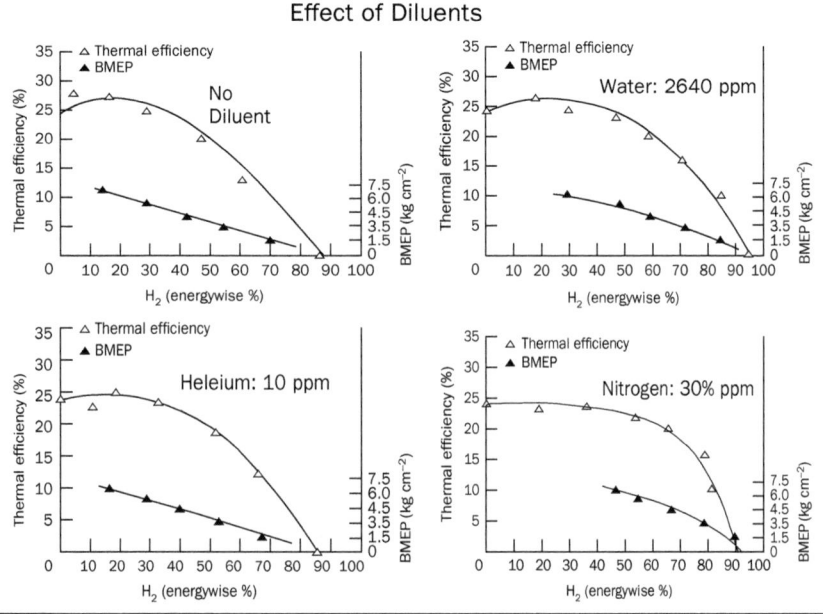

FIGURE 10.35   Effect of various diluents on performance parameters

The use of hydrogen in combination with several other fuels are being actively pursued in several parts of the world. Most prominent among the other fuels has been compressed natural gas. Many researchers in several parts of the world have been working on hydrogen blended to CNG (HCNG). Combustion and emission characteristics have improved substantially in HCNG operated engine and vehicles. However, a detailed discussion on the subject is beyond the scope of this chapter. Relevant and useful information is available in literature [80–86]. Hydrogen-blended CNG fuelled automobiles vehicle with various proportions of hydrogen blended with CNG have been very widely demonstrated in Vancouver (Canada), Washington D.C., Pennsylvania, Phoenix, Las Vegas (USA), New Delhi (India), and several other places.

In addition to CNG, a host of other fuels such as ethanol, LPG, biodiesel, DME, and biogas are also being tested in a laboratory scale, in combination with hydrogen. There are several other applications also in turbine, marine application, and railways in which hydrogen has been successfully used and efforts are being made to have wider applications of such systems.

Apart from the regular configuration of the SI and the CI engines as discussed above, hydrogen is also being investigated for use in homogeneous charge compression ignition (HCCI) engines. This is also a useful technique to have lean operation of hydrogen engine and bring down the NOx level emissions. High level of fuel efficiency (around 45 percent) has been reported by researchers. However, HCCI operation with a hydrogen engine is often restricted at higher loads because of maximum cylinder pressure and high rate of pressure rise. A detailed discussion on this topic has not been completed in this chapter.

10.13   Hydrogen-Fueled Vehicles

As far as the use of hydrogen in vehicles is concerned, there are two routes: fuel cell and internal combustion engines. Fuel cell is the most recent technology and several configurations of fuel cell vehicles have been developed and road tested. However, the technology of hydrogen vehicle development based on internal combustion engines has been a matter of interest spanning over a century. This is primarily because the IC engines have proven to be a versatile age-old prime mover used in vehicles.

Apart from researchers and industries all over the world, several automobile manufacturers have also taken interest and developed dedicated hydrogen-fuelled vehicles based on internal combustion engines [87–91]. Indian Institute of Technology (IIT), Delhi, in collaboration with Mahindra and Mahindra and Air products (USA) had developed a typical three-wheeler operated on hydrogen. This two-phase multiyear project was initially funded by UNIDO and subsequently continued by Ministry of New and Renewable Energy, Government of India. The typical three-wheeler vehicular engine of Mahindra and Mahindra was

FIGURE 10.36 Hydrogen fuel dispensing station

FIGURE 10.37 Bank of hydrogen gas cylinders

exhaustively tested in the laboratory in the Indian Institute of Technology, Delhi, and the operating parameters were optimized to ensure good performance and least emission characteristics of the vehicular engine. Adequate preventive steps care was taken and appropriate safety features were adopted to avoid any abnormal combustion symptoms such

Figure 10.38 Hydrogen fuelled three-wheeler

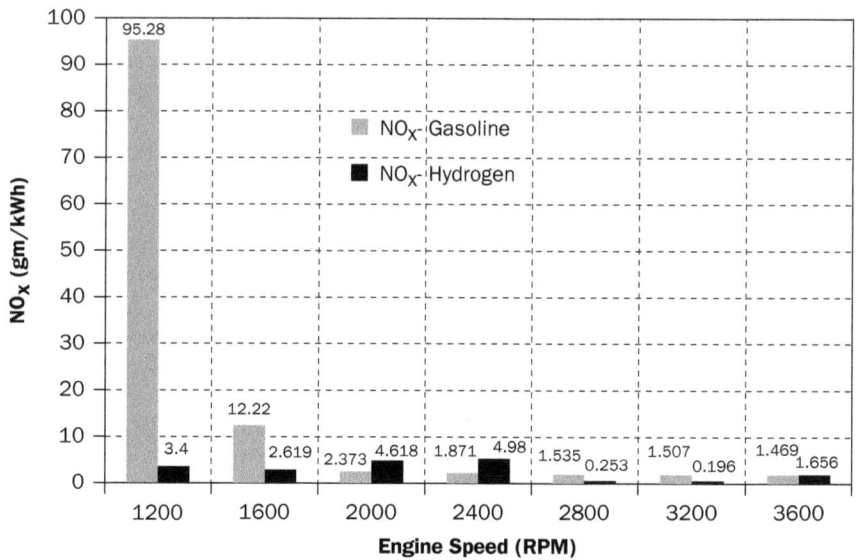

Figure 10.39 NOx emissions in mass concentration level

Figure **10.40**   Hydrogen-fuelled minibus **(Hydrogen bus developed by IIT Delhi and Mahindra & Mahindra, sponsored by MNRE)**

as backfire, high maximum pressure, and rapid rate of pressure rise. These optimized operating conditions generated during lab tests were subsequently transferred to the prototype vehicle that was tested in Pragati Maidan in New Delhi for long-term field tests. Such field operation for several years continuously has established the practical feasibility of hydrogen as a fuel for these types of vehicle. Figures 10.36 to 10.38 show the hydrogen fuelling system, the hydrogen cylinders, and the hydrogen vehicle [92]. Some results of the field tests show that NOx emission from these vehicles are negligible.

It has already been described that NOx is the only pollutant of concern in a hydrogen engine/vehicle. There are several techniques for controlling the NOx level. Figure 10.39 gives a comparative picture of NOx concentration level from the three-wheeler engine. As is evident from the figure NOx level is practically negligible in these hydrogen fuelled vehicles. Technical details of these tests on three wheelers have been encouraging [93].

The next step of practical vehicle development by the team (IIT Delhi and Mahindra and Mahindra) was taken up to develop a hydrogen operated minibus. The major objective was to develop the relevant technology for a vehicle with a multicylinder engine. These activities were also funded by MNRE. In the same sequence of technical activities, the typical minibus engine was tested in the laboratory in IIT under a wide operating range. Engine operating conditions for good performance and low emission characteristics were generated. Steps were taken in the test rig development to eliminate any possibility of backfire and other associated abnormal combustion phenomena. A specially designed timed manifold injection system was installed in the engine. Exhaust gas recirculation (EGR), retarded spark timing, and installation of catalytic converter were used to ensure smooth engine operation and control the exhaust emission of NOx.

This optimized configuration of the engine evolved in the lab after exhaustive tests has been installed in the prototype vehicle of Mahindra and Mahindra. The vehicle shown in Figure 10.40 is still being road tested. The test so far has given enough proof of hydrogen being used as a low-emission fuel for vehicle for public transportation system.

10.14   Conclusion

Hydrogen energy has the unique ability to save the planet from the devastating effects of environmental degradation. With an infinite source potential, it also has the distinctive feature to cater to the urgency created by the depletion of fossil fuels. An integrated study on all aspects of hydrogen energy starting from production, transmission and delivery, storage and usage is needed.

References

1. J. Bockris, T. N. Veziroglu, and D. Smith. *Solar Hydrogen Energy: the power to save the Earth.* London, UK: Optima, 1991.
2. J. Bockris and T. N. Veziroglu. A solar hydrogen energy system for environmental compatibility. *Environ. Conserv* 12(2), 105–115, 1985.
3. T. N. Veziroglu and F. Barbir. Hydrogen: The wonder fuel. *Int. J. Hydrogen Energy* 17(6), 391–404, 1992.
4. V. Agarwal, S. Patel, and K. K. Pant. H_2 production by steam reforming of methanol over $Cu/ZnO/Al_2O_3$ catalysts: Transient deactivation kinetics modeling. *Applied Catalysis A: General* 279(1–2), 155–164, 2005
5. A. Mastalir, B. Frank, A. Szizybalski, H. Soerijanto, A. Deshpande, M. Niederberger, R. Schomäcker, R. Schlögl, and T. Ressler. Steam reforming of methanol over $Cu/ZrO_2/CeO_2$ catalysts: A kinetic study. *Journal of Catalysis* 230(2), 464–475, 2005.
6. J. Hagen. Industrial Catalysis—*A Practical Approach.* 2nd ed., Weinhiem, Germany: Wiley-VCH Verlag GmbH & Co. KGaA, 2006.
7. G. Hoogers. *Fuel Cell Technology Handbook,* Boca Raton, FL: CRC Press, 2003.
8. G. J. Stiegel and M. Ramezan. Hydrogen from coal gasification: An economical pathway to a sustainable energy future. *International Journal of Coal Geology* 65, 173–190, 2006.

9. J. M. Beer. Combustion technology developments in power generation in response to environmental challenges. *Progress in Energy and Combustion Science* **26**, 301–327, 2000.

10. C. Higman and M. van der Burgt. *Gasification*, 2nd ed. Oxford, UK: Elsevier Science, 2008.

11. Rezaiyan and Cheremisinoff. *Gasification Technologies: A Primer for Engineers and Scientists*. Boca Raton, FL: CRC Press, 2005.

12. K. Nagaoka, T. Eboshi, Y. Takeishi, R. Tasaki, K. Honda, K. Imamura, K. Sato. Carbon-free H_2 production from ammonia triggered at room temperature with an acidic RuO2 /γ-Al2O3catalyst. *Science Advances* **3**(4), e1602747, 2017.

13. W. Rice. Hydrogen production from methane hydrate with sequestering of carbon dioxide. *International Journal of Hydrogen Energy* **31**(14), 1955–1963, 2006.

14. Havva Balat Elif Kırtay. Hydrogen from biomass: Present scenario and prospects. *International Journal of Hydrogen Energy* **35**, 7416–7426, 2010.

15. P. C. Hallenbeck and J. R. Benemann. Biological hydrogen production; fundamentals and limiting processes. *International Journal of Hydrogen Energy* **27**(11–12), 1185–1193, November–December 2002.

16. Y. Miura. Hydrogen production by biophotolysis based on microalgal photosynthesis. *Process Biochemistry* **30**(1), 1–7, 1995.

17. V. Utgikar and T. Thiesen. Life cycle assessment of high temperature electrolysis for hydrogen production via nuclear energy. *International Journal of Hydrogen Energy* **31**(7), 939–944, 2006.

18. K. Zeng and D. Zhang. Recent progress in alkaline water electrolysis for hydrogen production and applications. *Progress in Energy and Combustion Science* **36**, 307–326, 2010.

19. F. Barbir. PEM electrolysis for production of hydrogen from renewable energy sources. *Solar Energy* **78**(5), 661–669, May 2005.

20. S. Abanades and G. Flamant. Solar hydrogen production from the thermal splitting of methane in a high temperature solar chemical reactor. *Solar Energy* **80**(10), 1321–1332, October 2006.

21. A. Steinfeld. Solar hydrogen production via a two-step water-splitting thermochemical cycle based on Zn/ZnO redox reactions. *International Journal of Hydrogen Energy* **27**(6), 611–619, 2002.

22. M. Michael, K. H. Leung, and D. Y.C. Leung. Technological development of hydrogen production by solid oxide electrolyzer cell (SOEC). *International Journal of Hydrogen Energy* **33**(9), 2337–2354, May 2008.

23. G. Amikam, P. Nativ, and Y. Gendel. Chlorine-free alkaline seawater electrolysis for hydrogen production. *International Journal of Hydrogen Energy* **43**(13), 6504–6514, March 2018.

24. J. D. Holladay, J. Hu, D. L. King, and Y. Wang. An overview of hydrogen production technologies. *Catal Today* **139**(4), 244–260, 2009.

25. A. Demirbas. Hydrogen Sulfide from the Black Sea for Hydrogen Production. Part A: Recovery, *Utilisation and Environmental Effects* **31**(20), 1866–1872, 2009.
26. N. Kumar and D. Das. Continuous hydrogen production by immobilized Enterobacter cloacae IIT-BT 08 using lignocellulosic materials as solid matrices. *Enzyme and Microbial Technology* **29**(4–5), 280–287, September 2001.
27. J. Schwartz, H. Lim, and R. Drnevich. INTEGRATED CERAMIC MEMBRANE SYSTEM FOR HYDROGEN PRODUCTION Final Report for the Period February 2003–June 2010. Report PREPARED FOR THE UNITED STATES DEPARTMENT OF ENERGY Under Cooperative Agreement No. DE-FC36-00GO10534.
28. H. L. Chen, H. M. Lee, S. H. Chen, Y. Chao, and M. B. Chang. Review of plasma catalysis on hydrocarbon reforming for hydrogen production—Interaction, integration, and prospects. *Applied Catalysis B: Environmental* **85**(1–2), 1–9, December 2008.
29. B. Y. Mujid and S. Kazimi. Efficiency of hydrogen production systems using alternative nuclear energy technologies. *International Journal of Hydrogen Energy* **31**(1), 77–92, January 2006.
30. R. S. Irani. Hydrogen storage: High-pressure gas containment. *MRS Bull.* **27**(9), 680–682, 2002.
31. ISO 7866:1999(E), Gas cylinders—Refillable seamless aluminium alloy gas cylinders—Design, construction and testing, International Standards Organization, http://www.iso.org/iso/iso_catalogue/catalogue_tc/catalogue_detail.htm?csnumber=14795.
32. ISO 9809-1:1999(E), Gas cylinders—Refillable seamless steel gas cylinders—Design, construction and testing—Part 1: Quenched and tempered steel cylinders with tensile strength less than 1 100 MPa, International Standards Organization http://www.iso.org/iso/iso_catalogue/catalogue_tc/catalogue_detail.htm?csnumber=17683.
33. J. Wolf. Liquid-hydrogen technology for vehicles. *MRS Bull.* **27**(9), 684–687, 2002.
34. Y. M. Park. Literature research on the production, loading, flow, and heat transfer of slush hydrogen. *International Journal of Hydrogen Energy* **35**(23), 12993–13003, December 2010.
35. G. Sandrock. Applications of hydrides. In *Hydrogen Energy System—2 Production and Utilization of Hydrogen and Future Aspects*, Y. Yurum, ed., NATO ASI Series E 295, p. 253. Amsterdam, the Netherlands: Kluwer, 1995.
36. G. Sandrock. A panoramic overview of hydrogen storage alloys from a gas reaction point of view. *J. Alloys Compd.* 293–295:877–888, 1999.
37. R. C. Bowman Jr. and B. Fultz. Metal hydrides i: Hydrogen storage and other gas-phase applications. *MRS Bull.* **27**(9), 688–693, 2002.

38. G. Sandrock and R. C. Bowman Jr. Gas-based hydride applications: Recent progress and future needs. *J. Alloys Compd.* **356/357**, 794–799, 2003

39. J. Weitkamp, M. Fritz, and S. Ernst. Zeolites as Media for Hydrogen Storage. ht. J. *Hydrogen Energy* **20**(12), 967–970, 1995.

40. Federal Institute of Material and Testing. Hydrogen safety, German HYDROGEN ASSOCIATION, Brussels 18–21, 2002.

41. Colorant flame WWW

42. Fact Sheet Series No 1.008 HYDROGEN safety, National Hydrogen Association, Washington D.C.

43. A. Zuettel, A. Borgschulte, and L. Schlapbach (Eds.). *Hydrogen as a future Energy Carrier* Berlag, German y: Wiley-VCH, 2008, Chapter 4.

44. M. R. Swain. Proceedings of the 2001 DOE Hydrogen Energy Program Review, NREL/CP-570-30535, U S Dept of Energy, Washington DC.

45. L. Raymond, Ed. Hydrogen Embrittlement Testing, ASTM STP 543, ASTM, West Conshohocken, PA, 1972.

46. NACE Glossary of Corrosion-Related Terms. NACE International, 2002.

47. E. A. Rudolf and C. W. Hastings. Hydrogen: A commercial Fuel for Internal Combustion Engine and other Purposes. *Journal of the Institute of Fuel* **VI**(29), 277–290, 1933.

48. M. Oehmichen. Wasserstoff als Motortreibmittel. Deutsche Kraft Fahart-Forschung, Heft 68,VDI,Verlag GmbH, Berlin, 1942.

49. H. R. Ricardo. Further Note on Fuel Research. Proc. *Institute of Auto Engr.* **XVIII**(I) 327–341, 1924.

50. K. H. Weil. The Hydrogen IC Engine- Its Origin and Future in the Emerging Energy-Transportation-Environment Systems. *SAE* 729212, 1972.

51. W. J. D. Escher. Hydrogen-Fueled Internal Combustion Engine, A Technical Survey of Contemporary U.S Projects. *ETA Report* PR-51, 1975.

52. L. M. Das. Hydrogen Engines: A view of the past and look into the future. *International Journal of Hydrogen Energy* **15**(6) 425–433, 1990.

53. S. Furuhama. State of the art and future trend of Hydrogen Fuelled Engines. *JSAE Review* 4, 1981.

54. W. Cecil. On the Application of Hydrogen Gas to Produce a Moving Power in Machinery; with a Description of and Engine which is Moved by the pressure of Atmosphere upon a Vacum caused by Explosion of Hydrogen Gas. Cambridge Philosophical Society, Vol. I, 1822.

55. C. M. White, R. R. Steeper, and A. E. Lutz. The hydrogen-fueled internal combustion engine: a technical review. *International Journal of Hydrogen Energy* **31**, 1292–1305, 2006.

56. S. Verhelst and T. Wallner. Hydrogen-fueled internal combustion engines. *Progress in Energy and Combustion Science* **35**, 490–527, 2009.

57. S. Verhelst. Recent progress in the use of hydrogen as a fuel for internal combustion engines. *International Journal of Hydrogen Energy* **39**, 1071–1085, 2014.

58. L. M. Das. Fuel induction techniques for hydrogen operated engine. *International Journal of Hydrogen Energy* **15**, 833–842, 1990.

59. R. Sirens. Influence of Injection Parameters on the Efficiency and Power Output of a Hydrogen Fuelled Engine, ICE Technical Conference, *ICE* **36**(1), 2001.

60. C. A. McCarley. A study of factors influencing Thermally induced Backfire in Hydrogen-fuelled Engines and Methods of Backfire Control. IECEC, 1981.

61. K. S. Varde and G. A. Frame. Development of a High Pressure Injection for SI Engines and Results of Engine Behaviour. 5th World Hydrogen Energy Conference, 1984.

62. K. S. Varde and G. A. Frame. A Study of Combustion and Engine Performance Using Electronically Hydrogen Fuel Injection. Paper presented at World Hydrogen Energy Conference-IV, Pasadena, California, 1982.

63. H. B. Mathur and L. M. Das. Performance Characteristics of Hydrogen Fuelled SI Engine Using Timed Manifold Injection. *International Journal of Hydrogen Energy* **16**, 115–127, 1991.

64. L. M. Das. Studies of timed manifold injection in hydrogen operated spark ignition engine: performance, combustion and exhaust emission characteristics. Ph.D. dissertation. Department of Mechanical Engineering Indian Institute of Technology, Delhi, India, 1986.

65. L. M. Das. Exhaust Emission Characterization of Hydrogen-Operated Engine System: Nature of Pollutants and their Control Techniques. *International Journal. Hydrogen Energy* **16**(11), 1991.

66. L. M. Das. Hydrogen Oxygen Reaction Mechanism and its Implication to Hydrogen Engine Combustion. *Intrnl. J. Hydrogen Energy* **21**(8), 1996.

67. K. Koyanagi, M. Hiruma, and S. Furuhama. Backfire in Hydrogen engines. *SAE Special Publication* **10**.4, 271/942035.

68. L. M. Das. Abnormal Combustion in Hydrogen Engine: causes and Remedies. *Hydrogen Energy Progress*, Proceeding of the 8th World Hydrogen Energy Conference, **3**, 1379–1397, 1990.

69. R. O. King and M. Rand. The Oxidation, Decomposition, Ignition and Detonation of Fuel Vapours and Gases XXVII, The Hydrogen Engine. *Canadian Journal of Technology* **33**, 1955.

70. R. O. King, W. A. Wallace, and B. Mahapatra. The Oxidation, Decomposition, Ignition and Detonation of Fuel Vapours and Gases V, The Hydrogen Engine and Detonation of the End by the Igniting Effect of Carbon Nuclei Formed by Pyrolysis of Lubricating Oil Vapour. *Canadian Journal of Technology* **34**, 1957.

71. R. O. King, S. V. Hayes, A. B. Allan, R. W. P. Anderson, and B. J. Wacker. The Hydrogen Engine: Combustion Knock and Related Flame Velocity. *Trans. Eng. Inst. Canada* (EIC) **2**(4), 1958.

72. M. Ikegami et al. A study on hydrogen-fuelled diesel combustion. *Bulletin of the JSME* **23**(181), 1980.

73. M. Ikegami. A study of hydrogen-fuelled compression ignition engine. *International J of Hydrogen Energy* **7**(4), 1982.

74. H. S. Homan. An Experimental Study of reciprocating Internal Combustion Engines Operated on Hydrogen. Ph.D. Thesis, Cornell University, 1978.

75. A. B. Welch and J. S. Wallace. Performance characteristics of a hydrogen fuelled diesel engine with ignition assistance. *Final Report*, NRCC, Report No.DSS, Contract File No.24SU.31155-2-2664. Serial No. ISU 82-00340, 1986.

76. S. Stranislaw. Hydrogen Combustion in a Compression Ignition Diesel Engines, *International J of Hydrogen Energy* **34**, 2009.

77. H. B. Mathur, L. M. Das, and T. N. Patro. Hydrogen fuel utilization in CI engine powered end utility systems. *International Journal of Hydrogen Energy* **17**, 369–374, 1992.

78. T. N. Patro. Utilization of Hydrogen fuel in small utility CI engine and its effect on Engine Performance, emission and combustion characteristics. Ph.D. Thesis, Indian Institute of Technology, Delhi, New Delhi, India, 1990.

79. H. B. Mathur, L. M. Das, and T. N. Patro. Hydrogen fuelled diesel engines: Performance improvement through charge dilution techniques. *International Journal of Hydrogen Energy* **18**, 421–431, 1993.

80. L. M. Das and M. Polly. Experimental evaluation of a Hydrogen Added Natural Gas (`HANG) Operated SI engine. *SAE Paper* No 2005-26-29, January 2005.

81. S. S. Sandhu, M. K. G. Babu, and L. M. Das. Effect of hydrogen supplementation on the performance, combustion and emission characteristics of a natural gas fuelled S.I. engine. *Int. J. Alternative Propulsion* **2**(2), 181–196, 2012.

82. S. S. Sandhu, M. K. G. Babu, and L. M. Das. Investigations of emission characteristics and thermal efficiency in a spark-ignition engine fuelled with natural gas–hydrogen blends. *International Journal of Low Carbon Technologies* **6**(4), December 2011.

83. R. Mathai, R. K. Malhotra, K. A. Subramanian, and L. M. Das. Comparative evaluation of performance, emission, lubricant and deposit characteristics of spark ignition engine fuelled with CNG and 18% hydrogen-CNG. *International Journal of Hydrogen Energy* **37**, 6893–6900, 2012.

84. S. S. Sandhu, M. K. G. Babu, and L. M. Das. Investigations of emission characteristics and thermal efficiency in a spark-ignition engine fuelled with natural gas–hydrogen blends. *International Journal of Low Carbon Technologies* **6**(4), December 2011.

85. K. Duk Sang. Development of hydrogen-compressed natural gas blend engine for heavy duty vehicles. *International Journal of Automotive Technology* **18**(6), 1061–1066, December 2017.

86. R. Mehra, H. Duan, R. Juknelevicius, and F. Ma. Progress in hydrogen enriched compressed natural gas (HCNG) internal combustion engines - A comprehensive review. *Renewable and Sustainable Energy Review* **80**, 1458–1498, 2017.

87. Lapez et al. The design, development,validation and delivery of the Ford H2 ICE 450 Shuttle Bus Symposium on Hydrogen Internal Combustion Engine pp. 20–33, BMW Hydrogen 8 Rescue guidelines, 2006.

88. T. Wallner et al. Fuel economy and emission evaluation of a BMW hydrogen 7 monofuel demonstration vehicle. *Int. J. Hydrog. Energy* **33**, 7607–7618, 2008. http://dx.doi.org/ 10.1016/j.ijhydene.2008.08.067

89. R. Gopalkrishnan, M. Throop, et al. Engineering the Ford H2 IC Engine Powered E-450 Shuttle Bus. *SAE Technical Paper* 2007-01-4095.

90. S. Jilakara, S. Saravanan, K. Jayakrishnan, L. M. Das, et al. An Experimental Study of Turbocharged Hydrogen Fuelled Internal Combustion Engine. *SAE International Journal of Engines* 8(1), 314–325, January 2015.

91. L. M. Das. Hydrogen-fuelled Internal Combustion Engine, *Compendium of Hydrogen Energy* Chapter 7, Volume 3. Hydrogen Energy Conversion edited by F. Babir, A. Basile, and T. Nejat. Elsevier Publishing, pages 177–216.

92. S. Natarajan, M. Abraham, M. Rajesh, G. Subash, and L. M. Das. DelHy 3w-Hydrogen-fuelled HYalfa Three wheelers, *SAE Technical Paper* 2013-01-0224, 2013.

Questions

1. What is the lower flammability limit of hydrogen?

   a. 75%

   b. 24%

   c. 4%

2. Which of the following method is used for ammonia production?

   a. heating nitrogen and mixing with hydrogen

   b. reducing nitrogen catalytically by hydrogen

   c. hydrogen and nitrogen subjected to high pressure for a long time

3. Which of the following method is adopted for commercial hydrogen production?

   a. coal gasification and steam reforming of natural gas

   b. thermal decomposition of methane

   c. partial oxidation of hydrocarbons

   d. any of the above

4. There are some methods suggested below to use sunlight for hydrogen production without conversion to electric. With which of these do you agree?

   a. by way of using concentrated energy to reach higher temperature level, say above 2000°C and adopt thermochemical cycles to produce hydrogen

   b. using concentrated solar energy to generate very high temperature to split methane

   c. by the use of photoelectrochemical materials based on nanotech

   d. any of the above

5. Indicate which flame is least visible from among the following:

   a. methane

   b. propane

   c. ethyl alcohol (d) hydrogen

6. Which of the following is the most toxic?

   a. gasoline

   b. diesel

   c. methane

   d. hydrogen

7. In what form does hydrogen usually exist?

   a. gas

   b. component of compounds

   c. individual atoms

   d. diatomic molecules

8. What are the physical states in which hydrogen can exist?

   a. solid

   b. liquid

   c. gaseous

   d. plasma

   e. all of the above

9. What material is used for storage of compressed hydrogen gas at high pressure?

   a. steel cylinder

   b. titanium tube welded at the ends

   c. plastic cylinder

   d. aluminum core encapsulated in carbon fiber

10. Identify the most critical problem for hydrogen distribution system in liquid or gaseous form.

    a. compression

    b. liquefaction

    c. storage

    d. permeability

    e. pipelines

11. a. Explain the difference between photovoltaic and photocatalytic modes of hydrogen production.

    b. Explain major advantages of storing hydrogen in compressed gas form.

    c. Discuss the main disadvantages of storing hydrogen in liquid for a long period.

    d. Find out the approximate specific energy ratio for liquid hydrogen and gasoline at normal conditions of temperature and pressure.

    e. Compare the highest gravimetric density (hydrogen mass to the mass of storage) of the different modes of hydrogen storage types: compressed hydrogen storage, liquid hydrogen storage, and reversible metal hydride storage.

    f. In an open space, hydrogen flame is of lesser threat as compared to a flame from gasoline combustion. Do you agree? Why/Why not?

12. Evaluate the stoichiometric air fuel ratio for hydrogen combustion with air.

Relevant Definition of Energy/Work Units

Btu British thermal unit. Heat energy necessary to raise the temperature of 1 lb of water 1°F.

cal or gcal Calorie or gram calorie. Heat energy required to raise the temperature of 1 mL of water 1°C (from 15 to 16°C).

electron volt 1.6×10^{-12} erg $= 1.6 \times 10^{-19}$ J $= 23.06$ kcal/mol. Energy gained by an electron passing through a potential of 1 V.

foot · pound ft · lb. Work energy needed to raise 1 lb to a height of 1 ft $= 0.138$ kg · m.

force Correct force definition can be obtained from the second law of Newton stating that the inertia is disturbed by unbalanced force, which causes acceleration on a body directly proportional to the force (F) and inversely proportional to the mass of the body $F = K\, mf$ ($F = Kmf$ where m is mass and f is acceleration). If all are reduced to unity, the unit of force becomes pound foot per second per second (poundal) or gram centimeter per second per second (dyne).

joule Work energy to raise 1 kg to a height of 10 cm $= 0.1$ kg · m $= 0.74$ ft · lb.

joule (electrical) 0.239 cal. Energy developed when 1 C of electrons (10.364×10^{-6} mole) passes through a potential of 1 V.

kcal/einstein Energy of a mole of a photon (einstein) of wavelength (in μ) 28589.7 μ^{-1} kcal/mol.

power Rate of doing work, $P = W/t$.

$$J/s = W \text{ or ft} \cdot lb/s$$

$$\text{or horsepower} = 550 \text{ ft} \cdot lb/s \text{ or } 33{,}000 \text{ ft} \cdot lb/min.$$

So, $1\,W = 10^7\,ergs/s = 1\,J/s = 0.239\,cal/s$

$1\,hp = 550\,ft \cdot lb/s = 33{,}000\,ft \cdot lb/min$

$1\,hp = 746\,W = 178\,cal/s$

$1\,kW = 1000\,W = 1.34\,hp$

$1\,kWh = 3.6 \times 10^6\,J = 860\,kcal/h = 3413\,Btu/h$

$1\,ft \cdot lb/s = 1.356\,W = 0.324\,cal$

quantum   Wavelength $\mu eV = 1239.8\,\mu^{-1}$

Wave number $cm^{-1}\,eV = 1239.8 \times 10^{-4}\,cm^{-1}$

units $1\,erg = 1\,dyne \times 1\,cm\;(dyne = 1\,g \cdot cm\,s^{-2})$

$1\,J = 10^7\,ergs = 0.74\,ft \cdot lb = 0.239\,cal$

$1\,ft \cdot lb = 1.3549\,J$

Problems
and Answers

1. What is aviation gasoline?

Aviation gasoline is a complex mixture of relatively volatile hydro-carbons with small additives blended to form a fuel suitable for use in aviation reciprocating engines. Fuel specifications are provided in ASTM Specification D910 where naphtha is generally used for blend-ing or compounding into finished aviation gasoline.

2. What is global warming and solar greenhouse?

Global warming indicates the increase of average temperature in the earth, which may be due to increase of carbon dioxide and other greenhouse gases released by the burning of fossil fuels and other human activities contributed in the earth.

Solar greenhouse is an enclosed space that provides the required environment for growth and production of plants under adverse cli-matic conditions. Its design for plant growth includes light intensity, temperature, humidity, and air movement.

3. Discuss classification of solar energy systems.

Solar energy systems are classified as follows:

1. Thermal chemical storage

2. Thermal energy storage may be of two types:

(i) Sensible heat storage

(a) Solids (b) Liquids (c) Solar ponds (d) Absorbents

(ii) Latent heat storage

3. Electric storage

(a) Battery storage   (b) Capacitor storage   (c) Inductor storage

4. Electromagnetic storage

5. Mechanical energy storage

$\downarrow$ $\downarrow$ $\downarrow$

(a) Pumped hydroelectric   (b) Compressed air (c) Fly wheel
     storage   storage   storage

4. What are solar ponds and solar furnaces?

Solar ponds are artificially designed ponds filled with salty water that maintains a definite constant concentration gradient. These ponds are used for cooling and storing solar energy. The limitation is that they require sunny climate, large area, availability of salt, and also availability of large amounts of water.

Solar furnaces are optical equipment to get high temperature by concentrating solar radiations onto a specimen. Their advantages are simple working, availability of high heat flux, heating without contamination, and easy control of temperature.

5. What is maximum efficiency of a solar cell?

Maximum efficiency (η_{max}) of a solar cell is defined as the ratio of electric power output to incident solar radiation:

$$\eta_{max} = \frac{V_{max}\, I_{mp}}{I_s\, A_s}$$

$V_{max}\, I_{mp}$ = Voltage and current maximum power output
I_s = Incident solar flux
A_s = Cell' area

6. What is barrel?

Barrel is a unit of volume equal to 42 U.S. gallons.

7. What is catalytic cracking?

The decomposition of alkanes at high temperature (500°C to 800°C) in the absence of air is called cracking. Cracking generally takes place in the presence of catalyst to make less temperature, which is called catalytic cracking.

$$Al_2O_3 + SiO_2 \text{ (catalyst)}$$

Kerosene (C_{10}–C_{16}) $\rightarrow$ Gasoline (C_4–C_7)

Boiling range   Boiling range

180–250°C   40–200°C

8. What is reforming?

Petrol is produced from gasoline fraction by reforming. This involves converting straight chain alkanes into branched alkanes and aromatic hydrocarbons. The two reforming processes are isomerization and aromatization.

a) Isomerization: In this process, straight chain alkanes (from gasoline fraction) are converted into branched chain isomers, which are carried in presence of aluminum chloride at 100°C and 50 atmospheric pressure as, for example, n-octane can be converted into iso-octane. Branched chains have higher octane number than straight chains.

$$CH_3(CH_2)_6\,CH_3 \quad \xrightarrow[\text{100°C/50 atm}]{AlCl_3} \quad CH_3-\underset{\underset{CH_3}{|}}{\overset{\overset{CH_3}{|}}{C}}-CH_2-\underset{\underset{CH_3}{|}}{CH_2}-CH_3$$

Octane   Isooctane

Octane number = - 19   Octane number = 100

b) Aromatization: In this process straight chain alkane (from naptha and gasoline fractions) are converted into aromatic hydrocarbons. The reaction is carried in the presence of platinum supported alumina at 500°C as, for example, n-hexane are converted into benzene.

$$CH_3\,CH_2\,CH_2\,CH_2\,CH_2\,CH_3 \quad \xrightarrow[\text{500°C}]{Pt/Al_2O_3} \quad \bigcirc$$

n- Hexane

(octane number = 25)   (Octane number = 106)

9. Which fuel is commonly available in both caking and non-caking forms?

Bituminous coal is the most commonly used form of coal, as it is available in both caking and non-caking forms.

10. What is LPG?

LPG is liquefied petroleum gas where a group of hydrocarbons-based gases (ethane, ethylene, propane, propylene, butane, isobutene, butylene, and iso butylene) derived from crude oil. For convenience of transportation these gases are liquefied through pressurization. LPG is perfectly colorless. Its properties are –42°C boiling point, –188°C freezing point, heavier than air density, flame temperature, 470°C auto ignition temperature, and –104°C flash point.

11. What is the function of a fuel cell? What are the output voltage of common fuel cells?

A fuel cell is used for conversion of chemical energy into electrical energy through a chemical reaction of positively charged hydrogen ions with oxygen or another oxidizing agent. The standard emf of the hydrogen-oxygen fuel cell is 1.23 V. This type of cell operates efficiently in the temperature range 343 K to 413 K. The output voltage of hydrogen-oxygen fuel cell, carbon-oxygen fuel cell, and methane-oxygen fuel cell is –1.23 V, –1.02 V, and –1.06 V.

12. How do microbial fuel cells generate power?

Microbes grow in microbial fuel cells and they can extract electrons from their food sources such as organic materials and feed them into an electrical circuit to generate power.

13. What is motor gasoline?

Motor gasoline is a complex mixture of volatile hydrocarbons with additives in small quantity to make the blended fuel suitable for engines. Motor gasoline is designated as ASTM specification D-4814, which has the characteristic boiling point range 122–158°C.

14. What is gas hole?

Gas hole is a blend of finished motor gasoline containing alcohol at a concentration of 10 percent or less by volume. Data have gas hole at least 2.7 percent oxygen by weight and is intended for sale inside carbon monoxide nonattainment areas.

15. What is heavy oil?

After distillation fuels remaining after the light oil distilled off during the refinery process are called heavy oil; virtually all petroleum used in stream electric power plants is heavy oil. Its density or specific gravity is higher than that of light crude oil. It is highly viscous oil that cannot easily flow to production wells under normal reservoir conditions. Heavy crude oil is generally categorized in two ways as first those having over 1 percent sulphur with aromatics and asphaltenes and second those having less than 1 percent sulphur with aromatics, naphthalene, and resins.

16. What is light oil?

Lighter fuel oils are distilled off first during the refining process. Light oil has low density that flows freely at room temperature. It has low viscosity, low specific gravity, and high API gravity due to the presence of a high proportion of light hydrocarbon fraction. All petroleum used in internal combustion and gas turbine engines is light oil.

17. What is low ash engine oil?

Low ash engine oil is an engine lubricant for older diesel and gasoline engines designated as a lower ash straight grade oil, which is used to protect against sludge, varnish, engine wear, acid accumulation, and oil consumption.

18. What are the essential parts of a diesel engine? What are the functions of diesel engines?

a) Engine b) Engine starting system c) Fuel system d) Air intake and exhaust systems e) Cooling system f) Lubricating system g) Governing system

A diesel engine is an internal combustion engine in which ignition of the fuel is caused by the elevated temperature of the air in the cylinder due to the mechanical compression (adiabatic compression).

19. What are the classifications of internal combustion engines? Discuss differences between petrol and diesel engines.

The internal combustion engine has been classified according to cycle of operation, cycle of combustion, mode of operation, uses, fuel employed, speed of engine, method of cooling the system and governing system, and lastly by valve arrangement.

a) Cycle of operation

(i) Two-stroke cycle engine (ii) Four-stroke cycle engine

b) Cycle of combustion

(i) Combustion at constant volume (Otto cycle) (ii) Combustion at constant pressure (Diesel cycle engine) (iii) Duel combustion (combustion partly at constant volume and partly at constant pressure)

c) Arrangement of cylinder

(i) Horizontal engine (ii) Vertical engine (iii) V-type engine (iv) Radical engine

d) According to their uses

(i) Stationary engine (ii) Portable engine (iii) Marine engine (iv) Automobile engine (v) Aeronautical engine

d) Fuel supply

(i) Oil engine (ii) Petrol engine (iii) Gas engine (iv) Kerosene engine

e) According to speed of engine

(i) Low speed engine (ii) High speed engine (iii) Medium speed engine

f) Methods of ignition

(i) Spark ignition (ii) Compressed ignition

g) Methods of cooling the system

(i) Air-cooled engine (ii) Water-cooled engine

h) Methods of governing

(i) Hit and miss governed engine (ii) Quality governed engine (iii) Quantity governed engine

i) Methods of valve arrangement

(i) Over head valve engine (ii) L-head type engine (iii) T-head type engine (iv) F-head type engine

k) Number of cylinders

(i) Single-cylinder engine (ii) Multi-cylinder engines

Differences between petrol and diesel engines

Petrol engine	Diesel engine
1. It works on Otto cycle.	1. It works on diesel cycle.
2. Air petrol mixture is sucked in the engine cylinder during suction stroke.	2. Only air is sucked during suction stroke.
3. Spark plug is used.	3. Employs an injector.
4. Power is produced by spark ignition.	4. Power is produced by compression ignition.
5. The compression ratio is higher than diesel engine.	5. The compression ratio is lower than petrol engine.
6. Thermal efficiency up to 25 percent.	6. Thermal efficiency up to 40 percent.
7. It occupies less space but more running cost.	7. It occupies more space but less running cost.
8. Fuel value is costlier than diesel.	8. Fuel value is cheaper than petrol.
9. It is light in weight and petrol as volatile is more dangerous. It is less dependable.	9. It is heavy in weight but is not dangerous as it is nonvolatile. It is more dependable.
10. Pre-ignition is possible.	10. Pre-ignition is not possible.
11. The power developed in a petrol engine is low due to lower compression ratio. They are used in cars and motorcycles.	11. The power developed in diesel engine is high due to higher compression ratio. They are used in heavy duty vehicles like trucks, buses, and heavy machinery.

20. What is synthetic petrol?

Synthetic petrol is a liquid fuel obtained from coal, natural gas, oil shale, or biomass. It may also refer to fuels derived from other solids such as plastics or rubber waste. It was first produced using Fischer-Tropsch process, which is a catalyzed chemical reaction in which synthesis gas (syngas), a mixture of carbon monoxide and hydrogen, is converted into liquid hydrocarbons of various forms.

21. What are 2- and 4-stroke engines? Compare 2-stroke and 4-stroke engines.

2-stroke engine

It performs in two ways: (1) compression stroke and (2) combustion stroke. In that process fuel that is added in the engine is pulled from the carburetor and it is pressurized. As the piston travels down the cylinder a vacuum is created, opening the exhaust port and letting the pressurized fuel mixture into your cylinder. Next the piston travels up the cylinder to the spark plug, which gives off a spark for igniting the fuel and sends the piston back down the cylinder, opening the exhaust port. As the piston travels back up the cylinder some of the exhaust goes out of the exhaust port on the engine and new fuel is introduced into the cylinder.

4-stroke engine

It has (i) intake stroke, (ii) compression stroke, (iii) combustion stroke, and (iv) power stroke.

(i) The first stroke opens the intake valve and the piston travels down the cylinder, pulling the fuel and air mixture from the carburetor. (ii) Then the piston travels up the cylinder, closing the intake valve. Right before the piston reaches top dead center (TDC) the spark plug gives off a spark to ignite the fuel mixture. The piston's journey to the top of the piston is known as the compression stroke. (iii) The fuel ignites (combustion stroke). (iv) It will force the piston back down the cylinder. This stroke is the one that gives the engine its power, which is known as the power stroke.

Differences between 2- and 4-Stroke Engines

2-stroke engine	4-stroke engine
1. Engines are cheaper and are simple for manufacturing.	1. Engines are expensive due to lubrication and valves and are tough to manufacture.
2. It has poor lubrication and results in more wear and tear. It also has high torque.	2. Less wear and tear occurs due to lubrication and less torque.
3. It is an engine of lower thermal efficiency.	3. It is an engine of higher thermal efficiency.
4. It has one revolution of the crankshaft during one power stroke.	4. It has two revolutions of the crankshaft during one power stroke.
5. Engines are basically lighter and are noisy.	5. Engines are basically heavier because its flywheel is heavy and less noisy.
6. It generates more smoke and shows less efficiency.	6. It generates less smoke and shows more efficiency.
7. It uses ports for fuel outlet and inlet.	7. It uses valves for outlet and inlet of a fuel.
8. It requires more lubricating oil as some oil burns with the fuel.	8. It requires less lubricating oil.

22. How do you control sugar degradation during storage?

It can be controlled by keeping biomass as dry as possible and keeping it in cold storage. In the case of liquid sugars, sterile storage is one key, temperature is another, as well as in-time production.

23. What kinds of biomass can be used to generate biofuels? What are limitations of utilizing biomass?

Generally different verities of biomass such as crop wastes, forestry residues, purpose-grown grasses, woody energy crops, algae, industrial wastes, non-recyclable municipal solid waste, urban wood waste, and food waste are used to produce biofuels.

Limitations of utilizing biomass

a) Low concentration of energy, i.e., small percentage of sunlight is utilized to convert biomass by plant

b) Relatively expensive energy conversion

c) Relatively low concentration of biomass per unit area of land and water

24. What is energy plantation?

When land plants are grown purposely for their fuel value by capturing solar radiation is called energy plantation.

25. How will you determine iodine number of vegetable oil?

Iodine reacts with the double bond of unsaturated fatty acid and the amount of iodine that reacts with fat will indicate the amount of the fat. The iodine number is defined as the weight of iodine absorbed by 100 g of the fat.

Materials required
(i) Fat (ii) Wij's solution (iii) Potassium iodide (10 percent) (iv) Sodium thiosulphate (0.1 N) (v) Starch indicator

Procedure
(i) Preparation of Wij's solution

Wij's solution is prepared by dissolving 7.9 g of pure iodine trichloride in 100 ml of glacial acetic acid in a beaker by warming on a water bath. In another flask dissolve 8.7 g of resublimed iodine in a second portion of warm 100 cc glacial acetic acid. Mix the two solutions in a 1000 ml measuring flask and fill up the solution to the mark with glacial acetic acid. Store the Wij's solution in a well-stopper amber bottle.

(ii) Determination of iodine number

Weigh 0.1–0.2 g of oil or fat in a 250 ml iodine flask and add 15 cc of chloroform or carbon tetrachloride to dissolve it. Along with this carry out a blank experiment, omitting the fat. Add 25 cc of Wij's solution to the flasks containing the fat and the blank. Close the mouth of the flasks tightly and keep the flasks in the dark for about 30 minutes, after adding 25 ml of 10 percent solution of potassium iodide at the neck of the flasks. Then dilute the mixture with 50–60 ml of water. Titrate it immediately against 0.1 N sodium thiosulphate solution. Shake vigorously and again titrate until the yellow color almost disappears. Add starch solution and titrate again until the blue color disappears.

Let V_b and V_w be the number of cc of 0.1 N sodium thiosulphate required for 'W' g of fat.

So the equivalent of thiosulphate is equivalent to the iodine that reacts.

$$\text{Iodine number} = \frac{(V_b - V_w) \times 10^{-3} \times 127}{W \times 10}$$

where 127 is the equivalent weight of iodine.

26. Determine oxidation stability of vegetable oils by Rancimat method.

The oxidation processes that slowly take place in fats at ambient temperature are known as *auto-oxidation*. They start reacting with fatty acids in multistage processes, leading to a variety of decomposition products such as peroxides, alcohols, aldehydes, and different carboxylic acids.

In Rancimat method, the sample is exposed to high temperature (50–200°C), and the volatile oxidation products are transferred to the measuring vessel by air stream and are absorbed in the measuring water solution. When the conductivity of the measuring solution is recorded continuously, an oxidation curve is obtained whose point of inflection is known as the *induction time*, which provides the characteristic value of the oxidation stability.

Apparatus required
Rancimat apparatus attached with computer.
Procedure

1. Check initially Rancimat apparatus is correctly attached with computer and then switch both the machines.

2. Sample vessel and measuring vessels should be cleaned; otherwise, it hampers the reaction conditions.

3. Before starting the measurement, check the required temperature and flow of air.

4. Place the weighed sample (approximately 2.0–2.5 g) and cover the vessel with caps fitted with screw and tubes.

5. Place the measuring vessel containing 60 ml of distilled water in the Rancimat apparatus.

6. Attach the outlet tube of sampling vessel in the measuring vessel to absorb the oxidized products in water.

7. Place the exhaust tube of the measuring vessel in a hood as the products of oxidation are very irritant.

8. After the required time as programmed on a computer, take the induction result from the computer. See below figure.

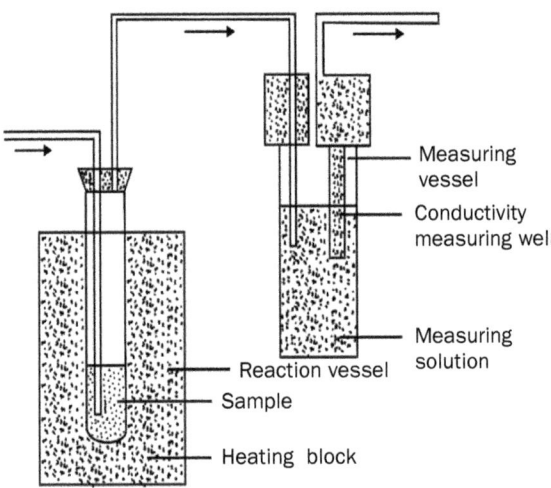

Schematic measuring arrangement of Rancimat apparatus

27. What are problems associated with hydrogen use? What are common methods of hydrogen storage?

Problems associated with hydrogen

a) Difficult to store, since it is highly explosive, so there is lack of safety and management.

b) Effective in energy utilization

c) Commercial production of hydrogen in cheap cost

Common methods of storage hydrogen

a) Compressed gas storage

b) Solid stable storage

c) Liquid storage

28. What are the costs of hydrogen storage?

The cost of hydrogen storage is listed as follows:

a) The cost of a container to store hydrogen

b) The cost of maintenance of low temperature in the storage of liquid hydrogen

c) The compression of gas hydrogen to high pressure

d) The compression and liquification cost of hydrogen

e) the weight of hydrogen tank increases the weight of vehicle, thereby increasing specific fuel consumption when vehicle is running

f) The cost of transportation of gas cylinder

29. For a hydrogen fuel cell find the following factors:

a) Electromotive force

b) Cell efficiency

c) Electric work output per mole of hydrogen consumed per mole of hydrogen consumed and per mole of hydrogen produced

Given $\Delta G_{25°C} = 235.4 \times 10^3$ kJ/(gm mole); $\Delta H_{25°C} = -284 \times 10^3$ kJ (kg mole)

a) Hydrogen–oxygen fuel cell reaction is as follows:

$$H_2 + \frac{1}{2}O_2 = H_2O$$

$$E \text{ (Electromotive force)} = \frac{\Delta G}{\pi F}$$

$$= \frac{235.4 \times 10^3}{2 \times 956500} = 1.22 \text{ V}$$

b) Cell efficiency $= \dfrac{\Delta G}{\Delta H}$

$$= \frac{235.4 \times 10^3}{284 \times 10^3} \times 100 = 82.7\%$$

c) Electric work output per mole $= \Delta W = n\, FE$

But $E = \dfrac{\Delta G}{nF}$

Hence $\Delta W = \Delta G = 235.4 \times 10^3$ kJ (kJ mole) of hydrogen

1 mole of water generated for 2 moles of hydrogen atoms and 1 mole of oxygen atoms

ΔW per mole of water $= \Delta W$ per more of hydrogen

$$= 235.4 \times 10^3 \, kJ \, (kg \, mole) \, H_2O$$

Heat transfer to surrounding is given $\Delta Q = \Delta H - \Delta W = \Delta H - \Delta G$

$$= -284 \times 10^3 + 234 \times 10^3$$

$$= \quad kJ \, (kg \, mole)$$

30. How can the performance of a solar cell be increased? What is the area of PV cells needed to generate electric power to run (i) a computer using 1200 W and (ii) a heater using 600 W?

Assuming the solar radiation of 300 J/m^2 and per unit time during day light, the efficiency of PV to be 20 percent.

The performance of a solar cell can be maximized by the following methods:

a) The efficiency of solar energy conversion depends on open electric voltage (V_{oc}) and solar generated current (I_{sc}) and also V_{oc} depends on the ratio of I_{sc} and I_o (reverse leakage current). So it depends on V_{oc} and I_{sc}.

b) It depends on low series resistance

c) It depends on high shunt resistance, which can be increased by preventing any leakage occurring perimeter of the cell

d) It depends also on optimum solar cell area.

i) Power output $= \pi \times$ solar power $= 0.20 \times 300 = 60$ W

$$\text{Area of PV cells for computer} = \frac{1200}{60} = 20 \, m^2$$

As Power required $= 1200$

ii) Power required $= 600$ W

$$\text{Area required to heater} = \frac{600}{60} = 10 \, m^2$$

Total area $= 20 + 10 = 30 \, m^2$

31. How can energy be increased from wind power by turbine?

The wind energy extracted from wind power by turbine can be increased by using the following methods:

i) Large size blades

ii) Variable speed drive

iii) Tilt control of blades

iv) Stall control by shifting of blades

v) Eddy current braking system to maintain speed

vi) Pitch control

vii) Yaw control (system of wind turbine responsible for the orientation of wind turbine rotor toward the wind)

viii) High power components used in the circuit

32. What is rotor solidity? What are its implications?

a) A windmill has rotor of 5 m with 50 blades. Each blade has a width of 0.40 m. Find solidity.

b) Find the maximum power output a turbine given wind speed is 6m² and density is 1.2 kg/m² and rotor diameter 50 m.

Rotor solidity is the area of the rotor disk that is actually occupied blade area. The greater the solidity of a rotor, the slower it needs to turn to intercept the wind.

$$\text{Solidity} = \frac{\text{Projected area of blades}}{\text{Swept area}} = \frac{50 \times 0.40 \times 100}{\Pi \times 5^2} = 25.4\,\%$$

$$\text{Area of rotor} - \frac{\pi\, d^2}{4} = \frac{3.14 \times 50^2}{4} = 1570\ \text{m}^2$$

Power of window $(P_w) = 1/2 \times$ density $\times$ area $\times$ wind speed
$$= 1/2 \times 1.2\ 1570 \times 6^2 = 33.9\ \text{kW}$$

Maximum power output $(P_{/T}) = 16/27 \times P_w = 16/27 \times 33.9 = 20.1\ \text{kW}$

33. Minimum ignition energy of hydrogen is much lower as compared to that of other conventional fuels. Therefore, hydrogen can be ignited in an engine with much lower energy spark. But practically, this has not been so. What could be the reason?

Hydrogen's low ignition energy quite often enables the energy concentrated at the hot spots of the combustion chamber to initiate combustion. Such a situation gives rise to preignition/backfire.

34. What are the main causes of backfire?

    (a) Hot spots on the spark plug and cylinder wall

    (b) Hot particles in the cylinder

    (c) Cross talk between spark plug wires

    (d) Mixing of exhaust gases with the incoming charge

35. Answer briefly.

(i) Lower power output of a hydrogen engine is relatively lesser than a corresponding gasoline engine.

This is mainly due to the lower volumetric energy density of hydrogen.

(ii) How is hydrogen's property of wider flammability limit of benefit to a vehicle/engine?

The wide range of flammability limit makes it possible to operate hydrogen engines far from the stoichiometric ratio value. Such a situation ensures lean mixture operation, greater fuel economy, and lesser NOx emission due to lower combustion temperature.

(iii) Why is it difficult to achieve compression ignition of hydrogen?

Auto ignition temperature of hydrogen is about 576 degrees; thus, very high compression ratios will be required to achieve the ignition temperatures.

(iv) Thermal efficiency of a hydrogen engine is much higher compared to that of a gasoline engine.

Hydrogen engines can have unthrottled operation using quality governing. Furthermore, hydrogen engines can run on higher compression ratios as compared to conventional gasoline engines. Hydrogen has a flame temperature of about 2100 deg. Celsius. The higher flame temperature results in higher thermal efficiency for an engine cycle operating with hydrogen.

36. Oxides of nitrogen is the pollutant of concern in a hydrogen vehicle. Suggest three methods to control the NOx concentration level.

    a. Operating at lower equivalence ratio

    b. Adopting exhaust gas recirculation

    c. Use of three-way catalytic converters (TWC)

37. From among the temperatures of 576°C, 540°C, and 490°C, identify the respective auto ignition temperatures of hydrogen, methane, and propane.

    Hydrogen 585°C   Methane –540°C Propane –490°C

38. Why does hydrogen combustion result in lower output?

Heat of combustion of hydrogen is 120 MJ/kg or 8.5 MJ/L (Liquid). Heat of combustion per unit mass is high, but the low mass density makes the corresponding per unit volume low. Thus when hydrogen is drawn into a given cylinder size at given conditions, it gives lower power output per unit volume of fuel.

39. A person can be closer to a hydrogen fire than a gasoline fire without being burned.

Hydrogen fires burn very rapidly and radiate very little heat, and thus are relatively short-lived. If liquid hydrogen does leak in a crash, it will evaporate and disperse exceedingly fast. On the other hand, gasoline will puddle and remain a fire hazard for much longer until the entire amount evaporates.

40. Storage of hydrogen as a compressed gas in metallic pressure vessels contributes significantly to the total weight of the system. This leads to low gravimetric efficiency. What steps are being adopted to achieve high volumetric and gravimetric efficiency?

Composite polymer pressure vessels in combination with increased working pressure up to 68.94 MPa are used to achieve high volumetric and gravimetric efficiency.

41. With the present level of technology please suggest the type of storage for (a) short range road vehicles; (b) medium range vehicles; (c) off-road construction; (d) mining equipment; (e) rail locomotives; (f) marine vehicles.

(a) compressed gas (b) metal hydrides and microsphere (c) metal hydrides (d) metal hydrides (e) liquid hydrogen (f) liquid hydrogen

42. Briefly explain the effect of exhaust gas recirculation (EGR) on the combustion and exhaust emission characteristics of a hydrogen engine.

With an EGR system, a portion of the exhaust gases back into the intake manifold. As far as the combustion behavior of the engine is concerned, the introduction of exhaust gases brings down the temperature of the hot spots. This, in turn, reduces the possibility of backfire and pre-ignition. Lowering the peak combustion temperature also brings down the NOx concentration level.

43. What is the difference between black hydrogen and "green hydrogen"?

Black hydrogen refers to hydrogen produced through processes that result in pollution or greenhouse gas emissions. Black: derived from fossil fuels (coal, oil, natural gas) or nuclear power; created by processes involving pollution or greenhouse gas emissions. "Green hydrogen" refers to hydrogen produced through processes that have zero emissions of carbon dioxide, or no net emissions of carbon dioxide. Hydrogen derived from plants; created from renewable energy.

44. What is a flame arrestor in the context of hydrogen fuel? Where is it normally installed in a hydrogen operated system?

Flame arrestor is a system for suppressing explosions inside hydrogen containing system. Normally these are installed upstream and downstream of possible ignition sources in piping systems containing hydrogen gases.

45. Excessive hydrogen leakage into open atmosphere does not result in any explosion. Justify your answer.

Low molecular weight of hydrogen results in high molecular diffusivity such that hydrogen diffuses 3 to 8 times faster than air.

46. Compounds such as mercaptans and thiophanes that are used to scent natural gas may not be added to hydrogen for fuel cell use.

They contain sulphur that would poison the fuel cells.

47. Hydrogen is not toxic. But under certain circumstances, it has been observed to have caused asphyxiation, frostbite, and hypothermia. Explain the stage of hydrogen and the situation under which such problems do occur.

Hydrogen, being colorless and odorless, is not detectable by human senses even in very low levels of concentration. In some situations, it dilutes the concentration level of oxygen in air and thus the air becomes oxygen-deficient. Inhalation of cold gas or vapor causes respiratory problems and the subsequently problem gets aggravated to asphyxiation.

Hydrogen has a low boiling point: 20.3 K at sea level. In the event that any liquid hydrogen splashes on the skin or in the eyes, it causes hypothermia or frostbite burns.

48. Diatomic hydrogen is the most stable form of hydrogen, but what are the difficulties it often causes when provided with enough energy?

When provided with enough energy it can dissociate into hydrogen atoms, which are small enough to permeate a container surface. This permeation gives rise to embrittlement, thereby causing structural damage.

49. Fuels with an Ocatne number over 100 have more resistance to autoignition. Write down the autoignition temperatures of hydrogen, propane, and methane in decreasing order.

Hydrogen 130 +, Methane 125, and Propane 105

50. Identify the metal hydride, chemical hydride, and the complex hydride from the following:

(a) MgH_2 (b) $NaAlH_4$ (c) $LiBH_4$ (d) $NH_3 BH_3$

Metal hydrides: MgH_2, Complex hydrides: $NaAlH_4$, and Chemical hydrides: $LiBH_4$, $NH_3 BH_3$

51. Lower heating value for hydrogen is about 120 MJ/kg whereas the higher heating value is 142 MJ/kg. Why such a difference?

The large difference in higher and lower heating value can be explained as H_2O is the sole combustion product of hydrogen.

52. Kitchen model stoves were tested for leakage using hydrogen, methane, and propane. No fans were supplied to prevent convection and there was a simple vent. Twelve minutes after the leak it was observed that in one case there was no cloud, in the second case there was a small combustible cloud, and the third case demonstrated a large cloud. Identify the effect of leakage with respect to the three gases and comment on the safety characteristics.

Hydrogen simply diffused up straight through the vent. No combustible cloud: least danger.

Methane: small combustible cloud, fairly dangerous.

Propane: large cloud, posing greatest danger.

53. Estimate the air fuel ratio for stoichiometric combustion of hydrogen with air.

The theoretical or stoichiometric combustion of hydrogen and oxygen is given as:

$$2H_2 + O_2 = 2H_2O$$

Moles of H_2 for complete combustion $= 2$ moles

Moles of O_2 for complete combustion $= 1$ mole

Because air is used as the oxidizer instead of oxygen, the nitrogen in the air needs to be included in the calculation:

Moles of N_2 in air $=$ Moles of $O_2 \times$ (79% N_2 in air / 21% O_2 in air)

$= 1$ mole of $O_2 \times$ (79% N_2 in air / 21% O_2 in air)

$= 3.762$ moles N_2

Number of moles of air $=$ Moles of $O_2 +$ moles of N_2

$= 1 + 3.762$

$= 4.762$ moles of air

Weight of O_2 = 1 mole of O_2 × 32 g/mole

   = 32 g

Weight of N_2 = 3.762 moles of N_2 × 28 g/mole

   = 105.33 g

Weight of air = weight of O_2 + weight of N (1)

    = 32g + 105.33 g

    = 137.33 g

Weight of H_2 = 2 moles of H_2 × 2 g/mole

   = 4 g

Stoichiometric air/fuel (A/F) ratio for hydrogen and air is:

A/F based on mass: = mass of air/mass of fuel

    = 137.33 g / 4 g

    = 34.33:1

A/F based on volume: = volume (moles) of air/volume (moles) of fuel

      = 4.762 / 2

      = 2.4:1

The percent of the combustion chamber occupied by hydrogen for a stoichiometric mixture:

% H_2 = volume (moles) of H_2/total volume (2)

     = volume H_2/(volume air + volume of H_2)

     = 2 / (4.762 + 2)

     = 29.6%

54. Indicate if the following statements are true (T) or false (F).

(i) Certain metals (such as martensitic and ferritic) are more susceptible to embitterment than highly alloyed austenitic steels. (T)

(ii) Carbonaceous steels are not usually used at conditions of high temperature as hydrogen can react with carbon in the steel to form CH_4 bubbles in the metal. (T)

(iii) When used as vehicle fuel, the low density of hydrogen necessitates that a large volume of hydrogen be carried to provide an adequate driving range. (T)

(iv) Even an invisible spark or static electricity discharge from the human body may have enough energy to cause ignition. (T)

(v) Hydrogen flames are very pale blue and are almost invisible in daylight due to the absence of soot. (T)

(vi) For a given liquid spillage volume, gasoline fires will last longest and hydrogen the shortest. Both fuels burn at about same flame temperatures. (T)

(vii) Hydrogen fires burn very rapidly and radiate very little heat, and thus are relatively short lived. (T)

(viiii) As compared to gasoline, propane, and methane, hydrogen can accelerate to its detonation velocity in a much shorter distance. (T)

(ix) Cryogenic burns (frostbites) often result from contact with extremely cold liquid hydrogen. (T)

(x) Over long distances, trucking liquid hydrogen is more economical than trucking gaseous hydrogen. (T)

55. The graph below shows variation of thermal efficiency with equivalence ratio for gasoline and hydrogen. What conclusions are evident from this graph?

Variation of Thermal Efficiency with Equivalence ratio

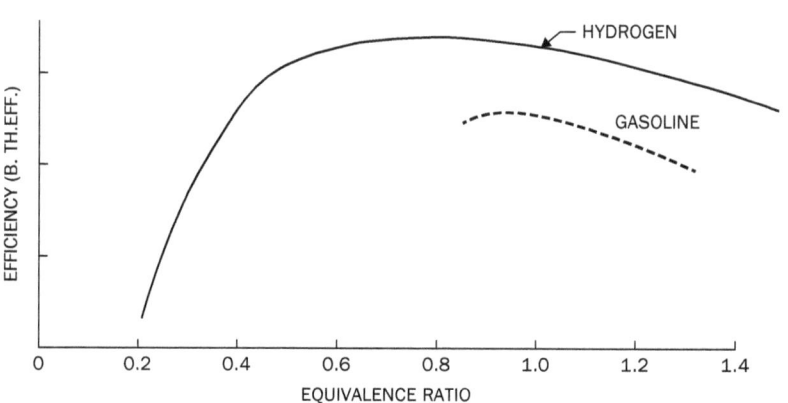

(a) Hydrogen is combustible over a much wider range of equivalence ratio (from about 0.2 to beyond 1.4) whereas gasoline has a much shorter range (equivalence ratio of 0.7 to approximately 1.3).

(b) Over the entire range of operation, thermal efficiency of hydrogen is much higher than that of gasoline.

Index

Page numbers followed by "f" are to figures; page numbers followed by "t" are to tables.

01 14
J